Testing Methods for Seed-transmitted Viruses: Principles and Protocols

Testing Methods for Seed-transmitted Viruses: Principles and Protocols

S.E. Albrechtsen

Danish Seed Health Centre for Developing Countries
Thorvaldsensvej 40, DK-1871 Frederiksberg C, Denmark

formerly
Danish Government Institute of Seed Pathology for
Developing Countries

CABI Publishing

CABI Publishing is a division of CAB International

CABI Publishing
CAB International
Wallingford
Oxfordshire OX10 8DE
UK

Tel: +44 (0)1491 832111
Fax: +44 (0)1491 833508
E-mail: cabi@cabi.org
Website: www.cabi-publishing.org

CABI Publishing
875 Massachusetts Avenue
7th Floor
Cambridge, MA 02139
USA

Tel: +1 617 395 4056
Fax: +1 617 354 6875
E-mail: cabi-nao@cabi.org

A catalogue record for this book is available from the British Library, London, UK.

ISBN 0 85199 016 9

Library of Congress Cataloging-in-Publication Data

Albrechtsen, S.E. (Sven Erik)
Testing methods for seed-transmitted viruses: principles and protocols / S.E. Albrechtsen.
p. cm.
Includes bibliographical references and index.
ISBN 0-85199-016-9 (alk. paper)
1. Seeds--Testing. 2. Seed pathology. 3. Virus diseases of plants. I. Title.
SB117.A35 2005
632'.8--dc22

2005003343

Typeset by AMA DataSet Ltd, UK.
Printed and bound in the UK by Biddles Ltd, King's Lynn.

Contents

Preface

The yield losses and reduced crop quality caused by plant virus diseases have gained increasing recognition in recent years. The use of virus-resistant cultivars appears to be one of the best control strategies, but is not always possible or effective. Outbreaks of new diseases due to new variants of viruses and viroids continue to occur. Several important plant viruses are, in addition to their sap and vector transmission, also transmitted through seed. Seed transmission of plant pathogens plays an important role for the early outbreak of crop diseases and for the survival of inoculum from one crop season to the next. And seeds are instrumental in an effective worldwide spread of a range of diseases through international exchange of seed. One of the key principles of seed-transmissible viruses is that their elimination by seed treatment is virtually impossible.

In the tropical and subtropical zones, where a great variety of virus-susceptible crop species are grown and virus vector pressure continues year-round, effective management of viral diseases is particularly important. And effective management of seed-transmitted viruses depends upon the use of healthy planting material, through seed quality control and by plant quarantine monitoring and testing.

Tests of seed for viral infection are crucial for providing adequate supplies of virus-free seeds or seed with very low infection rates to avoid intra-national and international dispersal, leading to potentially high losses. Until 1970–1980, large-scale testing of sowing material for viruses was impracticable, but a number of suitable techniques have since been (and are being) developed.

The primary aim of this book is to provide principles and protocols for a number of well-established routine detection methods for viruses and viroids in seeds of, primarily, field and vegetable crops, methods that should also be applicable for the detection of many other viruses and viroids.

In the introductory Part I, an attempt is made to provide a 'mini-review' of important aspects of seed-transmitted viruses, especially their economic importance, location in seeds, seed transmission and strategies for their control.

Chapters 4, 5 and 6 in Part II describe principles and stepwise protocols for the detection methods, and also contain information on method optimization and variations, and discuss plusses and drawbacks of different methods. Further included are practical notes, interpretation of data, notes on pitfalls and trouble shooting. The chapters contain both techniques manageable in modestly equipped laboratories and those requiring more sophisticated facilities.

The last chapter provides philosophies and outlines for setting up seed-health testing for viruses, including seed sampling, group testing, statistical analyses of test results, tolerance levels and standardization of seed-health assays.

A list of seed-transmitted viruses and viroids can be found in Appendix 1. A few recipes for reagents and buffers appear in the protocols, but most of them are placed in Appendix 2, and suppliers of equipment and materials are indicated in Appendix 3. Some manufacturers and sources are mentioned in the text but this does not infer that other comparable makes and brands may not be equally satisfactory for use.

The author's experience of virus seed pathology was obtained during 30 years of appointment, the final 12 years as Associate Professor, at the Danish Government Institute of Seed Pathology for Developing Countries (DGISP), an institute funded by the Danish Development Assistance (DANIDA). In 2004 the institute was changed to a centre, the Danish Seed Health Centre for Developing Countries (DSHC) under the Royal Veterinary and Agricultural University, Copenhagen, Denmark, and DANIDA. The book is based on methods and materials used in teaching and training students in seed virology courses held regularly in Denmark as well as in the developing countries, for which the author was responsible during the last 16 years of his appointment. The materials used are updated on the basis of consultations of the latest and most relevant literature. The book is primarily technical, and thus such topics as, for example, systematic descriptions of individual viruses are not included; for details on individual viruses or for further information on plant virology, the reader is referred to other sources (books and websites) specified in the text.

According to the numerous students from developing countries who have come for studies at DGISP over the years, as well as seed pathologists met abroad, and others, this book has been needed for a long time.

To the author's knowledge, there is no such book in existence, targeted towards use in seed-health testing for viruses and viroids and at the same time a source of general information on seed-borne viral diseases and their management. It is hoped that the book will find a use both by laboratory personnel at seed quality control and plant quarantine agencies and by others working with seed-health testing for viruses.

Acknowledgements

The author expresses his thankfulness to Dr S.B. Mathur, former Director of DGISP, for his continuous encouragement to write this book and for his help in structuring the book.

My sincere gratitude goes to Dr R.O. Hampton, former Professor at the Oregon State University, USA, who kindly read the entire manuscript with meticulous care and provided numerous and highly constructive suggestions, which enabled me to strengthen and improve the text to an ultimate extent.

I am indebted to Dr B. Albrektsen, Copenhagen, Denmark, for his critical perusal and linguistic corrections of a part of the manuscript. For critical comments and helpful suggestions for Chapter 6, I owe a debt of gratitude to Dr M. Nicolaisen, Danish Institute of Agricultural Sciences, Flakkebjerg, Denmark. I am grateful to Dr S.J. Roberts, Plant Health Solutions, Warwick, UK, for valuable and substantial help with Section 7.2 – I consider Dr Roberts as co-author of that section.

My sincere thanks to the present and former staff at DGISP/DSHC and to DANIDA for all help and support. Among the staff, I am especially indebted to Ms L. Hasle, Laboratory Technician, whose assistance in and inspiration for the virology training programme during 13 years have contributed to improvement of the protocols. Thanks also to Ms H. Westh, Secretary, for her constant help with providing reprints, etc., and Mr M. Ragab, photographer, for assisting with pictures.

I wish to acknowledge with thanks the individuals and the editors of journals, as well as private companies, who provided several of the illustrations and tables that appear in this book. Thanks are also due to the scientists who kindly provided requested reprints and those mentioned under personal communications.

Last but not least I am greatly indebted to my wife, Marianne, for her patience and forbearance with me throughout the entire writing period. I thank her for her untiring assistance in typing and retyping the manuscript as well as for keeping track of a comprehensive selection of reprints and help with making the subject index.

1 Introduction

Viruses and viroids are unique pathogens, differing completely from other pathogens, such as fungi and bacteria. The virus pathogen is a particle consisting of nucleic acid encapsidated in a protein coat. The nucleic acid (the genome) of the majority of plant viruses is single-stranded RNA; only a relatively few species have a genome of double-stranded RNA or of DNA. The viroid particle is the smallest plant pathogen known; it consists of a single circular strand of RNA without a protein coat. Potato spindle tuber disease is one of several important diseases caused by viroids. Viruses as well as viroids have no protein-synthesizing system, but depend totally on that of the host for their replication (Hull, 2002).

Passage of viral inoculum from diseased plants to their offspring via seeds was long thought to be a rare phenomenon. Today, seed transmission is known to occur for about one-seventh of the known viruses in one or more of their hosts (Hull, 2002), and the number is still increasing. Three important effects of seed transmission are: (i) direct injury to the crop and/or indirect injury, as even a low incidence of infected seeds sown results in numerous randomly scattered foci of inoculum, facilitating early secondary spread in the crop by vectors; (ii) survival of virus inoculum from one crop season to the next; and (iii) several viruses and viroids have been, and undoubtedly still are, disseminated worldwide through exchange of seed having undetected infection.

1.1. Seed-transmitted Viruses and Viroids

Many species of viruses are known to occur in seeds harvested from infected plants without being transmitted to the offspring, i.e. they are seed-borne but not seed-transmitted. As a general rule, only those able to infect the embryo are seed-transmissible to the next generation. A notable exception is some members of the genus *Tobamovirus* that do not enter the embryo, but are stable enough to survive in or on seeds and from there infect the offspring. Hull (2002) listed a number of virus genera, indicating the number of seed-transmitted viruses for each. Among these viruses, 31 belong to the cryptoviruses (*Alphacryptovirus*, *Betacryptovirus*), appearing to be of no economic importance and solely transmitted through seed and pollen. Genera containing relatively high proportions of seed-transmitted members are (no. of seed-transmitted/no. of members): *Alfamovirus* (1/1), *Comovirus* (6/15), *Cucumovirus* (3/3), *Ilarvirus* (8/17), *Nepovirus* (17/40), *Potyvirus* (16/179), *Sobemovirus* (4/14), *Tobamovirus* (7/17) and *Tobravirus* (3/3), whereas for genera like *Dianthovirus*, *Luteovirus*, *Marafivirus* and *Tenuivirus* no seed-transmitted members are reported (Hull, 2002). In the large family *Geminiviridae* (102 members), no member is assumed to be seed-transmitted (see Section 3.1.2). Out of the 28 viroids known (Steger and Riesner, 2003), ten are reported to be seed-transmitted (Mink, 1993; Singh *et al.*, 2003). A list of seed-transmitted viruses and viroids and their assignment to genera and families is given in Appendix 1. Thirty-four economically important seed-transmitted viruses and one viroid, with their principal hosts, are listed in Table 4.1.

1.2. The Development of the Science

Transmission through seed was first suspected for tobacco mosaic in tomato by Westerdijk in 1910 (quoted from Bos, 1977) and later evidenced for *Bean common mosaic virus* (BCMV, *Potyvirus*) in bean (Reddick and Stewart, 1918, 1919, quoted from Bos, 1977), *Cucumber mosaic virus* (CMV, *Cucumovirus*) in wild cucumber (Doolittle and Gilbert, 1919, quoted from Bos, 1977), *Lettuce mosaic virus* (LMV, *Potyvirus*) in lettuce (Newhall, 1923, quoted from Bos, 1977) and *Soybean mosaic virus* (SMV, *Potyvirus*) in soybean (Kendrick and Gardner, 1924, quoted from Bos, 1977). Until 1951, only eight seed-transmitted viruses were known (Smith, 1951, quoted from Mink, 1993), but 6 years later about 20 viruses seed-transmitted in about 40 plant species were listed by Crowley (1957, quoted from Bennett, 1969). In 1964, 36 seed-transmitted viruses in 63 host species were reported (Fulton, 1964, quoted from Bennett, 1969) and in 1969 the recognized number was 47 (Bennett, 1969). Phatak (1974) listed 85 and Agarwal and Sinclair (1988) estimated as many as 156 viruses as being seed-transmitted.

Since 1951, several reviews on seed-transmitted viruses have been published, e.g. Bennett (1969), Bos (1977), Neergaard (1977), Mandahar (1981), Mink (1993) and Johansen *et al.* (1994). In his critical review of the literature, Mink (1993) found a number of viruses erroneously listed as seed-transmitted and concluded that 108 viruses (excluding cryptoviruses) and seven viroids could be regarded as transmissible through the seeds of one or more hosts. Since then, the list has changed due to new findings and changed taxonomy, etc. The list in Appendix 1 is updated accordingly (but is probably not complete). Full virus names of those that are abbreviated in the following text can be found in Table 4.2, together with their genera.

The increasingly recognized importance of seed transmission in plant virus ecology has led to the strengthening of seed-health testing for viruses in certification and quarantine agencies internationally. There is also an increasing awareness of seed transmission of important viruses in many weed hosts, seeds of which could provide a perpetual inoculum reservoir in the field, until such weeds are controlled.

1.2.1. Seed-health testing for viruses

Seed-health testing is a science as well as a technology which is used in screening seed lots for 'planting value' at seed testing stations and in plant quarantine laboratories, where seeds are tested against disease agents that are dangerous to countries or regions (Mathur, 1995).

The earliest method used for seed-health testing for viruses was examination of seedlings raised from the seeds, the growing-on test. This test was first proposed by Fajardo (1930, quoted from Bos, 1977) for testing of bean seed lots for BCMV. Later the method was proposed for *Barley stripe mosaic virus* (BSMV, *Hordeivirus*) in barley by McKinney (1954, quoted from Bos, 1977) and LMV by Rohloff (1962, 1967). Hampton *et al.* (1957, quoted from Bos, 1977), Rohloff (1966) and Phatak (1974) emphasized that optimal quality and quantity of light are critical factors for optimal virus-symptom development in growing-on tests. Symptom development in infected seedlings is, however, also a function of the virus or virus strain and the host species.

Thus, Lister and Murant (1967, quoted from Bennett, 1969) reported that symptomless infection in seeds was characteristic of seeds infected with nematode-transmitted viruses (see also Section 4.1). Because of such latency or of faint symptoms, and also the fact that symptoms caused by other factors may resemble virus symptoms, it was early realized that growing-on tests must be supplemented by confirmative biological or serological tests. Still, a growing-on test has the indisputable advantage of revealing seeds carrying seed-to-plant-transmissible virus (Chapter 2). Quantz (1957, 1962, quoted from Bos, 1977) reported the detection of BCMV, first from bean leaves by inoculating detached primary leaves of the hypersensitive bean cultivar 'Top Crop', followed by incubation on moist filter paper in Petri dishes for 3 days at 32°C under constant artificial light, and later, in 1962, the same method was effective for inoculum from bean seeds, pre-germinated on moist filter paper for 3–4 days. A highly sensitive biological test for LMV in lettuce seed was described, first by Pelet and Gagnebin (1963) and Pelet (1965) and later by Marrou and Messiaen (1967), by which one infected seed out of a group of 700 was detected by inoculation of dry seed triturated in buffer on to young *Chenopodium quinoa* plants (see also Section 4.4.3).

One of the first to develop serological techniques for virus indexing of seed was Hamilton (1965). Embryos separated from barley seeds were crushed onto small filter-paper discs, which, when placed on agar gel near another disc soaked with antiserum to BSMV, caused a visible precipitate in the gel if the embryos were BSMV-infected. A detergent was added to the agar, preventing virus aggregation and dissociating the virus into more rapidly migrating antigens, thus increasing test sensitivity. Virus-specific bands could be observed after about 48 h. Slack and Shepherd (1975, quoted from Bos, 1977) later developed an improved agar gel test, the radial-diffusion technique, for BSMV for cleaning of germplasm seed of wheat and barley. In this test, small leaf pieces (1 mm) from germinated seeds were embedded by pressing them into a gel containing diluted antiserum, a detergent and sodium azide. The antiserum was prepared against pyrrolidine-degraded virus protein, since antiserum raised against intact virions reacts poorly to detergent-degraded virions. After 24–36 h incubation, the reactions could be observed with a stereomicroscope. Up to 500–600 specimens per 9 cm diameter agar plate could be detected simultaneously and the test had the advantages of directly showing the percentage of infection, and that virus-free seedlings could be grown out, thus preserving the germplasm material. Both tests for BSMV have been used routinely for commercial seed certification in Montana, USA. Relatively few other sensitive and reliable serological tests suited for virus indexing of seed were developed up to the mid-1970s (Bos, 1977). Among those was the latex agglutination test, developed by Bercks (1967) and Bercks and Querfurth (1969). This test has a relatively high sensitivity compared to gel diffusion tests and was reported as usable, for example, for routine seed-health testing for BSMV in barley (Lundsgaard, 1976).

Electron microscopy (EM) (Section 7.1) is an invaluable tool for virus diagnosis, but is not suited for routine seed-health testing, whereas the highly sensitive immunosorbent EM (ISEM), developed by Derrick (1973), is occasionally used to detect viruses in seeds or to verify results of other detection methods. The principles of some of the early serological methods and ISEM are briefly described in Section 5.5.

In contrast to the situation two to three decades ago, highly sensitive and reliable methods for virus detection, well suited for testing of seeds, are available today. Among antibody-based methods, the enzyme-linked immunoassays (EIAs) have become the principal ones, being relatively simple to use, high in sensitivity and reliability and suited for large-scale testing and partial automation. The nucleic acid-based (NAB) methods, especially the polymerase chain reaction (PCR) with its high specificity and extremely high sensitivity, are increasingly used in plant virus detection, including seed-health testing for viruses. Although simple in principle and with obvious advantages, PCR still lacks simplicity of use and suitability for large-scale testing compared with the EIAs. However, intense effort in many places is under way to improve and simplify molecular detection techniques.

1.3. Economic Importance of Seed-transmitted Viruses

When plants become infected with virus, they are infected for life, and the infection normally results in crop losses. Such losses can be substantial for annual crops, but may obviously have even more fatal consequences for perennial crops, such as fruit trees. About 90% of all food crops in the world are propagated by seeds (Maude, 1996), and losses caused by seed-borne pathogens,

Table 1.1. Reported yield losses from seed-borne viruses.

Virus	Crop	Per cent yield loss
Bean common mosaic	French bean	35–98[a]
	Mung bean	31–75[b]
Bean yellow mosaic	Broad bean	≤ 59[a]
Broad bean stain	Lentil	14–61[c]
Cucumber mosaic	Lupin	25–42[d]
Lettuce mosaic	Lettuce	≤ 30[a]
Pea seed-borne mosaic	Pea	11–36[e]
Peanut mottle	Groundnut	20–72[a]
*Peanut stripe**	Groundnut	6–79[a]
Soybean mosaic	Soybean	48–99[f]
Tomato mosaic	Tomato	5–50[g]
Zucchini yellow mosaic	Cucurbit	0–99[a]

*Strain PSt of BCMV.
Sources: [a]quoted from Shukla *et al.* (1994); [b]Kaiser and Mossahebi (1974); [c]Makkouk and Kumari (1990); [d]Bwye *et al.* (1994); [e]Khetarpal and Maury (1987); [f]Tu (1989); [g]quoted from Walkey (1991).

including viruses, therefore, are of great significance. Economic losses from seed-transmitted viruses are both direct losses in reduced yield or quality of crops, and indirect losses as costs of control measures. As previously mentioned, even traces of virus-infected seeds sown can, in the presence of efficient vectors, lead to early epidemic spread in the field, resulting sometimes in 100% crop infection. Thus, at a seeding rate of, for example, 250,000 seeds per hectare (25 seeds/m^2), which is not very high, seed with as little as 0.1% infected seeds results in 250 infected plants/ha. Immediately after emergence, these randomly scattered plants function as reservoirs of inoculum. Examples of yield losses caused by seed-transmitted viruses are shown in Table 1.1.

Two famous examples are the high losses in barley caused by BSMV in Montana, USA, estimated to be more than US$30 million during 1953–1970 (Carroll, 1983), and the great losses in lettuce fields in California in the 1950s due to 'June yellows', caused by LMV (Grogan, 1980). In both of these cases, the problems were solved by using seed free of or nearly free of virus. A more recently reported case is the losses to the Australian dairy industry due to infection of pastures with *Subterranean clover mottle virus* (SCMoV, *Sobemovirus*), which were estimated to be AUS$31 million per year (Jones *et al.*, 2001). Seed transmission of *Zucchini yellow mosaic virus* (ZYMV, *Potyvirus*), a devastating pathogen of cucurbits causing yield losses up to 99% (Table 1.1), was not considered to play an important role in its epidemiology, as it was only found seed-transmitted in traces, if at all, until 2002, where up to 5% of the seeds of *Cucurbita pepo* var. *styriaca* (oil pumpkin) were reported to carry transmissible virus (Riedle-Bauer *et al.*, 2002).

In the case of mixed infections, drastic reductions in yield may occur due to synergistic effects. Cowpea stunt disease, caused by simultaneous infection of the blackeye cowpea mosaic (BlCM) strain of BCMV and CMV, resulted in an 84% reduction in number of seeds and 87% reduction in seed weight of cowpea. In the case of single infections of cowpea, these figures were only 19 and 14%, respectively, for CMV and 11 and 3% for BCMV-BlCM (Pio-Ribeiro *et al.*, 1978).

As previously mentioned, two of the important risks connected with seed transmission are: (i) survival of diseases from one crop season to another; and (ii) seed as a vehicle for the introduction of diseases into new areas via the seed trade. In contrast to most other seed-transmitted diseases, viruses constitute special problems. Viral infections often either show faint symptoms or are difficult to diagnose by visual inspection in the field. Testing of seeds for viruses is demanding in both cost and labour, compared to, for example, testing for fungi. And, although many attempts have been made, no treatment has eliminated viruses from seed without reducing seed viability (exception: disinfection of tomato and pepper seeds for tobamoviruses; see Section 4.4.3).

Indirect losses include costs of farmers' control measures and costs of seed certification and quarantine inspection. One of the characteristics of plant disease diagnostics is the 'low value added' to samples being indexed (Candresse *et al.*, 1998). The extent of control measures and their costs, therefore, has to be weighed against the value of the crop and the risk of economic loss (risk/benefit ratio). There are examples of effective control of viruses through the use of certified seeds, such as in the previously mentioned serious attacks of BSMV and LMV, where losses were considerably lowered or the virus nearly eradicated by relatively modest economic inputs.

Especially in developing countries, farmers commonly use their own saved grains as seeds for the following crop. By this practice, not only may seed-transmitted viral infection be perpetuated in annual crop species, but also the amount of inoculum may build up year by year. On the other hand, there are also examples of less virus incidence or susceptibility in farmers' seed than in seed from breeding stations, as was found for *Pea seed-borne mosaic virus* (PSbMV, *Potyvirus*) in lentil seed on one occasion in Nepal (Joshi *et al.*, 2004). In breeding and seed production programmes, breeders' seed and nuclear stocks should be free of virus before being multiplied into foundation and commercial stocks. Moreover, it is less costly to control viruses in the small quantities of seed in the early phases of breeding programmes than later when the seed is multiplied into large quantities.

1.4. Testing of Seeds for Viruses: Why?

Drastic losses in crop yield may in many cases be prevented or yield losses reduced by use of seed with no or low virus incidence. Assessing virus incidence in commercial seed production fields by field inspection alone is usually insufficient. Diseased plants may not be noticed in dense plant stands or because of inconspicuous symptoms in many virus–host combinations, and diagnosis based on field symptoms is difficult and often impossible. In quarantine, apparently healthy plants may have latent infection, the presence and identity of which can only be revealed by detection and diagnosis. Use of detection methods with high specificity is also needed in quarantine, for detecting and excluding viruses or strains of a virus. Seed-health testing is required in germplasm collections to eliminate viruses from incoming wild relatives, landraces or genetically improved genotypes, and for detecting viruses introduced during germplasm regeneration. Thus, detection and diagnosis of viruses by reliable, specific and sensitive techniques are required in seed production, in gene banks and in quarantine.

The degree of freedom from seed-transmissible viruses needed to minimize disease in the field varies according to the virus, the crop species and the field conditions, especially the presence of effective vectors, in the cropping area.

1.4.1. Seed certification

The tolerance limits for seed-transmitted viruses required for seed quarantine are distinct from those required for seed certification. The purpose for certifying seed is to provide growers with seed of optimal quality and, if seed-health testing is included, to ensure that the certified seed contains seed-transmissible pathogens at an acceptably low level through testing of representative samples. An acceptable level of inoculum, however, remains debatable. Stace-Smith and Hamilton (1988) noted that hard data for inoculum thresholds for seed-transmitted viruses are woefully lacking. According to Jones (2000) the situation 12 years later remains essentially the same. In contrast to the advances in testing procedures, surprisingly little effort has been made for relating test results and field spread of pathogens and the associated yield losses from different rates of infection. Or, as expressed by the same author: 'There is need for a better balance between research resources devoted to plant virus epidemiology and those allocated to cellular and molecular virology. This is particularly evident in the study of seed-borne virus diseases.'

A realistic tolerance limit for the level of virus-infected seeds in a seed lot can only be set on the basis of thorough field experiments running over a number of years and at different sites in a region, and in which seeds with different infection levels are sown in randomized plots (see also Chapter 3).

The famous field trials in Salina Valley, California, are among the few such trials ever conducted. It was found there that, unless lettuce seed lots had zero infected seeds per 30,000 seeds planted (< 0.003%), serious outbreaks of LMV still occurred in the crop (Grogan 1980). For the same crop and virus, a 0.1% level of infected seeds (Marrou and Messiaen, 1967) is considered acceptable in Europe and most other regions where vector pressure is less and where lettuce is not grown all year round (as it is in Salina Valley). Jones (2000) reports the determination of reliable inoculum thresholds in field trials at different sites over 5, 7 and 7 years, respectively, for CMV in lupin (*Lupinus angustifolius*), CMV in subterranean clover (*Trifolium subterraneum*) and *Alfalfa mosaic virus* (AMV, *Alfamovirus*) in burr medic (*Medicago polymorpha*) in Western Australia.

For most growers in developing countries, however, availability of certified seed is rather limited. In some of these countries, 90% or more of the total seed sown had been saved from the previous season's harvest. In Tanzania, farm-saved seed constituted about 96% of crop seeds planted in 2002 (Manento *et al.*, 2002). In such cases certified seed is either not affordable or not easily available due to limited poor-country infrastructure. Farmers are unable to test for virus infection in seed, but there are simple measures by which they could produce seed free of, or with low, virus incidence (Section 3.2.1).

1.4.2. Quarantine

Most embryo-transmitted viruses remain viable in seeds for years and often as long as the seeds remain viable (Bennett, 1969; Bos, 1977); thus, seeds are an obvious means of dispersing viruses over long distances by trade and exchanges of seeds. Several seed-transmitted viruses have undoubtedly been spread worldwide in this way, especially those attacking legumes (Hampton, 1983), e.g. SMV in soybean (Goodman and Oard, 1980) and PSbMV in pea (Hampton and Braverman, 1979). Even when a virus is known to exist in a country or region, care should be taken not to introduce new and more severe strains of the virus. As pointed out by Mathur (1995), quarantine agencies should maintain constant contact with research institutes on the status of important seed-transmitted diseases, including the occurrence of new diseases or new strains of disease agents.

A new serious pathogen of groundnut, peanut stripe virus (later identified as a strain of BCMV), was introduced to USA in the 1980s through seed germplasm imported from China (Demski *et al.*, 1984). The same pathogen was intercepted in Australia (Persley *et al.*, 2001) from imported groundnut seed held in post-entry quarantine. Exclusion of diseases carried in seeds is impossible by simply testing a representative sample of a seed lot, as done in certification. Any required zero tolerance is possible only for small seed lot quantities, such as crop germplasm.

Several legume germplasm lines imported from large germplasm seed banks were found to be infected in Australian post-entry quarantine (Jones, 1987). In this case, among lines introduced in 1978–1981, 29 of 302 (9.6%) of *Vigna* spp. and 54 of 309 lines (17.5%) of *Glycine max* and *Phaseolus* spp. were infected with seed-transmissible viruses. Most of the infected seedlings expressed obvious viral symptoms, and identities were confirmed by bioassay, serology and electron microscopy. In Australia, imported germplasm seeds of annual crops are planted in vector-free greenhouses and only seeds from the virus-free plants are released. Jones emphasizes the need for disease-free seed stocks in germplasm collections to enable countries with less sophisticated quarantine systems to import disease-free seeds, intended for improving crop production and not for introducing new pathological problems. Removal of exotic viruses from germplasm by this means inevitably causes a delay in the release of seeds. Non-destructive testing of the seeds could have shortened this time. However, this is possible only in a few virus–host combinations (see Sample preparation, Section 5.2.2).

In the 1970s it was realized that US germplasm collections of pea (*Pisum sativum*) had been a major inoculum reservoir of PSbMV, resulting in a number of PSbMV outbreaks in North American breeding programmes. Also, lines and cultivars of pea from seed companies worldwide were discovered in these years to be infected with this virus (Hampton, 1983).

Among pea accessions introduced to the US Department of Agriculture (USDA) Plant Introduction (PI) *Pisum* collection, accessions originating in northern India had a remarkably high frequency (138 of 472) of PSbMV-infected seed lots. At the same time, sources of immunity were also found in a high number of pea PI accessions (33) from northern India. In comparison, none of 500 PI accessions from developed countries was found to contain PSbMV-immune germplasm. The geographical concentration of both PSbMV inoculum and PSbMV-immune germplasm in northern India is unique, being an association probably not occurring in any other region of the world. It has, therefore, been hypothesized that northern India represents an original epicentre of PSbMV, from which the virus was disseminated to other parts of the world in infected *Pisum* seeds (Hampton, 1986).

The successful elimination of this devastating pathogen from a large number of US germplasm seed accessions can serve as a case history (Hampton *et al.*, 1993). In 1976 a working group was organized by three US research laboratories with the purpose of removing PSbMV from pea germplasm, which at that time was underutilized by breeders due to the virus risk. From each seed lot a number of seeds were planted in vector-free greenhouses. Emerging plants with virus symptoms were discarded, and a minimum of 24 phenotypically representative, healthy-looking plants were selected per seed lot as mother plants for seed production. These plants were enzyme-linked immunosorbent assay (ELISA)-tested one or more times during their growth, and seeds harvested from virus-free plants were deposited in germplasm collections as virus-free accessions. During 1988–1991 this work resulted in elimination of PSbMV from 2700 PI accessions of *P. sativum*. According to the authors, the method used led to only minimal losses of genetic diversity in the germplasm accessions.

Similar efforts for establishing virus-free accessions of germplasm seed have been made, e.g. in the USA for cowpea (Gillaspie *et al.*, 1995) and in Brazil for groundnut (Pio-Ribeiro *et al.*, 2000). Unfortunately, plantings with virus-free seeds, in regions with widespread incidence of viruses that are both seed-borne and insect-borne, may also quickly succumb to virus infection and serious crop losses.

Quarantine for seed-transmitted viruses is also discussed in Sections 3.2.2 and 7.3.

References

Agarwal, V.K. and Sinclair, J.B. (1988) *Principles of Seed Pathology*, Vols I and II. CRC Press, Boca Raton, Florida.

Bennett, C.W. (1969) Seed transmission of plant viruses. *Advances in Virus Research* 14, 221–261.

Bercks, R. (1967) Methodische Untersuchungen über den serologischen Nachweis pflanzenpathogener Viren mit dem Bentonit-Flockungs-Test, dem Latex-Test und dem Bariumsulfat-Test. *Phytopathologische Zeitschrift* 58, 1–17.

Bercks, R. and Querfurth, G. (1969) Weitere methodische Untersuchungen über den Latextest zum serologischen Nachweis pflanzenpathogener Viren. *Phytopathologische Zeitschrift* 65, 243–256.

Bos, L. (1977) Seed-borne viruses. In: Hewitt, W.B. and Chiarappa, L. (eds) *Plant Health and Quarantine in International Transfer of Genetic Resources.* CRC Press, Boca Raton, Florida, pp. 39–69.

Bwye, A.M., Jones, R.A.C. and Proudlove, W. (1994) Effects of sowing seed with different levels of infection, plant density and the growth stage at which plants first develop symptoms on cucumber mosaic virus infection of narrow-leafed lupins (*Lupinus angustifolius*). *Australian Journal of Agricultural Research* 45, 1395–1412.

Candresse, T., Hammond, R.W. and Hadidi, A. (1998) Detection and identification of plant viruses and viroids using polymerase chain reaction (PCR). In: Hadidi, A., Khetarpal, R.K. and Koganezawa, H. (eds) *Plant Virus Disease Control.* APS Press, St Paul, Minnesota, pp. 399–416.

Carroll, T.W. (1983) Certification schemes against barley stripe mosaic. *Seed Science and Technology* 11, 1033–1042.

Crowley, N.C. (1957) Studies on the seed transmission of plant virus diseases. *Australian Journal of Biological Sciences* 10, 449–464.

Demski, J.W., Reddy, D.V.R., Sowell, G. and Bays, D. (1984) Peanut stripe virus – a new seed-borne potyvirus from China infecting groundnut (*Arachis hypogaea*). *Annals of Applied Biology* 105, 495–501.

Derrick, K.S. (1973) Quantitative assay for plant viruses using serologically specific electron microscopy. *Virology* 56, 652–653.

Doolittle, S.P. and Gilbert, W.W. (1919) Seed transmission of cucurbit mosaic by the wild cucumber. *Phytopathology* 9, 326–327.

Fajardo, T.G. (1930) Studies on the mosaic disease of the bean (*Phaseolus vulgaris*). *Phytopathology* 20, 469–494.

Fulton, R.W. (1964) Transmission of plant viruses by grafting, dodder, seed and mechanical inoculation. In: Corbett, M.K. and Sisler, H.D. (eds) *Plant Virology.* University of Florida Press, Gainesville, Florida, pp. 39–67.

Gillaspie, A.G., Jr, Hopkins, M.S., Pinnow, D.L. and Hampton, R.O. (1995) Seedborne viruses in preintroduction cowpea seed lots and establishment of virus-free accessions. *Plant Disease* 79, 388–391.

Goodman, R.M. and Oard, J.H. (1980) Seed transmission and yield losses in tropical soybeans infected by soybean mosaic virus. *Plant Disease* 65, 913–914.

Grogan, R.G. (1980) Control of lettuce mosaic with virus-free seed. *Plant Disease* 64, 446–449.

Hamilton, R.I. (1965) An embryo test for detecting seed-borne barley stripe mosaic virus in barley. *Phytopathology* 55, 798–799.

Hampton, R.E., Sill, W.H. and Hansing, E.D. (1957) Barley stripe mosaic virus in Kansas and its control by a greenhouse seed lot testing technique. *Plant Disease Reporter* 41, 735–740.

Hampton, R.O. (1983) Seed-borne viruses in crop germplasm resources: disease dissemination risks and germplasm-reclamation technology. *Seed Science and Technology* 11, 535–546.

Hampton, R.O. (1986) Geographic origin of pea seedborne mosaic virus: an hypothesis. *The Pisum Newsletter* 18, 22–26.

Hampton, R.O. and Braverman, S.W. (1979) Occurrence of pea seedborne mosaic virus in North American pea breeding lines, and new virus-immune germplasm in the Plant Introduction collection of *Pisum sativum. Plant Disease Reporter* 79, 631–633.

Hampton, R.O., Kraft, J.M. and Muehlbauer, F.J. (1993) Minimizing the threat of seedborne pathogens in crop germ plasm: elimination of pea seedborne mosaic virus from the USDA-ARS germ plasm collection of *Pisum sativum. Plant Disease* 77, 220–224.

Hull, R. (2002) *Matthews' Plant Virology*, 4th edn. Academic Press, London.

Johansen, E., Edwards, M.C. and Hampton, R.O. (1994) Seed transmission of viruses: current perspectives. *Annual Review of Phytopathology* 32, 363–386.

Jones, D.R. (1987) Seedborne diseases and the international transfer of plant genetic resources: an Australian perspective. *Seed Science and Technology* 15, 765–776.

Jones, R.A.C. (2000) Determining 'threshold' levels for seed-borne virus infection in seed stocks. *Virus Research* 71, 171–183.

Jones, R.A.C., Fosu-Nyarko, J., Jones, M.G.K. and Dwyer, G.I. (2001) Subterranean clover mottle virus. *AAB Descriptions of Plant Viruses* No. 387. Available at: http://www.dpvweb.net

Joshi, S., Shrestha, S.K., Albrechtsen, S.E. and Mathur, S.B. (2004) *Pea seed-borne mosaic virus* in lentil seed samples of Nepal. In: Proceedings of the Fourth National Conference on Science and Technology, 23–26 March 2004. Royal Nepal Academy of Science and Technology (RONAST), Kathmandu, Nepal (in press).

Kaiser, W.J. and Mossahebi, G.H. (1974) Natural infection of mungbean by bean common mosaic virus. *Phytopathology* 64, 1209–1214.

Kendrick, J.B. and Gardner, M.W. (1924) Soybean mosaic virus: seed transmission and effect on yield. *Journal of Agricultural Research* 27, 91–98.

Khetarpal, R.K. and Maury, Y. (1987) Pea seed-borne mosaic: a review. *Agronomie* 7, 215–224.

Lister, R.M. and Murant, A.F. (1967) Seed transmission of nematode-borne viruses. *Annals of Applied Biology* 59, 49–62.

Lundsgaard, T. (1976) Routine seed health testing for barley stripe mosaic virus in barley seeds using the latex-test. *Zeitschrift für Pflanzenkrankheiten und Pflanzenschutz* 83, 278–283.

Makkouk, K.M. and Kumari, S.G. (1990) Variability among 19 lentil genotypes in seed transmission rates and yield loss induced by broad bean stain virus infection. *LENS Newsletter* 17, 31–32.

Mandahar, C.L. (1981) Virus transmission through seed and pollen. In: Maramorosch, K. and Harris, K.F. (eds) *Plant Diseases and Vectors: Ecology and Epidemiology.* Academic Press, New York, pp. 241–292.

Manento, B.B., Mtolera, C.C., Mwakitwange, F.F., Mallya, J., Horombo, L.M.N., Wakangwale, M., Ngwediagi, P.S. and Maingu, T. (eds) (2002) *Seed Technology.* Technical Handbook No. 8, Department of Training Institutes, Instructional Materials Development Unit/ASPS, Ministry of Agriculture and Food Security, Dar es Salaam, Tanzania.

Marrou, J. and Messiaen, C.M. (1967) The *Chenopodium quinoa* test: a critical method for detecting seed transmission of lettuce mosaic virus. *Proceedings of the International Seed Testing Association* 32, 49–57.

Mathur, S.B. (1995) Some aspects of seed pathology that deserve immediate attention. *Indian Journal of Mycology and Plant Pathology* 25, 13–24.

Maude, R.B. (1996) *Seedborne Diseases and Their Control – Principles and Practice.* CAB International, Wallingford, UK.

McKinney, H.H. (1954) Culture methods for detecting seed-borne virus in Glacier barley seedlings. *Plant Disease Reporter* 38, 152–162.

Mink, G.J. (1993) Pollen and seed-transmitted viruses. *Annual Review of Phytopathology* 31, 375–402.

Neergaard, P. (1977) *Seed Pathology.* Macmillan Press, London.

Newhall, H.G. (1923) Seed transmission of lettuce mosaic. *Phytopathology* 13, 104–106.

Pelet, F. (1965) Dosage du virus de la mosaïque de la laitue par indexage de la graine sur *Chenopodium quinoa* Willd. *Revue Horticole Suisse* 38, 7–10.

Pelet, F. and Gagnebin, F. (1963) Production de graines de laitue sans virus. *Revue Horticole Suisse* 36, 22–38.

Persley, D.M., McMichael, L. and Spence, D. (2001) Detection of *Peanut stripe virus* in post-entry quarantine in Queensland. *Australasian Plant Pathology* 30, 377.

Phatak, H.C. (1974) Seed-borne viruses – identification and diagnosis in seed health testing. *Seed Science and Technology* 2, 3–155.

Pio-Ribeiro, G., Wyatt, S.D. and Kuhn, C.W. (1978) Cowpea stunt: a disease caused by a synergistic interaction of two viruses. *Phytopathology* 68, 1260–1265.

Pio-Ribeiro, G., Andrade, G.P., Cruz, S.M., Santos, R.C., Reddy, D.V.R. and Andrade, A.A.M.X. (2000) Virus eradication from peanut germplasm based on serological indexing of seeds and analyses of seed multiplication fields. *Fitopatologia Brasileira* 25, 42–48.

Quantz, L. (1957) Ein Schalentest zum Schnellnachweis des Gewöhnlichen Bohnenmosaikvirus (*Phaseolus* virus I). *Nachrichtenblatt des Deutschen Pflanzenschutzdienstes* 9, 71–74.

Quantz, L. (1962) Zum Nachweis des Gewöhnlichen Bohnenmosaikvirus im Bohnensamen mit Hilfe des Schalentests. *Nachrichtenblatt des Deutschen Pflanzenschutzdienstes* 14, 49–54.

Reddick, D. and Stewart, V.B. (1918) Varieties of beans susceptible to mosaic. *Phytopathology* 8, 530–534.

Reddick, D. and Stewart, V.B. (1919) Transmission of the virus of bean mosaic in seed and observations on thermal death-point of seed and virus. *Phytopathology* 9, 445–450.

Riedle-Bauer, M., Suarez, B. and Reinprecht, H.J. (2002) Seed transmission and natural reservoirs of *Zucchini yellow mosaic virus* in *Cucurbita pepo* var. *styriaca*. *Journal of Plant Diseases and Protection* 109, 200–206.

Rohloff, I. (1962) Entwicklung einer Laboratoriumsmethode zur kurzfristigen Untersuchung von Salatsamen (*Lactuca sativa* L.) auf Befall mit Salatmosaikvirus (SMV). *Gartenbauwissenschaft* 27, 413–436.

Rohloff, I. (1966) Weitere Ergebnisse der Prüfung von Leuchtstoffröhren auf ihre Eignung für den Salatvirustest nach der Klimakammermethode. *Gemüse* 2, 38–40.

Rohloff, I. (1967) The controlled environment room test of lettuce seed for identification of lettuce mosaic virus. *Proceedings of the International Seed Testing Association* 32, 59–63.

Shukla, D.D., Ward, C.W. and Brunt, A.A. (1994) *The Potyviridae*. CAB International, Wallingford, UK.

Singh, R.P., Ready, K.F.M. and Nie, X. (2003) Biology. In: Hadidi, A., Flores, R., Randles, J.W. and Semancik, J.S. (eds) *Viroids*. CSIRO Publishing, Collingwood, Victoria, Australia, and Science Publishers, Enfield, New Hampshire, pp. 30–48.

Slack, S. and Shepherd, R.J. (1975) Serological detection of seed-borne barley stripe mosaic by a simplified radial-diffusion technique. *Phytopathology* 65, 948–955.

Smith, K.M. (1951) A latent virus in sugarbeets and marigolds. *Nature* 167, 1061.

Stace-Smith, R. and Hamilton, R.I. (1988) Inoculum thresholds of seedborne pathogens: viruses. *Phytopathology* 78, 875–880.

Steger, G. and Riesner, D. (2003) Molecular characteristics. In: Hadidi, A., Flores, R., Randles, J.W. and Semancik, J.S. (eds) *Viroids*. CSIRO Publishing Collingwood, Victoria, Australia, and Science Publishers, Enfield, New Hampshire, pp. 15–29.

Tu, J.C. (1989) Effect of different strains of soybean mosaic virus on growth, maturity, yield, seed mottling and seed transmission in several soybean cultivars. *Journal of Phytopathology* 126, 231–236.

Walkey, D. (1991) *Applied Plant Virology*. Chapman & Hall, London.

Westerdijk, J. (1910) Die Mosaikkrankheit der Tomaten. *Mededelingen Phytopathologisch Laboratorium Willie Comelin Scholten, Amsterdam* 1, 20.

2 Seed Transmission of Viruses

In most cases there is a high degree of protection of seed embryos against invasion by viruses affecting the mother plant. However, a number of viruses are able to pass from one generation to the next through seed. This phenomenon is more common in some plant families than in others. Thus, for some reason, two of three virus–host combinations resulting in documented viral transmission through seeds involve legume hosts (Hampton *et al.*, 1993). Of 105 seed-transmitted viruses presented by Mink (1993), 36 (34.3%) were seed-transmitted in leguminous hosts. Of 32 presented by Johansen *et al.* (1994), 14 (43.8%) were seed-transmitted in leguminous hosts. In contrast, very few viruses are seed-transmitted in members of *Poaceae*, for example. Some aspects of the location of viral inoculum in seeds, how it passes from the plant to the seeds and from seeds to the plant, and factors that affect seed transmission are discussed in the following sections.

2.1. Location of Inoculum in the Seed

Unlike other seed-transmitted pathogens, such as fungi, viral inoculum can, with very few exceptions, survive only if it enters the seed embryo. Infection of other parts of the seed early in seed development is likely to occur with most viruses that move from cell to cell through cytoplasmic strands (Bennett, 1969; Bos, 1977). During seed maturation, however, viral inoculum located outside the embryo generally becomes inactivated.

The very few instances where viruses survive outside the embryo are seeds contaminated with members of the tobamoviruses, e.g. *Cucumber green mottle mosaic virus* (CGMMV) and *Tomato mosaic virus* (ToMV) (see Section 4.4.3.). The only virus of other genera reported to survive outside the embryo is *Southern bean mosaic virus* (SBMV), belonging to *Sobemovirus* (Crowley, 1959;

McDonald and Hamilton, 1972; Uyemoto and Grogan, 1977, all quoted from Johansen *et al.*, 1994). For virus acronyms, see full names in Table 4.2.

It should be noted that embryo-transmissible viruses can also occur in an inactivated state in seed parts outside the embryo, e.g. endosperm and testa of seeds, both with and without embryo infection. For example, up to 50% of mature seeds of a pea seed lot infected with PSbMV have been shown to contain inactivated but detectable virus in the seed-coats, but only 2–3% of these seeds carried transmissible virus (Sample preparation, Section 5.2.2).

This phenomenon is common for many seed-transmitted viruses in many hosts and has an important bearing on seed-health testing, because inactivated (non-infectious) viruses are readily detectable by serological and NAB methods. Testing of whole seeds by these methods can, therefore, lead to a great over-estimation of the rate of seed transmission in a seed lot (Sample preparation in Section 5.2.2). There are only a few virus–host combinations for which the result of a whole-seed assay for viruses corresponds to the actual rate of seed transmission. Five of these cases, and probably the only ones known, are LMV in lettuce seed, *Peanut mottle virus* (PeMoV, *Potyvirus*) and the peanut stripe strain of BCMV in groundnut seed, SMV in soybean and CMV in lupin (Sample preparation, Section 5.2.2).

Most viruses located in the embryo remain viable as long as the seeds are viable (Bennett, 1969; Bos, 1977). Among the virus–host combinations listed by Bennett (1969), up to 14-year-old seed (*Chenopodium murale*) contained active virus (*Sowbane mosaic virus* (SoMV, *Sobemovirus*)). One virus, BCMV, was isolated from bean seeds after 30 years of storage (Pierce and Hungerford, 1929, quoted from Bos, 1977). Viable inoculum of *Potato spindle tuber viroid* (PSTVd, *Pospiviroid*) in 21-year-old true potato seed has also been reported (Section 4.5).

2.2. Plant-to-seed and Seed-to-plant Transmission

The review of Johansen *et al.* (1994) on seed transmission of viruses is, to the author's knowledge, the latest one that discusses several factors and theories related to how viruses pass from mother plant to seed and from seed to seedling. Of the two events, the virus passage from plant to seed embryo appears to be the most complex, or, as expressed by Maule and Wang (1996) in their review, 'a lesson in biological complexity'.

2.2.1. Plant-to-seed transmission

The ability to invade generative tissues (embryo, pollen) from vegetative tissue seems confined to certain virus genera, such as *Potyvirus*, *Nepovirus*, *Cucomovirus*, cryptoviruses, as well as to viroids. One of the characteristics for many of these pathogens is that they are relatively unstable outside the host, whereas others, such as the extremely stable and highly contagious *Tobacco mosaic virus* (TMV,

Tobamovirus), cannot reach the embryo although present in huge amounts in vegetative tissue throughout the host (Bos, 1999). This is probably true for some other tobamoviruses. No TMV has been detected in the embryo of tomato (Broadbent, 1965) or of *Arabidopsis thaliana* (de Assis Filho and Sherwood, 2000). This most probably applies also to ToMV, which was earlier considered a tomato strain of TMV. Another characteristic of seed-transmissible viruses, apart from the cryptoviruses, is that practically all are present in relatively high concentrations in vegetative host cells and are readily sap-transmissible from plant to plant, accounting for the easy detectability of most of these viruses.

In the exceptional group, cryptoviruses (family *Partitiviridae*, comprising the two genera, *Alphacryptovirus* and *Betacryptovirus* (Appendix 1)), 31 members are so far identified (Fauquet and Mayo, 1999; Van Regenmortel *et al.*, 1999), which cause no or very mild symptoms in their hosts and are considered of no economic importance. Their virions are low in concentration in host tissue, none of the viruses is sap- or graft-transmissible and no invertebrate vector is known; they appear to spread entirely and with high efficiency through pollen and seed (Hull, 2002).

Among viroids, Mink (1993) listed seven as being seed-transmitted. Since then an additional three grapevine viroids (Appendix 1) have been reported to be seed-transmitted (Wan Chow Wah and Symons, 1997, 1999).

Despite progress in the study of virus seed transmission in recent years, the reason that some viruses are transmitted through seeds and others are not is still not fully understood. Among perceived obstacles are difficulties associated with virus replication and movement in reproductive tissues, anatomically very difficult to investigate, as well as the lack of technologies needed for defining seed transmission limitations (Maule and Wang, 1996).

The present knowledge of seed transmission mechanisms appears to include the following:

1. (a) Most viruses are readily sap-transmissible, indicating an ability to invade parenchymatous tissue.

(b) Viruses transmitted by certain types of vectors are more often seed-transmitted than those transmitted by other types of vectors. Thus, viruses transmitted by leafhoppers and those transmitted by aphids in a persistent manner (e.g. most members of *Luteoviridae*) are not seed-transmitted, whereas those transmitted by nematodes, beetles and, in a non-persistent manner, by aphids may be seed-transmitted (Bennett, 1969; Stace-Smith and Hamilton, 1988).

2. There are two ways in which viruses can infect the developing embryo: by indirect invasion, i.e. by infection of the gametes before fertilization, or by direct embryo invasion, i.e. after fertilization. For many virus–host interactions, both modes of embryo infection may result in maximal seed transmission (Johansen *et al.*, 1994).

3. Virus survival during seed transmission requires its immunity to host ploidy changes (diploid to haploid to diploid) and its intact movement through both vegetative and reproductive tissues. If established in the embryo the virus must remain stable during seed maturation and storage and finally be activated during or after seed germination (Wang and Maule, 1992).

4. The difference between viruses in their capacity to enter meristematic tissue and remain viable, or in their incapacity to do so, may determine the specificity of seed transmission (Bos, 1999).
5. The earlier the mother plant becomes infected, the higher the level of seed transmission, whereas mother plant infection after flowering reduces the probability of transmission through seed. For a single exception to this generality, see Section 2.2.3.
6. The distribution of virus-infected seeds in infected mother plants tends to be erratic, at least for early infected plants, e.g. LMV in lettuce and some viruses in legumes, where distribution within pods was completely random (Bos, 1977; Hull, 2002).

Direct embryo invasion

Although most experimentation appears to suggest indirect entry of viruses into the embryo, evidence for direct embryo invasion during some stage of embryogenesis has also been reported. How certain viruses are able to reach the embryo is not fully understood. The apparent early separation of the egg and embryo sac from the maternal tissue should impede direct viral invasion of the embryo, because the separation results from the disappearance of the plasmodesmata and the formation of a callus layer and later a thick pectocellulosic wall (Johansen *et al.*, 1994). If established in the embryo, the virus must further survive the subsequent changes mentioned in **3** above.

The most recent studies of direct embryo invasion are those of Wang and Maule (1992, 1994, 1997), Maule and Wang (1996) and Roberts *et al.* (2003), which involved PSbMV in pea. For the virus strains and pea cultivars used in the study, seed transmission was assumed to occur because of direct embryo invasion from maternal tissue, because no pollen transmission was observed and no detectable virus occurred in pea ovules prior to fertilization (Wang and Maule, 1992). By studying the early stages of embryo development and virus locations in two pea cultivars, one permissive and one non-permissive of seed transmission, using techniques such as light microscopy, electron microscopy, *in situ* hybridization, RNA analysis and immuno-staining, they attempted to locate the 'window' through which virus could pass into the embryo.

Based on these studies, Wang and Maule (1994) proposed that the virus uses the embryonic suspensor as the route for the direct invasion of the embryo. The suspensor (Fig. 2.1) is a transient structure that provides nutritional and positional support for the embryo early in its development but later degenerates as the cotyledons expand. This degeneration effectively closes this hypothetical window for further direct invasion of the embryo. What supported the hypothesis that the suspensor functions as the window was the observation of virus accumulation in the suspensor and the fact that the embryo always became infected at the same primary location: the contact point between the suspensor and the embryonic radicle (Fig. 2.1). However, while plasmodesmata are present between the cells of the suspensor and between it and the embryo (Fig. 2.1), no such connections seemed to occur between the maternal tissue and the endosperm or between the endosperm and the suspensor (Wang and Maule, 1994; Maule and Wang, 1996).

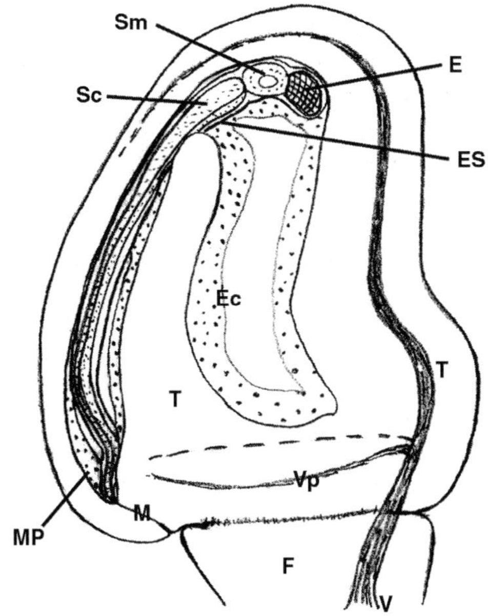

Fig. 2.1. Sketch of an immature pea showing the relative organization of the major structures. **E**, embryo; **Ec**, endosperm cytoplasm; **ES**, embryonic sheath; **F**, funiculus; **M**, micropyle; **MP**, micropyle pocket; **Sm**, suspensor middle cell; **Sc**, suspensor columnar cells; **T**, testa; **V**, main vascular strand; **Vp**, peripheral vascular strands. From Roberts *et al.* (2003) with permission from authors and the copyright holder © Protoplasma.

It appeared from these studies that the only possible entrance for virus into the suspensor was at the suspensor base in the micropylar region (Fig. 2.1), but how the virus or its genome could pass the boundaries from the infected testa to the suspensor via the endosperm without plasmodesmata remained unclear. Additional studies of cell and tissue ultrastructure were undertaken, particularly around the micropyle of young developing seeds, from plants infected with a seed-transmissible or a non-seed-transmissible PSbMV isolate (Roberts *et al.*, 2003). These studies supported the hypothesis of the suspensor base region as the entry point for the virus by revealing that symplastic connections apparently existed at the testa–endosperm boundary wall (although not directly observable, but indicated by the presence and positioning of viral cylindrical inclusion bodies (CIs), which are believed to play a role in the cell-to-cell viral movement). When closely examining the endosperm–suspensor boundary at the base of the suspensor, discontinuities (something like a kitchen colander) in the suspensor sheath wall were observed at certain seed development stages.

Virus accumulation and translocation were different for the two isolates studied, and in the case of the host that was non-permissive of seed transmission it was concluded (Wang and Maule, 1994) that the chance of virus accessing the base of the suspensor before the degeneration of the suspensor would be very low. This is probably because of the observed lower extent of PSbMV accumulation and/or spread from the vascular strands (Fig. 2.1) into the testa tissue compared to that of the seed transmission-permissive cultivar. The vegetative tissues of the two cultivars, however, were equally susceptible, suggesting that resistance genes in the non-permissive cultivar probably function in the testa tissue, resulting in the reduced extent of virus persistence and spread (Wang and Maule, 1994).

Although appearing plausible, the conclusions drawn from these comprehensive studies are, as also stated by the authors, still hypothetical. As previously mentioned, these processes are very difficult to investigate because, as noted by Johansen *et al.* (1994), literally hundreds of complex cellular/biochemical factors could interact with virions or viral RNA moving from infected maternal cells towards the ovule and a subsequently developing embryo.

Indirect embryo invasion

Evidence for indirect embryo invasion has been provided by ultrastructural studies (Johansen *et al.*, 1994). Virus has been observed in the megaspore mother cell and egg or in the pollen mother cells and pollen. Thus, *Tobacco rattle virus* (TRV, *Tobravirus*) has been reported to be present in pre-meiotic pollen mother cells and, later, in pollen, and AMV has been detected in the cytoplasm of lucerne pollen. Both viruses are transmitted via virus-infected pollen (Johansen *et al.*, 1994).

The high rate of seed transmission of *Tobacco ringspot virus* (TRSV, *Nepovirus*) in soybean is apparently related to the capacity of TRSV to invade meristematic tissue and infect the megaspore mother cell (Yang and Hamilton, 1974, quoted from Johansen *et al.*, 1994). In his review, Carroll (1981) reported his and co-workers' studies on the seed transmission of BSMV, demonstrating that the passage of BSMV from plant to seed was seemingly determined by the ability of this virus to invade male and female reproductive meristems very early in their development, thereby infecting the embryo indirectly. Wang and Maule (1997) found that embryos of pea plants infected with *Pea early-browning virus* (PEBV, *Tobravirus*) were uniformly infected with the virus from the earliest stages of embryo development, and they detected PEBV in the egg cell and pollen grains, indicating gametic transmission into the embryo.

De Assis Filho and Sherwood (2000) demonstrated the seed transmission of *Turnip yellow mosaic* (TYMV, *Tymovirus*) in *A. thaliana*, and studies of the seed transmission showed that indirect embryo invasion occurred. Crosses between healthy and infected plants resulted in TYMV passage to seed both from: (i) infected female × infected male; (ii) infected female × healthy male; and (iii) healthy female × infected male, indicating an indirect embryo invasion to occur, at least in (iii), but the rate of transmission to seeds in (iii) was only about 1/15 of that in (i) and (ii). It was also shown that, when seed embryos were infected via pollen (iii), the mother plant remained free of infection. This phenomenon has also been reported for other virus–host systems (e.g. Johansen *et al.*, 1994).

It is thus evident that some viruses can infect floral meristems and, hence, the gametes, but details of the mechanism are unknown (Hull, 2002).

Cases where viruses have been suggested to move from pollen-infected embryos into healthy maternal tissue are rare (Johansen *et al.*, 1994) and also complicated by the fact that virus carried in pollen can be transmitted in other ways (see below).

Pollen transmission and seed transmission

Pollen may be involved in transmission of viruses and viroids, either: (i) in their passage from plant to seed (vertical transmission); or (ii) in their spread from plant to plant (horizontal transmission); or (iii) in both.

VERTICAL TRANSMISSION. The involvement of pollen in plant-to-seed transmission has been shown for several viruses and viroids. This also includes the cryptoviruses, which are vertically transmitted but probably not horizontally (Mink, 1993). Experimental crosses among carrier and non-carrier parents demonstrated that cryptoviruses are transmitted through either pollen or ovules to approximately half or more of the progeny; when both parents are carriers, all progeny carry the virus (Mink, 1993).

The ilarvirus *Prunus necrotic ringspot virus* (PNRSV) has been detected in seedlings grown from nectarine (*Prunus persicae*) seeds and seedlings from healthy plants of several *Prunus* species after hand pollination with pollen from infected plants, strongly suggesting that fertilization could be a method for virus transmission (Mink, 1992). Aparicio *et al.* (1999) found large amounts of PNRSV in the cytoplasm of infected nectarine pollen, but none in the generative cell. They speculated, therefore, whether the virus could reach the egg cell either: (i) by infection of the sperm cells during or after mitosis, but before fertilization; or (ii) by being transmitted by the vegetative cytoplasm within the pollen, containing large amounts of the virus, and which is the origin of the pollen tube that carries the male sperm cells within it.

Of the ten viroids reported to be seed-transmitted (see Section 1.1 and above) at least five are also reported to be pollen-transmitted (Mink, 1993; Wall and Randles, 2003).

The frequency of virus transmission to seedlings through pollen is generally less than through ovules (Mink, 1993). Virus-infected pollen often cannot compete in tube growth with healthy pollen, as demonstrated for TRSV in soybean (Hamilton, 1985); still, pollen-to-seed transmission plays a significant role for a number of viruses, probably not least in self-pollinated plants (Johansen *et al.*, 1994).

HORIZONTAL TRANSMISSION. According to the review of Mink (1993), the number of viruses earlier reported to be transmitted via pollen from plant to plant have, after re-examination, been reduced (in 1993) to less than ten, including four members of *Ilarvirus*, three of *Nepovirus*, one of *Sobemovirus* and one of *Idaeovirus*. However, the hypothesis that the ilarviruses PNRSV and *Prune dwarf virus* (PDV) are transmitted from pollen to mother plants during fertilization is not probable after reports on thrips-assisted pollen transmission. Sdoodee and Teakle (1987) were the first to report that an ilarvirus, *Tobacco streak virus* (TSV), could be transmitted from plant to plant by thrips feeding on leaves of a healthy plant that were covered with virus-carrying pollen from another plant. Later, others of the viruses mentioned above were found to be transmissible in this same way (Mink, 1993), e.g. SoMV (Hardy and Teakle, 1992).

The role of honeybees in virus transmission through pollen has also been studied. The primary mechanism of plant-to-plant transmission of *Blueberry shock virus* (BlShV, *Ilarvirus*), which is seed-transmitted at low levels in blueberry, appeared to be the transfer of virus-contaminated pollen by honeybees from flowers of infected plants to flowers of healthy plants (Bristow and Martin, 1999). Okada *et al.* (2000) demonstrated that pollinating bumblebees could transmit TMV from infected to healthy tomato plants grown in greenhouses (the virus was identified as TMV, not ToMV).

According to Mink (1993) three of the five pollen-to-seed-transmitted viroids mentioned above were also shown to invade the mother plants, demonstrating that horizontal transmission can occur as a direct result of pollination. The mechanism by which these viroids move from the reproductive tissues to vegetative tissues was not known (Mink, 1993). The recent review on viroids (Hadidi *et al.*, 2003) appears to contain little additional information on seed and pollen transmission of viroids.

2.2.2. Seed-to-plant transmission

The infectivity of ToMV externally carried on tomato seed and CGMMV on cucumber seed has been reported to decline rapidly after 7 months to 1 year of seed storage (Bos, 1977). The extent of seed-to-plant transmission for viruses established in the embryo depends, first, on their survival during seed storage and, secondly, on their reactivation during or after seed germination. Thus, if viruses survive the physiological changes of seed maturation, most of those located in the embryo retain their infectivity for years, and often as long as the seeds remain viable. As previously mentioned (Section 2.1), infective BCMV was isolated from 30-year-old bean seeds. For transmission of virus from embryo to seedling, the presence of detectable virus in the embryo axis is normally the key determinant (Johansen *et al.*, 1994). However, not all seeds with detectable virus in the embryo axis produce infected progeny, as found, for example, by Varma *et al.* (1992) for *Blackgram mottle virus* (BMoV, *Carmovirus*) in *Vigna mungo*, which was also shown in the work of de Assis Filho and Sherwood (2000). In the latter case, a study of TYMV transmission through seeds of *A. thaliana* indicated slightly lower incidences in progeny from two infected seed lots (2.6 and 20.6%) than in the embryo of the seeds (5.4 and 24.5%).

The presence of infective virus in seeds seemed not to affect seed germination, according to Johansen *et al.* (1994) and other reports, e.g. on PSbMV in pea (Hampton 1972), SMV in soybean (Irwin and Goodman, 1981), BCMV in bean (Raizada *et al.*, 1990), PEBV in pea (Wang and Maule, 1997), TYMV in *A. thaliana* (de Assis Filho and Sherwood, 2000), and the blackeye cowpea mosaic strain of BCMV in catjang bean and yardlong bean (Hao *et al.*, 2003). A lower germination rate of seeds from infected plants may be due to disease effects on the mother plant rather than the presence of virus in the seed (Johansen *et al.*, 1994). Kazinczi and Horváth (1998) reported that SoMV, seed-transmitted in *Chenopodium* spp., caused a reduced viability of seeds from infected plants. The viability of seeds from TMV-infected *Solanum nigrum* plants was also significantly reduced compared to that of seeds from healthy plants (Kazinczi *et al.*, 2002).

Symptoms on seedlings raised from infected seeds vary substantially, ranging from severe to mild or complete absence, depending on the virus, virus strain, host species and cultivar and environmental conditions. Symptoms may develop in the cotyledons of emerging seedlings, as reported for some virus–host combinations (Bennett, 1969), but symptoms are often absent in cotyledons, becoming evident first at the primary leaf stage (Fig. 4.5). In some cases, symptoms may not occur until later stages of seedling development, as observed in seedlings grown from BCMV-infected mungbean (*Vigna radiata*) seeds, where virus symptoms at times did not appear until the second or third trifoliate leaf (Kaiser and Mossahebi, 1974). In seedlings from seed infected with Urdbean leaf crinkle virus (ULCV, unassigned), typical crinkle symptoms appeared first at the third trifoliate leaf stage (Pushpalatha *et al.*, 1999, and others).

2.2.3. Factors affecting seed transmission

The occurrence and extent of seed transmission depends on several factors, first of all virus species, virus strain, host species and cultivar, but also the time of mother plant infection and environmental conditions.

Virus and virus strain

As previously mentioned (Chapter 1 and Section 2.2.1), members of certain virus genera are more often seed-transmitted than others, while members of some genera have never been found to be seed-transmitted. A large number of the 40 *Nepovirus* members, at least 17, are able to pass through seeds to the next generation and many of them at frequencies up to 100% in certain hosts. Despite these high rates of seed transmission, nepoviruses are generally considered less harmful than, for instance, potyviruses. Thus, those known to be transmitted via nematodes are not very efficiently spread in the field compared to potyviruses, which spread via aphids and often cause high yield losses, and whose seed-transmitted members are also often seed-transmitted at high frequencies (Table 4.1).

Two important viruses affecting bean (*Phaseolus vulgaris*) are *Bean yellow mosaic virus* (BYMV, *Potyvirus*) and BCMV. Of these, only BCMV is transmitted through the seed of this host, whereas seed transmission of BYMV occurs in other of its host species, e.g. broad bean, lupin, pea and lentil.

Mixed infections may, apart from inducing a synergistic reduction of crop yield, also influence the rate of seed transmission. The double infection of cowpea with the non-seed-transmissible *Cowpea chlorotic mottle virus* (CCMV, *Bromovirus*) and the seed-transmissible SBMV resulted in 20% seed transmission of SBMV compared to the 12% observed for single infection of SBMV (Kuhn and Dawson, 1973). (Today, the cowpea-infecting strain of SBMV is considered a distinct virus, *Southern cowpea mosaic virus* (SCPMV)).

Seed transmission of TYMV in *A. thaliana* doubly infected with TYMV and the non-seed-transmissible TMV was more than doubled (70%) compared to TYMV alone (31%) (de Assis Filho and Sherwood, 2000). An opposite effect of double infection was reported for PEBV, which was seed-transmitted at a lower rate in faba bean when simultaneously inoculated with *Broad bean true mosaic virus* (BBTMV, *Comovirus*) than when inoculated with PEBV alone (Boulton, 1996). Seed transmission of PSbMV, on the other hand, was completely blocked in pea plants co-inoculated with PEBV and PSbMV, whereas PEBV seed transmission was unaffected (Wang and Maule, 1997).

Virus strains differ greatly, sometimes, in their ability to pass through seeds. A soybean isolate of TSV was transmitted in soybean in up to 30% of seeds, whereas a tobacco isolate of the same virus had zero transmissibility in soybean (Ghanekar and Schwenk, 1974, quoted from Bos, 1977). An investigation of the viral genetic basis for differences in seed transmission of two TSV isolates by Walter *et al.* (1995) revealed many minor RNA species in the mild, infrequently seed-transmitted isolate, which were not detected in the severe, seed-transmitted isolate. The isolate NY of PSbMV is rarely seed-transmitted (Johansen *et al.*, 1996) and was used in the previously mentioned study on PSbMV transmission (Roberts *et al.*, 2003). Strains of CMV differed from 36% to 0 in seed transmissibility in *P. vulgaris* (Hampton and Francki, 1992).

Host genotype

Host species and host cultivar have a marked effect on seed transmission. As previously mentioned, in the family *Leguminosae*, particularly, many virus–host combinations result in seed transmission; in cowpea, for example, at least 12 viruses are seed-transmitted (Frison *et al.*, 1990).

Considerable variation has been found within pea cultivars for PSbMV, with five cultivars showing no seed transmission (Stevenson and Hagedorn, 1973). The movement of PSbMV in testa tissues of a selected non-permissive cultivar is reviewed in Section 2.2.1. Cultivar-specific resistance to seed transmission was reported in barley against BSMV (Carroll *et al.*, 1979) and in lucerne against AMV (Bailiss and Offei, 1990). Likewise, *Bean common mosaic necrosis virus* (BCMNV, *Potyvirus*) is not seed-transmissible in some bean cultivars, simply because of a lethal effect of the virus on these cultivars.

Seed transmission rates of *Broad bean stain virus* (BBSV, *Comovirus*) varied between 0.2% and 32% in 19 lentil genotypes (Makkouk and Kumari, 1990). A similar variation in seed transmission for BBSV in lentil genotypes (93) was found in a later study (Al-Khalaf *et al.*, 2002). The seed transmissibility of seven SMV strains in eight soybean cultivars differed widely (from 0 to 70%), but none of the cultivars were resistant to seed transmission by all strains (Tu, 1989).

Time of infection and environmental factors

It is a general rule for viruses transmitted through embryos that: (i) the earlier plants are infected, the higher percentage of seeds that will transmit the virus; and (ii) no transmission will occur in plants infected after flowering (Bos, 1977; Hull, 2002). A single exception from this general rule appears to be BSMV in barley, where the percentage of infected seeds rose steadily as the time of infection was delayed, reaching a maximum 10 days before heading. After this time the percentage declined (Eslick and Afanasiev, 1955, quoted from Hull, 2002).

Research by Tomlinson and Carter (1970, quoted from Bos, 1977) indicated that there were fewer infected seeds (3 to 21%) from CMV-inoculated *Stellaria media* plants than from plants arising from CMV-infected seeds (21 to 40%).

Among environmental factors, temperature during plant growth was shown to influence the seed transmission of AMV in lucerne. Rates of seed transmission were higher when plants were grown at 18 or 24°C than when grown at 29°C (Frosheiser, 1974, quoted from Johansen *et al.*, 1994). The same author found, on the other hand, that 5 years of storage of infected lucerne seed seemed not to affect seed transmission of AMV whether stored at 21–27°C, 4°C or −18°C. The effect of temperature during plant growth on seed transmission of different nepoviruses was shown to differ not only between viruses but also between strains of the same virus. It was suggested that, at least in some cases, viruses are adapted to seed transmission in the different climatic conditions where they occur (Hanada and Harrison, 1977; Johansen *et al.*, 1994).

Viral determinants

Some viruses, including CMV, BSMV, PEBV, PSbMV and two nepoviruses, have been studied to determine portions of their genome that influence seed transmissibility (Johansen *et al.*, 1994; Hull, 2002). For both of two nepoviruses (bipartite genome) (Hanada and Harrison, 1977) and for CMV (tripartite genome) (Hampton and Francki, 1992), RNA 1 was found to be implicated as the primary determinant for seed transmissibility (both quoted from Johansen *et al.*, 1994). RNA 1 was also shown to be the major determinant of seed transmissibility of the bipartite PEBV (Wang *et al.*, 1997). Pseudorecombinants between BSMV strains (tripartite genome) ND18 (efficiently seed-transmitted) and CV17 (poorly seed-transmitted) demonstrated that the primary determinant of seed transmissibility was RNA 3 (Edwards, 1995). The same report also presents results of an analysis of RNA 3, made for more precisely mapping regions of the segment affecting seed transmissibility.

Such determinants were also reported for the monopartite genome of PSbMV (Johansen *et al.*, 1996). By constructing hybrids between a highly seed-transmitted and a poorly seed-transmitted PSbMV isolate, it was shown that the 5′ non-translated region, the helper-component proteinase (HC-Pro) and the coat protein regions of the genome affected seed transmission, with the HC-Pro having the major influence. These studies have also demonstrated that virus replication and/or movement may strongly influence seed transmissibility (Johansen *et al.*, 1994).

References

Al-Khalaf, M., Makkouk, K.M. and Kasem, A.H. (2002) Seed transmission of broad bean stain virus in lentil with respect to genotype variability and seed size. *Arab Journal of Plant Protection* 20, 106–110 (Abstr.).

Aparicio, F., Sánchez-Pina, M.A., Sánchez-Navarro, J.A. and Pallás, V. (1999) Location of Prunus necrotic ringspot Ilarvirus within pollen grains of infected nectarine trees: evidence from RT-PCR, dot blot and *in situ* hybridisation. *European Journal of Plant Pathology* 105, 623–627.

Bailiss, K.W. and Offei, S.K. (1990) Alfalfa mosaic virus in lucerne seed during seed maturation and storage, and in seedlings. *Plant Pathology* 39, 539–547.

Bennett, C.W. (1969) Seed transmission of plant viruses. *Advances in Virus Research* 14, 221–261.

Bos, L. (1977) Seed-borne viruses. In: Hewitt, W.B. and Chiarappa, L. (eds) *Plant Health and Quarantine in International Transfer of Genetic Resources*. CRC Press, Boca Raton, Florida, pp. 39–69.

Bos, L. (1999) *Plant Viruses, Unique and Intriguing Pathogens – a Textbook of Plant Virology*. Backhuys Publishers, Leiden, The Netherlands.

Boulton, R.E. (1996) Pea early-browning tobravirus. *Plant Pathology* 45, 13–28.

Bristow, P.R. and Martin, R.R. (1999) Transmission and the role of honeybees in field spread of blueberry shock ilarvirus, a pollen-borne virus of highbush blueberry. *Phytopathology* 89, 124–130.

Broadbent, L. (1965) The epidemiology of tomato mosaic, XI. Seed transmission of TMV. *Annals of Applied Biology* 56, 177–205.

Carroll, T.W. (1981) Seedborne viruses: virus–host interactions. In: Maramorosch, K. and Harris, K.F. (eds) *Plant Diseases and Vectors: Ecology and Epidemiology*. Academic Press, New York, pp. 293–317.

Carroll, T.W., Gossel, P.L. and Hockett, E.A. (1979) Inheritance of resistance to seed transmission of barley stripe mosaic virus in barley. *Phytopathology* 69, 431–433.

Crowley, N.C. (1959) Studies on the time of embryo infection by seed-transmitted viruses. *Virology* 8, 116–123.

de Assis Filho, F.M. and Sherwood, J.L. (2000) Evaluation of seed transmission of *Turnip yellow mosaic virus* and *Tobacco mosaic virus* in *Arabidopsis thaliana*. *Phytopathology* 90, 1233–1238.

Edwards, M.C. (1995) Mapping of the seed transmission determinants of barley stripe mosaic virus. *Molecular Plant–Microbe Interactions* 8, 906–915.

Eslick, R.F. and Afanasiev, M.M. (1955) Influence of time of infection with barley stripe mosaic on symptoms, plant yield and seed infection of barley. *Plant Disease Reporter* 39, 722–724.

Fauquet, M.C. and Mayo, M.A. (1999) Abbreviations for plant virus names – 1999. *Archives of Virology* 144, 1249–1273.

Frison, E.A., Bos, L., Hamilton, R.I., Mathur, S.B. and Taylor, J.D. (eds) (1990) *FAO/IBPGR Technical Guidelines for the Safe Movement of Legume Germplasm*. Food and Agriculture Organization of the United Nations, Rome/International Board for Plant Genetic Resources, Rome.

Frosheiser, F.I. (1974) Alfalfa mosaic virus transmission to seed through alfalfa gametes and longevity in alfalfa seed. *Phytopathology* 64, 102–105.

Ghanekar, A.M. and Schwenk, F.W. (1974) Seed transmission and distribution of tobacco streak virus in six cultivars of soybeans. *Phytopathology* 64, 112–114.

Hadidi, A., Flores, R., Randles, J.W. and Semancik, J.S. (eds) (2003) *Viroids*. CSIRO Publishing, Collingwood, Victoria, Australia, and Science Publishers, Enfield, New Hampshire.

Hamilton, R.I. (1985) Soybean bud blight: seed transmission of the causal virus. In: Shibles, R. (ed.) *Proceedings of the World Soybean Research Conference III*. Westview Press, Boulder, Colorado, pp. 515–522.

Hampton, R.O. (1972) Dynamics of symptom development of the seed-borne pea fizzletop virus. *Phytopathology* 62, 268–272.

Hampton, R.O. and Francki, R.I.B. (1992) RNA-1 dependent seed transmissibility of cucumber mosaic virus in *Phaseolus vulgaris*. *Phytopathology* 82, 127–130.

Hampton, R.O., Kraft, J.M. and Muehlbauer, F.J. (1993) Minimizing the threat of seedborne pathogens in crop germplasm: elimination of pea seedborne mosaic virus from the USDA-ARS germplasm collection of *Pisum sativum*. *Plant Disease* 77, 220–224.

Hanada, K. and Harrison, B.D. (1977) Effects of virus genotype and temperature on seed transmission of nepoviruses. *Annals of Applied Biology* 85, 79–92.

Hao, N.B., Albrechtsen, S.E. and Nicolaisen, M. (2003) Detection and identification of the blackeye cowpea mosaic strain of *Bean common mosaic virus* in seeds of *Vigna unguiculata* sspp. from North Vietnam. *Australasian Plant Pathology* 32, 505–509.

Hardy, V.G. and Teakle, D.S. (1992) Transmission of sowbane mosaic virus by *Thrips tabaci* in the presence and absence of virus-carrying pollen. *Annals of Applied Biology* 121, 315–320.

Hull, R. (2002) *Matthews' Plant Virology*, 4th edn. Academic Press, London.

Irwin, M.E. and Goodman, R.M. (1981) Ecology and control of soybean mosaic virus. In: Maramorosch, K. and Harris, K.F. (eds) *Plant Diseases and Vectors: Ecology and Epidemiology*. Academic Press, New York, pp. 182–220.

Johansen, E., Edwards, M.C. and Hampton, R.O. (1994) Seed transmission of viruses: current perspectives. *Annual Review of Phytopathology* 32, 363–386.

Johansen, E., Dougherty, W.G., Keller, K.E., Wang, D. and Hampton, R.O. (1996) Multiple viral determinants affect seed transmission of pea seedborne mosaic virus in *Pisum sativum*. *Journal of General Virology* 77, 3149–3154.

Kaiser, W.J. and Mossahebi, G.H. (1974) Natural infection of mungbean by bean common mosaic virus. *Phytopathology* 64, 1209–1214.

Kazinczi, G. and Horváth, J. (1998) Transmission of sowbane mosaic *Sobemovirus* by seeds of *Chenopodium* species and viability of seeds. *Acta Phytopathologica et Entomologica Hungarica* 33, 21–26.

Kazinczi, G., Horváth, J., Takács, A.P. and Pribék, D. (2002) Biological decline of *Solanum nigrum* L. due to *Tobacco mosaic tobamovirus* (TMV) infection. II. Germination, seed transmission, seed viability and seed production. *Acta Phytopathologica et Entomologica Hungarica* 37, 329–333.

Kuhn, C.V. and Dawson, W.O. (1973) Southern bean mosaic virus in single and double infections in cowpea. *Phytopathology* 63, 1380–1385.

Makkouk, K.M. and Kumari, S.G. (1990) Variability among 19 lentil genotypes in seed transmission rates and yield loss induced by broad bean stain virus infection. *LENS Newsletter* 17, 31–32.

Maule, J.A. and Wang, D. (1996) Seed transmission of plant viruses: a lesson in biological complexity. *Trends in Microbiology* 4, 153–158.

McDonald, J.G. and Hamilton, R.I. (1972) Distribution of southern bean mosaic virus in the seed of *Phaseolus vulgaris*. *Phytopathology* 62, 387–389.

Mink, G.I. (1992) Ilarvirus vectors. *Advances in Virus Vector Research* 9, 261–281.

Mink, G.I. (1993) Pollen- and seed-transmitted viruses and viroids. *Annual Review of Phytopathology* 31, 375–402.

Okada, K., Kusakari, S.-I., Kawaratani, M., Negoro, J.-I., Okhi, S.T. and Osaki, T. (2000) *Tobacco mosaic virus* is transmissible from tomato to tomato by pollinating bumblebees. *Journal of General Plant Pathology* 66, 71–74.

Pierce, W.H. and Hungerford, C.W. (1929) A note on the longevity of the bean mosaic virus. *Phytopathology* 19, 605–606.

Pushpalatha, K.C., Prakash, H.S., Albrechtsen, S.E., Shetty, H.S. and Mathur, S.B. (1999) Transmission of Urdbean leaf crinkle virus through urdbean seeds. *Seed Research* 27, 112–115.

Raizada, R.K., Albrechtsen, S.E. and Lange, L. (1990) Detection of bean common mosaic virus in dissected portions of individual bean seeds using immunosorbent electron microscopy. *Seed Science and Technology* 18, 559–565.

Roberts, M.I., Wang, D., Thomas, C.L. and Maule, A.J. (2003) Pea seed-borne mosaic virus seed transmission exploits novel symplastic pathways to infect the pea embryo and is, in part, dependent upon chance. *Protoplasma* 222, 31–43.

Sdoodee, R. and Teakle, D.S. (1987) Transmission of tobacco streak virus by *Thrips tabaci*: a new method of plant virus transmission. *Plant Pathology* 36, 377–380.

Stace-Smith, R. and Hamilton, R.I. (1988) Inoculum thresholds of seedborne pathogens: Viruses. *Phytopathology* 78, 875–880.

Stevenson, W.R. and Hagedorn, D.J. (1973) Further studies on seed transmission of pea seedborne mosaic virus in *Pisum sativum. Plant Disease Reporter* 57, 248–252.

Tomlinson, J.A. and Carter, A.L. (1970) Studies in the seed transmission of cucumber mosaic virus in chickweed (*Stellaria media*) in relation to the ecology of the virus. *Annals of Applied Biology* 66, 381–386.

Tu, J.C. (1989) Effect of different strains of soybean mosaic virus on growth, maturity, yield, seed mottling and seed transmission in several soybean cultivars. *Journal of Phytopathology* 126, 231–236.

Uyemoto, J.K. and Grogan, R.G. (1977) Southern bean mosaic virus: evidence for seed transmission in bean embryos. *Phytopathology* 67, 1190–1196.

Van Regenmortel, M.H.V., Fauquet, C.M., Bishop, D.H.L., Carstens, E., Estes, M.K., Lemon, S., Maniloff, J., Mayo, M.A., McGeoch, D., Pringle, C.R. and Wickner, R.B. (eds) (1999) *Virus Taxonomy.* Seventh Report of the International Committee on Taxonomy of Viruses, Academic Press, New York.

Varma, A., Krishnareddy, M. and Malathi, V.G. (1992) Influence of the amount of blackgram mottle virus in different tissues on transmission through the seeds of *Vigna mungo. Plant Pathology* 41, 274–281.

Wall, G.C. and Randles, J.W. (2003) Coconut tinangaja viroid. In: Hadidi, A., Flores, R., Randles, J.W. and Semancik, J.S. (eds) *Viroids.* CSIRO Publishing, Collingwood, Victoria, Australia, and Science Publishers, Enfield, New Hampshire, pp. 242–245.

Walter, M.H., Wyatt, S.D. and Kaiser, W.J. (1995) Comparison of the RNAs and some physicochemical properties of the seed-transmitted tobacco streak virus isolate Mel 40 and the infrequently seed-transmitted isolate Mel F. *Phytopathology* 85, 1394–1399.

Wan Chow Wah, Y.F. and Symons, R.H. (1997) A high sensitivity RT-PCR assay for the diagnosis of viroids in grapevines in the field and in tissue culture. *Journal of Virological Methods* 63, 57–69.

Wan Chow Wah, Y.F. and Symons, R.H. (1999) Transmission of viroids via grape seeds. *Journal of Phytopathology* 147, 285–291.

Wang, D. and Maule, A.J. (1992) Early embryo invasion as a determinant in pea of the seed transmission of pea seed-borne mosaic virus. *Journal of General Virology* 73, 1615–1620.

Wang, D. and Maule, A.J. (1994) A model for seed transmission of plant virus: genetic and structural analyses of pea embryo invasion by pea seed-borne mosaic virus. *The Plant Cell* 6, 777–787.

Wang, D. and Maule, A.J. (1997) Contrasting patterns in the spread of two seed-borne viruses in pea embryos. *Plant Journal* 11, 1333–1340.

Wang, D., MacFarlane, S.A. and Maule, A.J. (1997) Viral determinants of pea early browning seed transmission in pea. *Virology* 234, 112–117.

Yang, R.F. and Hamilton, R.I. (1974) The mechanism of seed transmission of tobacco ringspot virus in soybean. *Virology* 62, 26–37.

3 Ecology, Epidemiology and Control

3.1. Ecology and Epidemiology

A prerequisite of the control of plant viruses is, as for other plant pathogens, knowledge of the ecology and epidemiology of individual viruses. *Ecology* describes the factors influencing the behaviour of an individual virus in a given physical situation (Hull, 2002). *Epidemiology* describes disease incidence as a function of time, virus and host, number of infection sources, type and number of vectors and distance from sources of infection (Bos, 1999).

The interplay between virus, host and environment can be rather complex for some viruses. Some aspects of seed transmission in the ecology and epidemiology of viruses are discussed below.

3.1.1. Viruses and virus–host interaction

Seed transmission plays a significant, though varying role in the ecology and epidemiology of viruses and viroids. For some viruses, hibernating in seeds is a prerequisite to survival from one season to the next. BSMV (see Table 4.2 for virus names in full) is an example of a virus depending nearly 100% on seed transmission for survival. This virus is carried in barley seed at high rates, but has no known vector, and no weed or wild grass comprises a significant reservoir of the virus (Stace-Smith and Hamilton, 1988). Another virus for which survival in seed is important is BCMV, which has a relatively narrow host range, but is seed-transmitted at high rates (up to 83% in French bean (Frison *et al.*, 1990)). Other viruses, such as CMV, are less dependent on survival in seed because CMV has one of the largest host ranges known among plant viruses. It therefore exists in many inoculum sources in or near crop fields.

PSTVd is transmitted through tubers and also through true potato seeds and is, therefore, of importance, together with a few other true-seed-transmitted potato viruses, in certification programmes that are dependent on propagations from true potato seeds. Among other primary hosts of PSTVd is tomato, in which it is also seed-transmitted (Diener and Raymer, 1971). Until the advent of sensitive detection methods for viroids, *Avocado sunblotch viroid* (ASBVd, *Avsunviroid*) has also been a problem in avocado plantations. A high percentage of seedlings from seeds of a diseased, but symptomless, parent tree can be infected but without symptoms. If infected seedlings are used as rootstocks, the scions budded or grafted on to them become infected with ASVd. When scions are also symptomless, the viroid can remain undetected and lead to serious losses (Stace-Smith and Hamilton, 1988).

New virus–host relationships involving seed transmission are continually being reported. One of the more recent cases is ZYMV, previously mentioned. Others are *Artichoke Italian latent virus* (AILV, *Nepovirus*) and *Artichoke latent virus* (ArLV, *Potyvirus*) in artichoke (Bottalico *et al.*, 2002), and *Olive latent virus 1* (OLV-1, *Necrovirus*) and *Cherry leafroll virus* (CLRV, *Nepovirus*) in olive (Saponari *et al.*, 2002).

Important reservoirs of inoculum apart from crop seeds are volunteer plants from previous crops and, not least, weeds in and near the crop. Many crop viruses are hosted by weeds, in which they may also be seed-transmitted. For instance, CMV is transmitted (at rates up to 40%) through seeds of *Stellaria media*, as first reported by Noordam *et al.* (1965, quoted from Bos, 1981), and may persist in the seeds buried in the soil for at least 21 months (Tomlinson and Walker, 1973, quoted from Hull, 2002). If 10^7 seeds were present per hectare, with 1% infected with CMV, then a 10% emergence of seedlings would give rise to about one infected seedling per m^2. Such conditions could, in the presence of aphid vectors, lead to a rapid build-up of CMV infection of crop plants (Hull, 2002). High rates of seed transmission of nematode-borne viruses (up to 100%) were found in a number of weed species by Lister and Murant (1967). *Plum pox virus* (PPV, *Potyvirus*) is probably not seed-transmitted in fruit trees (Myrta and Savino, 1998), but Slovakova *et al.* (2002) found that *Nicotiana* spp. inoculated with PPV could transmit the virus through seeds at rates up to 19%, and speculated that these annual species could serve as potential inoculum sources for the spread of virus to new plantations of stone fruit trees by aphid transmission. In a study in Egypt of five viruses, seed-transmitted in faba bean, at least five weed species, naturally occurring in faba bean fields, were shown to host one or more of these viruses (El-Hammady *et al.*, 2004). The importance of wild plants in virus ecology has been recognized for many years, and a lot of information was available at the time of the comprehensive review of Bos (1981). According to the same author (1999) there has been a growing awareness of these phenomena since 1981. Virus infection of wild hosts may often escape attention in the field, because viruses are very often asymptomatic in these hosts. This absence of symptoms is probably due to many hundreds or thousands of years of virus–host associations, leading to optimal adaptation of both viruses and hosts.

Viruses have been dispersed over long distances and for hundreds of years through the exchange of crop seeds and germplasm. Seeds, both of cultivated plants and their wild relatives, collected for gene banks may harbour undetected viruses, which may then be spread to growers and breeders worldwide.

Another risk is spread of viruses to healthy germplasm during seed regeneration, including infected germplasm of the gene banks (see also Section 7.3). A particular case of human influence on the introduction of new viruses to a region is that demonstrated in New Zealand (Matthews, 1991), one of the most geographically isolated countries and having diverse and modern agriculture and horticulture. Over the past 170 years, European immigrants have introduced a wide range of agricultural and horticultural crop plants into the country. Also a large number of weed species were inadvertently introduced. By 1990, 139 viruses had been recorded in New Zealand, all present in introduced species. Almost all of the viruses have been identified as those occurring elsewhere, mainly Europe or North America, and obviously most of these diseases could have arrived in vegetative tissues, e.g. tubers, corms, rootstocks and/or seed.

3.1.2. Vector transmission

Apart from being sap-transmissible, the majority of seed-transmitted viruses are also transmitted in the field by vectors. Exceptions are BSMV and tobamoviruses, which have no known vectors, and the cryptoviruses, which are neither sap- nor vector-transmitted. Also viroids have no known vectors (exception: PSTVd; see Aphid-transmitted viruses below). Vectors in descending order of general significance in the ecology and epidemiology of seed-transmitted viruses are: aphids, beetles, nematodes and fungi. Thrip transmission seems to play a role in only a few of the seed-transmitted viruses and that mainly in connection with infected pollen (Section 2.2.1); and whitefly transmission seems not to occur with seed-transmitted viruses, except in a single case, the carlavirus *Cowpea mild mottle virus* (CPMMV).

All members of a given virus genus are normally transmitted by only one type of vector and in one mode of transmission (Table 3.1).

Aphid-transmitted viruses

Aphids are the most important vectors of plant viruses, both in numbers of viruses they transmit and in their efficiency as transmitters. The latter is due to the fact that they multiply rapidly, often occur in high populations and have high mobility. Alfamo-, cucumo- and potyviruses are transmitted in a non-persistent manner, i.e. the aphids acquire the virus from parenchyma cells in a short time (seconds), persistence in the vector is very brief (usually minutes) and virus is often lost after the first probe on a new plant.

Among seed- and aphid-transmitted viruses, *Pea enation mosaic virus* (PEMV) constitutes an exception, being transmitted in a persistent (circulative) manner. The disease is caused by the obligate symbiosis between PEMV-1 (*Enamovirus*, family *Luteoviridae*) and PEMV-2 (*Umbravirus*). Some isolates contain a third, non-essential satellite RNA (Skaf and de Zoeten, 2000). One seed-transmitted viroid, PSTVd, is reported to be aphid-transmitted from plants that are co-infected with *Potato leafroll virus* (PLRV, *Polerovirus*). The PSTVd RNA was 'transencapsidated' by PLRV coat proteins (Singh *et al.*, 2003).

Table 3.1. Virus genera with important seed-transmitted members and their vectors.

Virus genus	Example	Vector	Mode of transmission
Alfamovirus	*Alfalfa mosaic virus*	Aphids	Non-persistent
Cucumovirus	*Cucumber mosaic virus*	Aphids	Non-persistent
Potyvirus	*Bean common mosaic virus*	Aphids	Non-persistent
Pospiviroid	*Potato spindle tuber viroid*	Aphids[a]	Persistent
Comovirus	*Cowpea mosaic virus*	Beetles	Semi-persistent
Sobemovirus (most)	*Southern bean mosaic virus*	Beetles	Semi-persistent
Carlavirus	*Cowpea mild mottle virus*	Whitefly[b]	Semi-persistent
Nepovirus	*Tobacco ringspot virus*	Nematodes	Semi-persistent
Tobravirus	*Pea early browning virus*	Nematodes	Semi-persistent
Ilarvirus	*Tobacco streak virus*	Thrips[c]	Pollen/mechanical
Pecluvirus	*Peanut clump virus*	Fungus	*In vivo* acquisition[d]

[a]Only from plants co-infected with *Potato leafroll virus.*
[b]Exception; carlaviruses normally aphid-transmitted.
[c]In thrips-assisted pollen transmission.
[d]Other fungus-transmitted viruses have *in vitro* virus–vector relationship (Campbell, 1996).

According to Hull (2002) 192 aphid species can transmit one or more viruses, and 275 viruses are reported to be transmitted by these species. In contrast to those persistently transmitted, of which none is known to be seed-transmitted (exceptions: PEMV and PSTVd + PLRV), the virus–vector specificity of non-persistently transmitted viruses is low. For example, SMV is reportedly vectored by 24 aphid species (Irwin and Goodman, 1981) and CMV by more than 60 species (Hull, 2002). The black bean aphid (*Aphis fabae*) transmits a high number of viruses, but a higher number, probably three times as many, are transmitted by the green peach aphid (*Myzus persicae*). The latter species is very polyphagous, attacking more than 400 plant species (Bos, 1981), and is an efficient virus transmitter.

Importantly, non-persistent transmission of viruses can only take place over short distances, i.e. from sources within or near the field, whereas persistently transmitted viruses can be transported over long distances.

Beetle-transmitted viruses

Numerous species belonging to Coleoptera, especially leaf beetles (Chrysomelidae) are known as virus vectors. About 30 viruses are reportedly transmitted by 48 species in this family (Hull, 2002). Members of four virus genera, including *Comovirus* and *Sobemovirus*, are transmitted by beetles belonging to Chrysomelidae as well as Coccinellidae (ladybird beetles), Curculionidae (weevils) and Meloidae (blister beetles). The manner of transmission is semi-persistent, as beetles acquire the virus quickly, there is no latency period and the retention (transmission) period varies from about 1 to 10 days or sometimes longer, depending on the beetle species. There is a substantial degree of specificity among viruses and vector beetles (Hull, 2002).

A comovirus that has caused great and increasing losses in soybean in the USA in recent years is *Bean pod mottle virus* (BPMV). The virus is transmitted by several leaf-feeding beetles, but is seed-transmitted in soybean only at low levels (< 0.1%) (Giesler *et al.*, 2002). The same authors suggest, therefore, that inoculum sources other than seeds may be involved.

Whitefly transmission

The majority, if not all, of the begomoviruses (family *Geminiviridae*) are transmitted by the whitefly *Bemisia tabaci* (and *Bemisia argentifolia*) in a persistent, circulative manner (Hull, 2002). Whiteflies are efficient transmitters and are vectors of important viruses, such as the devastating *Tomato yellow leaf curl virus* (TYLCV, *Begomovirus*). Probable seed transmission is reported for only one of the geminiviruses (Brunt *et al.*, 1996; Hull, 2002). However, according to a later report, no geminiviruses are believed to be transmitted through seed (Briddon, 2003).

The seed-transmitted CPMMV belongs to the genus *Carlavirus*, whose members are all aphid-transmitted in a non-persistent manner, except CPMMV, which is whitefly-transmitted but in a non-persistent way (Jeyanandarajah and Brunt, 1993; Naidu *et al.*, 1998).

Transmission by nematodes

Viruses of two genera, of which many are seed-transmitted, are transmitted by soil-dwelling nematodes. Tobraviruses are transmitted by *Trichodorus* and *Paratrichodorus* species and nepoviruses by *Xiphinema* and *Longidorus* species; however, only about one-third of the nepoviruses are transmitted by nematodes. For the greater portion of nepoviruses, there are no reports of virus vectors; some, however, are reportedly pollen-transmitted. TRSV is, as an exception, reportedly transmitted by nematodes, thrips and aphids (Brunt *et al.*, 1996; Hull, 2002). Transmission by nematodes somewhat resembles the semi-persistent transmission by arthropods, i.e. the virus is acquired quickly, there is no latency period and the retention period varies from 12 weeks to more than a year. Only relatively few of the several hundreds of nematode species are capable of transmitting viruses, and among them there is a rather high degree of virus–vector specificity (Bos, 1999; Hull, 2002).

Transmission by thrips

About 5000 species of thrips exist, but only ten species, all in the family Thripidae, are known to transmit plant viruses. Thrips are important as efficient transmitters of members of *Tospovirus*, but apparently of limited importance for seed-transmitted viruses. While tospoviruses are transmitted in a persistent, propagative manner (Hull, 2002), some members of *Carmovirus*, *Ilarvirus* and *Sobemovirus* are spread by a kind of mechanical transmission, via pollen by 'thrips assistance' (Section 2.2.1). The nepovirus TRSV is also reportedly thrips-transmitted in a non-specific way (Brunt *et al.*, 1996).

Transmission by fungi

Fungal transmission of viruses has been either proved or suggested for certain members of some virus genera and for all members of other genera, totalling about 30 viruses in ten genera (Campbell, 1996). Among these are the seed-transmitted viruses *Peanut clump virus* (PCV) and *Indian peanut clump virus* (IPCV), both pecluviruses, *Soil-borne wheat mosaic virus* (SBWMV, *Furovirus*) and *Melon necrotic spot virus* (MNSV, *Carmovirus*). The first three viruses are transmitted by *Polymyxa graminis* and the latter by *Olpidium bornavanus*. Both of these fungi, as well as three other known virus-transmitting fungi, are obligate plant root parasites. The terms 'persistent' and 'non-persistent' are not appropriate for transmission by fungal vectors because it is based on virus survival in *dormant* resting spores, in contrast to the usual definition based on retention of virus by *feeding* vectors. For some viruses an *in vitro* acquisition is involved, where the virions are adsorbed to the surface of zoospores but are not located within the resting spores. The virions of other viruses become located within the fungal resting spores by *in vivo* acquisition (Campbell, 1996).

For the role of pollen in horizontal transmission, see Section 2.2.1. Details on vectors and virus–vector relationships can be found in, for example, Campbell (1996), Bos (1999) and Hull (2002).

3.1.3. Environment and cultural practices

The environment influences both viroid– and virus–host interactions, but has greater effects on vector activities, especially on those of the airborne vectors. Temperature may differently influence the host and virus replication and virulence, whereas replication and symptom development are generally enhanced for viroids as temperature increases above 20°C, at least to 35°C, probably resulting in more prevalent viroid diseases in warmer field climates (Singh *et al.*, 2003).

Low temperatures, rainfall and strong winds reduce multiplication and movement of aphids. The direction and velocity of wind also influence the direction of their movement (their transmission of viruses is impeded by high-velocity winds). Nematode transmission of viruses is more pronounced in light than in heavy soils (Hull, 2002). Wet and heavy soils generally favour fungal transmission of viruses (Teakle, 1988). In pepper crops grown under plastic tunnels, 84% of virus-infected plants contained TMV, whereas 88% of virus-infected pepper plants grown outdoors were infected by viruses transmitted by aphids in a non-persistent manner. Thus, the two types of environment seemed to favour different viruses to prevail (Conti and Masenga, 1977, quoted from Hull, 2002).

Environmental conditions affecting the rate of seed transmission are dealt with in Section 2.2.3.

Tropical climates

The ecology and epidemiology of plant viruses in tropical and subtropical regions differ greatly from those in temperate regions. In warm climates as compared to temperate climates: (i) the vector pressure is greater, often continuing all year round; (ii) the number of vector species and types is higher; (iii) a greater diversity of crops is grown throughout the year; (iv) the wild flora is richer in species, often present throughout the year and potentially providing more inoculum sources; and (v) temperature-sensitive diseases like those induced by viroids cause more severe crop damage.

Thottappilly (1992) lists more than 100 viruses reported to infect more than 30 field and vegetable crops in Africa. Of these, at least 25 are seed-transmitted. The list, which also shows the diseases' geographical distribution in Africa and mode of transmission, would have been longer had fruit tree viruses been included. The incidence of viroids in Africa is probably far greater than the single viroid listed; however, according to more recent information (da Graca and van Vuuren, 2003), there are still relatively few, mainly viroids of citrus, potato and avocado, reported from Africa compared to the rest of the world.

The great crop diversity in the subtropics and tropics includes in particular many legume species: soybean, cowpea, French bean, groundnut, mung bean, bambara groundnut, broad bean, pea, lens, lupin, etc. that are attacked by several viruses, of which many are seed-transmitted, at least 12 in cowpea, for example (Frison *et al.*, 1990).

Due to the factors (i)–(v) mentioned above, the chance for severe virus epidemics is considerably higher in these zones than in the temperate zones. Research and management efforts are therefore particularly needed in these climates. The fact that most countries in the subtropics and tropics are developing countries unfortunately dictates that only scarce resources are available for the investigation and management of plant diseases in this large part of the world. This is particularly so for virus and viroid diseases, whose detection and investigation are more resource-demanding than those of, for example, fungal diseases.

Cultural practices

The ecology and epidemiology of viruses are considerably influenced by growers' cropping practices. As mentioned (Chapter 1), the sowing of seed with even traces of virus infection may lead to early outbreaks of epidemics. For aphid-borne viruses, if the vector population is high at the time of plant emergence and inoculum sources are present, e.g. seedlings from infected seed, early and therefore devastating epidemics may occur.

Planting of a new crop near an overwintering crop infected with viruses to which the new crop is susceptible may also lead to early disease attacks. For example BYMV-infected clover in pastures in the neighbourhood of new cultures of lupin served as inoculum sources for aphid-borne BYMV infection of lupin (Jones, 2001).

In warm climates, where two or three crops per year may be grown in the same field, virus inoculum may persist, facilitated by irrigation, from which the new crop becomes infected, especially if the same crop species is grown successively. For instance, growing tomato and pepper crops successively in the same field endangers the second crop to *Tobamovirus* attack. Apart from being seed-transmitted, these very stable viruses can persist in plant debris in the soil for several months. In Taiwan, at least 6–7 months were advised between tomato crops to avoid ToMV infection from the soil (Green *et al.*, 1987). Because ToMV also infects pepper, pepper plantings should not follow tomato crops. In contrast, the pepper-infecting tobamovirus *Pepper mild mottle virus* (PMMoV) does not infect tomato (Green and Kim, 1991). The losses caused by ToMV in tomato may be even heavier in case of co-infection with the devastating (but non-seed-transmitted) TYLCV, a virus that, as well as other begomoviruses, is becoming more and more common in the tropics and subtropics (Bos, 1999).

Growing the same crop year after year in a field, if susceptible to nematode- or fungal-transmitted viruses, may lead to the build-up of soil-borne inoculum, particularly nepo- and tobraviruses, which are highly seed-transmitted in different weed species; thus also leading to a flora of infected weeds in the field. Nematode and fungal vectors move slowly in the soil, but may be spread from point sources by movement of soil during soil ploughing or cultivation.

3.2. Control Strategies

Because the replication of viruses is intimately associated with the metabolism of the host, virtually no cure of virus-infected plants or virus-infected seed embryos is possible without damaging host tissues. There are reports on reduction of embryo-located viral inoculum in seeds by heat treatment (Section 3.2.2) but, generally, the only way to control plant viruses is by preventive means. General methods of preventive plant virus control are reviewed by, for example, Hadidi *et al.* (1998), Bos (1999) and Hull (2002). Although complete elimination of viruses is rarely possible, their damage to crops can be reduced even by the use of simple measures. Below, some ways to control or reduce the damage of seed-transmitted viruses are briefly discussed.

3.2.1. Control measures at farm level

Avoidance of inoculum

Obviously, use of virus-free seed is an effective way to avoid inoculum of a range of seed-transmitted viruses. Use of seed free of or with low virus incidence is especially useful to control viruses with relatively narrow host ranges, such as BCMV, SMV and comoviruses, whereas it is less effective for controlling CMV, for example. CMV has an extremely wide host range and therefore the chance of infection from inoculum sources in or near the field is generally much higher. Elimination from seed by means of chemical or heat treatment is possible for tobamoviruses on, for example, tomato and pepper seed (Section 4.4.3). As a rule, avoidance of all other seed-transmitted viruses is only possible by obtaining seed harvested from virus-free plants or from species or varieties with resistance to seed transmission, although elimination of such viruses from seed by heat treatment has been partially successful in some cases (Section 3.2.2).

Spatial isolation of new crops from older stands of the same plants or other plants that are virus-infected may protect the crop against viral inoculum.

An adequate crop rotation can prevent build-up of viruliferous soil-living vectors and, for example, tomato and pepper crop exposure to tobamoviruses persisting in plant debris.

Most seed-transmitted viruses are also mechanically transmissible from plant to plant, but probably none as easily as tobamoviruses; therefore, to avoid spread of these viruses in, for example, tomato and pepper cultures, hygienic precautions must be taken during pruning and picking, etc. (see also Section 4.4.3, Protocol II).

Early removal of inoculum sources in or near the crop, such as volunteer plants, infected crop plants and weeds, are also means of preventing epidemic build-ups. This should be repeated a number of times to be effective. However, roguing of infected plants is not always easy, such as in dense plant stands or in cases of certain virus–host combinations. For example, lentil with seed-transmitted BBSV, PSbMV and BYMV usually produces mild or non-observable symptoms (Makkouk *et al.*, 1997). As earlier mentioned, many weed species are symptomless carriers of viruses, and therefore total rather than selective weed control is advisable in sensitive crops. Weed control should begin before crop emergence, because control after crop establishment may force the vector populations on the weeds to move on to crops, thus enhancing virus spread (Bos, 1999).

Prevention of spread

Prevention or reduction of viruses spread non-persistently by aphids (the most common vector of seed-transmitted viruses) can be done either by trying to avoid aphids (e.g. planting locations), by chemical control or by combining both measures simultaneously.

Avoidance of vectors. There are a number of ways to avoid aphids, such as early or late planting to avoid peak populations during active plant growth, use of insect repellents (foils or coloured mulches), use of barrier strips around the field of taller, non-susceptible plants on which incoming viruliferous aphids may land and rid their mouthparts of virus by probing before reaching the crop (good effects reported), covering the plants with a thin film of mineral oil or a mixture of oil and pyrethroids, where the oil possibly protects by oil interference with attachment of virions to the vector's mouthparts (only practicable for high-value crops), placing coarse white nets (2–8 mesh) 0.5 m above and around plots or covering with propylene fleece, producing crops in screened cages, screen houses or greenhouses (essential for propagation of nuclear seed stocks). More preventive measures and details can be found in Bos (1999) and Hull (2002).

Chemical control of vectors. Use of insecticides can be required but this entails the risks of inducing pesticide resistance in vectors and of killing natural vector enemies. Use of traps to monitor the aphid population can help in administering the timing and number of spray applications. Insecticides seem to have the best effect against the spread of persistent viruses. To have any impact on the spread of non-persistent viruses, the spray must kill vectors quickly; if not, aphid vectors move from plant to plant, probing a number of plants before dying (Walkey, 1991; Bos, 1999; Hull, 2002). Synthetic pyrethroids with a fast knock-down effect have been developed and found promising for the control of non-persistently transmitted viruses (Bos, 1999). One or two applications of one of the newer generation pyrethroids, alpha-cypermethrin, was effective in controlling CMV in lupin on one occasion by suppressing virus incidence in experimental plots by up to 62%, which surpassed the effect of other insecticides (Jones, 2001). However, application of alpha-cypermethrin did not always effectively control CMV spread.

Use of resistant cultivars and cross protection

The simplest measure of control would be the use of cultivars with resistance to the principal crop viruses. Such cultivars are available for some crops, e.g. BCMV/bean, but cultivars with multiple resistances to several viruses are few, and fewer cultivars exist with multiple resistances to important diseases of both viruses, fungi and bacteria (Hull, 2002). Another measure is the use of virus-tolerant cultivars, wherever complete resistance does not exist. However, tolerance to viruses is rarely comparable to genetic resistance, because virus-infected tolerant plants can be an unnoticed reservoir of infection transmitted to other hosts (Walkey, 1991; Hull, 2002).

RESISTANCE TO SEED TRANSMISSION. Cultivars resistant to seed transmission or with low transmission rates, such as are known for BSMV/barley and PSbMV/pea, may also be available (see also Incorporation of virus resistance, Section 3.2.2).

CROSS PROTECTION. Cross protection, i.e. inoculating crop plants with a mild virus strain to protect against severe strains, has been practised as a control measure, but only where no other control measures have been possible. Cross protection has also been demonstrated to occur between viroids (Lecoq, 1998). Gonsalves and Garnsey (1989) discuss the five elements of cross protection: (i) selection of mild strains; (ii) preliminary evaluation; (iii) pilot tests; (iv) field evaluation of mild strains; and (v) integration of cross protection into crop management systems. They also discuss the advantages and risks of the cross protection method. Some of the hazards of 'vaccinating' the entire crop, as pointed out by these authors as well as Lecoq (1998) and Hull (2002), are: (i) the protecting strain may spread to other hosts in which it may have more severe effects; (ii) serious disease may result from mixed infection when an unrelated virus is introduced into the crop (synergistic effect); and (iii) the protecting virus may mutate into a more severe form or its genes may recombine with that of another virus, resulting in a new virus.

Production of virus-free seed at farm level

As previously mentioned, farmers in developing countries either have no access to or cannot afford to purchase certified seed; instead, to a large extent, they use seed from their own harvest to plant the new crop. Further, in these countries, certified seeds of many crops are often not available, because certification methods are not practised or do not include health certification, and, if they do, testing for virus is rarely included. This is particularly so for a number of grain and vegetable legumes, according to information from seed-agency officers met during the author's travels in developing countries.

SEED PRODUCTION AT FARMS. To provide farm-saved seed free of or with low virus incidence, seeds only from plants that appear healthy at least to the end of flowering should be harvested. Virus-free plants, however, are not distinguishable when the crop is harvest-mature. A simple method to select virus-free plants at the farm, especially for legume seed for the next crop, could be to arrange a part (approx. 10%) of the crop area, for example, in the corner of the field, for seed production only. This seed production plot should be surrounded by a non-host barrier strip (cereal), and plants in the plot should be inspected frequently from emergence to the end of flowering, kept free of vectors and weeds, and plants with virus-like symptoms, if discernible, should be removed at short intervals. Plants infected only after the end of flowering need not be removed. These measures, if properly undertaken, should result in a stand of healthy-appearing plants with low pathogen incidence at harvest and therefore with little seed infection. Seeds from this plot should be stored separately and used only for sowing the next crop, and the procedure should be followed each year or every crop season. The seed production plot could alternatively be placed inside a cereal field for better isolation from the legume grain production area.

Negative consequences of frequent use of insecticides in such a plot, both ecologically and in the form of possible chemical residues in seeds, would be limited since the area is of a limited size and the seeds are to be used for sowing only, not for food. Initially, farmers should be instructed in recognizing virus-diseased plants. When farmers remove symptom-bearing plants, some plants with vectored but non-seed-transmissible viruses may also be removed. However, such plants might produce poor seed anyway, and, further, may be fewer in number due to systematic control measures in the plot.

A joint project between DGISP and Egypt (Abdelmonem, 2003), involving a number of Egyptian farmers, extension service personnel and scientists, has been initiated to evaluate such a seed production system for broad bean (*Vicia faba*), a species known to host at least six important seed-transmitted viruses. Promising results from the project have already been demonstrated, according to initial data recently reported (A. Abdelmonem, Egypt, 2004, personal communication). A similar system for production of seeds with low virus incidence might also be applicable to local seed production 'collectives'. For example, projects on such collectives in developing countries as alternatives to more formalized seed production systems have been initiated in cooperation with DANIDA, Denmark.

Integrated disease management

According to the definition of Jacobsen (1997), integrated pest management (IPM) is 'a sustainable approach to managing pests by combining biological, cultural, physical, and chemical tools in a way that minimizes economic, health, and environmental risks'. Although the early definitions and philosophical basis of IPM belong to entomologists (Jacobsen, 1997), plant pathology IPM or integrated disease management (IDM) is based on the same principles and has been practised for control of virus diseases from the early beginning of plant virology (Bos, 1999).

As a more recent example, Jones (2001) uses as a model programme the integrated management of two viruses, CMV and BYMV, in lupin. Among the individual measures combined within the IDM strategy were: sowing seed stocks with minimal virus content, sowing cultivars with inherently low seed transmission rates, and isolation from neighbouring lupin crops (for CMV only); perimeter non-host barriers and avoiding fields with large perimeter : area ratios (for BYMV only); promoting early canopy development, generating high plant densities (both found to decrease vectored infection), adjusting row spacing, sowing early maturing cultivars, maximizing weed control and crop rotation (for both viruses). Recommendations to apply insecticide were included solely for spraying high-value seed multiplication crops (for CMV only) or virus-infected pastures next to crops (BYMV only).

The report concludes that the IDM strategies used significantly reduced the losses caused by the two viruses in lupin, that most of the measures can be adopted by farmers with only few labour demands and minimal disruption to their normal farming operations or extra expenses, and that similar approaches should be applicable for other non-persistently transmitted viruses when their ecology and epidemiology are known.

3.2.2. Control measures at research and technology level

Incorporation of virus resistance

Cultivars resistant to one or more viruses can sometimes be found by screening for resistance or can be obtained by breeding by incorporation of resistance genes from cultivated or wild relatives. For example, on the Ethiopian plateau, where barley has probably been grown for millennia, many barley cultivars have a high degree of resistance to *Barley yellow dwarf virus* (BYDV, family *Luteoviridae*) (non-seed-transmitted) and BSMV (Hull, 2002). A new barley variety, Mobet, with good agronomic qualities and high resistance to seed transmission of the Montana isolates of BSMV, was obtained through breeding, introducing a resistance gene from an Ethiopian barley called Modjo (Carrol *et al.*, 1983, quoted from Hull, 2002). There are other examples of virus resistance incorporated by breeding, but also many examples of viruses, including important seed-transmitted ones, for which no resistant host material is known. Thus, until about 1998, resistance genes were reported against only 30 of the more than 100 seed-transmitted viruses known (Maury *et al.*, 1998). For some crop plants and viruses, resistance has lasted for many years, while, for others, resistance-breaking strains have emerged in a short time.

Breaking of resistance can be difficult or impossible to avoid, even when in breeding programmes the plant material, as required, has been screened for resistance to all known strains of the virus in question. Hull (2002) concludes that the search for new sources of resistance and their incorporation into useful cultivars will be a continuing and very long-term process. Breeding for resistance to plant viruses has been reviewed by Khetarpal *et al.* (1998); the review includes a comprehensive list of genetic resistance reported in various host–virus combinations.

TRANSGENIC PLANTS. Biotechnology methods make it possible to introduce exogenous genes into a plant across species, genus and family boundaries. Protection against viruses can be obtained from three main types of sources: (i) natural resistance genes; (ii) genes derived from viral sequences (pathogen-derived resistance, PDR); and (iii) genes from various other sources that interfere with virus synthesis (Hull, 2002). Among the different forms of PDR, coat protein-mediated and replicase-mediated resistance seem to be most effective and best studied (Kaniewski and Lawson, 1998). Replicase-mediated virus resistance was reviewed by Palukaitis and Zaitlin (1997). Transgenic plants expressing virus-specific antibodies (plantibodies) are one of the alternatives to PDR strategies (reviewed by Martin, 1998). For example, Terrada *et al.* (2000) constructed recombinant antibodies against *Citrus tristeza virus* (CTV, *Closterovirus*) and suggested that these constructs may be effective in inducing resistance in transgenic citrus plants (see also Section 5.2.7).

Incorporation of resistance transgenes, if stably integrated, adequately expressed and heritable in the host, is an alternative for the many viruses for which no resistant cultivars or relatives are known, and is also a less labour- and time-intensive process than traditional resistance breeding. Introducing resistance genes from related species by breeding may also be hampered by the fact that these genes may have linkages to undesirable agronomic or culinary traits that are sometimes unbreakable, whereas transgenic virus resistance can be introduced into susceptible crop varieties without affecting such properties (Barker *et al.*, 1998).

Introducing new genes into plants can be performed in most crops by either an *Agrobacterium*-mediated method or DNA delivery via a particle gun projected into plant tissues that subsequently differentiate into whole plants (Kaniewski and Lawson, 1998).

There could be risks of using transgenically resistant plants (reviewed by, for example, Hull, 1998, 2002). One supposed risk is the often debated one about transferring transgenes to related or wild species through pollen. However, it can also be asked whether pollen transfer from crop plants containing virus-resistant transgenes poses a greater risk than pollen transfer from disease-resistant varieties produced by conventional plant breeding methods (Timmerman-Vaughan, 1998).

Transgenic protection/resistance against plant viruses is reviewed, for example, by Hadidi *et al.* (1998) (six chapters) and by Hull (2002), and methods of generating virus-resistant transgenic plants are described in Foster and Taylor (1998).

Virus-free seed through seed-health testing

Use of seed free of, or with low incidence of, seed-transmitted viruses is a control strategy that for a number of virus–host combinations can be very effective. However, like any other control measure, exclusive use of this method may often be insufficient, and it should be combined with other control measures included in Section 3.2.1.

SEED-HEALTH TESTING IN CERTIFICATION. Estimating the rate of seed-transmitted virus(es) in a seed lot by testing representative samples from the lot is today possible for a large number of viruses and viroids. The increased sensitivity of detection methods has facilitated the testing of large samples. Testing of large samples is required in order to detect low levels of seed transmission. Production of certified seed with a tolerably low incidence of virus requires that seeds be multiplied from a virus-free nuclear stock. A prerequisite for a healthy nuclear stock is that the starting seed lot is free from virus, i.e. the germplasm and the breeder's lines, and the nuclear stock has been monitored for health during its maintenance.

Production of absolutely virus-free certified seed is costly, if possible at all. If the price of doing so exceeds the expected financial losses from crop infection from partially infested seed, growers may prefer cheaper, partially infected seed, or to multiply their own seed for a few years from certified seed (Bos, 1999).

In Western Australia, where the only significant source of CMV in lupin field crops is infected lupin seed, a fee-based testing service for farmers' lupin seed stocks has been provided since 1988 (Wylie *et al.*, 1993; Jones, 2000). During the first years, seeds were tested by ELISA, later by PCR (and ELISA) and more recently by a semi-quantitative TaqMan-PCR (Jones, 2001); for the test procedures used, see Section 6.2.2. Farmers submit seeds representatively sampled according to instructions. The testing service provides them with an approximate % of CMV infection, which (in the case of infection) should not exceed 0.5% for low-risk zones or 0.1% for high-risk zones. These zones have been identified by field experiments (Section 1.4.1), which also revealed zones suited for high-value seed multiplication crops. The lupin seed testing service and recommended tolerance levels have been widely adopted by growers and have contributed greatly to the currently diminished occurrence of the CMV-induced disease in lupin crops in south-west Australia (Jones, 2000, 2001).

As mentioned in Section 1.4.1, setting realistic tolerance levels for virus in seed depends on the virus and host, and on the ecology and epidemiology of the virus in the geographical area where the seed is going to be grown. As emphasized by Stace-Smith and Hamilton (1988), Jones (2000) and others, setting realistic thresholds should be based on data generated from field trials (Section 1.4.1), and such data are generally lacking. There are difficulties connected with providing realistic data, because many conditions affect seed-borne disease development. Most tolerances are based on empirical information, but, in many cases, years of testing at these empirical levels have proved quite successful in controlling diseases from seed-borne inoculum (Maddox, 1998).

According to Maury *et al.* (1998) a classification of seed might be more practical and have fewer inconveniences than certification. It would inform the grower about the primary level of inoculum in different classes of seed lots and allow the grower to decide whether or not to purchase a given class of infected seed lot, if virus-free lots are not available or are too costly. Information collected on the field incidence of the disease for a few years could be related to the incidence of virus in seeds according to the class of seed sown, and could reveal whether the tolerance level should be readjusted. Thus, a classification system could help in setting a level of inoculum tolerance without the need for an expensive programme for generating epidemiological data.

General data on the ecology and epidemiology of SMV were collected by Irwin and Goodman (1981) and were subsequently used for the development of a simulation model (Ruesink and Irwin, 1986, quoted from Maury *et al.*, 1998). The model was flexible enough to be useful over a wide range of environmental and agro-climatic conditions. It could predict that, in geographical areas where vector intensity is low before flowering, a tolerance limit of 1% seed transmission of SMV could be considered acceptable, whereas, in areas with early and high vector intensity, the tolerance limit was established as 0.01%.

ROUTINE SEED-HEALTH TESTING. Testing of seed for viruses is a technology that can be performed by trained technical personnel, but, as for testing for other types of pathogens, tests should be supervised by a pathologist. Most of the techniques require laboratory facilities of a certain standard, and, in general, the higher the sensitivity of assay, the higher are the requirements for facilities and equipment. For all assay types, sample preparation, whether from seed or seedlings, is the most time- and labour-intensive part of the tests. To determine whether a seed lot originated from plants infected with seed-transmitted virus(es), a relatively quick test can often be made on whole seeds. A fast method (direct immunostaining assay (DISA)) for testing pepper seed for externally borne tobamoviruses is described in Section 5.4. Whatever method is used, the number of seeds tested must correspond statistically to the actual tolerance level (Chapter 7). If the test is negative for the virus(es) tested for, it is quite likely that the seed lot is virus-free or at least below the tolerance level. If the test is positive for one or more viruses known to be seed-transmitted, a new sample must usually be tested by a method that can reveal whether the seeds carry transmissible virus(es) and at what rate.

The result of a whole-seed test rarely corresponds to the actual virus transmission rate, except for five virus–host combinations, LMV/lettuce, PeMoV and BCMV-PSt (peanut stripe strain of BCMV) in groundnut, SMV/soybean and CMV/lupin, due to lack of detectable viral antigen in testae (Sample preparation, Section 5.2.2). To determine the rate of seed transmission for all other virus–host combinations, either seeds must be decorticated before testing or seedlings raised from the seeds must be tested. For some virus–host combinations, like BSMV/barley, either the embryos must be separated and tested or seedlings be raised and tested because of high titres of non-transmissible virus in the endosperm (Maury *et al.*, 1998).

According to the same authors there can be large variations of virus concentration among infected embryos (cotyledons + embryo axis) and also the distribution of infective virus in the cotyledons and the embryo axis. Good correlations have been found between the percentage of ELISA-positive embryos in a given seed lot and the percentage of infected seedlings raised from the same lot for SMV/soybean seed, PeMoV/groundnut seed and PSbMV/pea seed. In other cases, presence of detectable virus in the embryo corresponds less well to the actual seed transmission rate, as found for BCMV-BlCM in cowpea (Gillaspie *et al.*, 1993), where in some seeds detectable virus was present only in the cotyledons, and not in the axis, and then assumed not to be transmissible (see also Seed-to-plant transmission, Section 2.2.2).

Obviously, the rate of seed transmission is generally most precisely determined when the seed progeny is tested rather than the seed embryos. An example of rational seed-health testing by virus assay of seed progeny is testing seedlings from lentil seed by tissue blot immunoassay (TBIA), where the transversely cut surfaces of seedlings in bundles of 25 were simultaneously printed on the membranes (Section 5.4). As a further rationalization of the assay, mixed antisera against three lentil viruses were used (Makkouk *et al.*, 1997).

Viroids are not serologically detectable, due to lack of a protein coat, but require the use of biological, electrophoretic (Section 7.1) or molecular assays. Due to low viroid titres in seed tissue, such as PSTVd in true potato seed (TPS), detection of viroids in whole seeds may be difficult when using molecular hybridization tests (Section 6.1), according to Shamloul *et al.* (1997), who recommended the more sensitive PCR assay (Section 6.2). However, Borkhardt *et al.* (1994), using a hybridization assay, detected PSTVd in RNA extracts of whole TPS diluted up to 1 : 50 (in denaturing solution) when phenol-chloroform extraction of RNA (Section 6.2.3) was used.

Seed certification for viruses is reviewed by Maury *et al.* (1998). Set-up and interpretation of seed-health assays and international standardization of testing methods are discussed in Chapter 7.

Exclusion of seed-transmitted viral inoculum by quarantine. As previously mentioned, seed is an effective vehicle for transport of virus inoculum over long distances, and this has inadvertently caused the spread of plant virus diseases by humans in the past, as exemplified by New Zealand in Section 3.1.1. Today, the risk of spreading viruses remains substantial, due to extensive, worldwide movement of seed. Unlike arthropods, weeds, fungi or nematodes, viruses cannot be seen by quarantine station personnel by the naked eye or light microscope, nor can they be routinely eliminated by treating seed or plant tissues. An exception is externally borne tobamoviruses on tomato and pepper seed, which can be inactivated by chemical or heat treatment (Section 4.4.3). There have also been reports on reduction of the incidence of embryo-transmitted viruses in seeds by heat treatment, but which at the same time reduced seed viability. Dry heat treatment at 70°C eliminating BBSV from lentil seed is an example of infected-seed therapy, but it reduced seed germination by 40%. These experiments were carried out at the International Center for Agricultural Research in the Dry Areas (ICARDA), where similar rescue attempts were considered for other virus–host combinations in order to retrieve valuable virus-infected germplasm (Diekmann, 1997).

Attempts to reveal the presence of seed-transmitted viruses or viroids by subjecting the seeds to a growing-on test are not always successful, as many of these pathogens appear symptomless or have mild symptoms only. Additional tests of progeny in such cases are, therefore, required for their detection (see also Non-destructive testing of seeds, Section 5.2.2). As mentioned earlier, the required zero tolerance can only be practised for small seed lots, such as germplasm. Effective exclusion of exotic viral diseases by quarantine, as done, for example, in Australia (Section 1.4.2), is costly, requiring sophisticated facilities and specialist supervision. Not all countries have such a well-developed quarantine system. Developing countries dependent on a few crops tend to be stricter than countries with diverse economies, but, as facilities for virus exclusion are absent in most developing countries, rigid rules are often hard to carry out, if not impossible (Bos, 1999).

Quarantine is easier to practise in countries with few ports of entry, such as those surrounded by sea, than between countries separated by long boundaries. Quarantines can be performed either in the country after entry (post-entry quarantine) or in a quarantine station in a third country (intermediate quarantine) – for example a country where the risks of inoculum escape is low. Effective quarantine to exclude seed-transmitted viruses requires certain facilities. There should be growth facilities, apart from the usual quarantine facilities, where seedlings and plants can be grown in containment to maturity (Section 1.4.2). There should also be equipment and skilled personnel for detection and identification of not only virus and viroid species under quarantine in the country, but also new, virulent strains of those viruses and viroids already present in the country. Quarantine stations may obtain access to advanced detection tools, like electron microscopy and molecular techniques, by cooperating with plant pathological centres. In such cases, adequate precautions should be taken, in accordance with quarantine regulations, to prevent escape of inoculum during transportation and handling of materials to be tested.

Among other important tools for quarantine work are the *Technical Guidelines for the Safe Movement of Germplasm*, jointly issued by the Food and Agriculture Organization (FAO) and the International Plant Genetic Resources Institute (IPGRI), one of which is for legume germplasm (Frison *et al*., 1990). Another useful guide is the *Crop Protection Compendium* (CPC) published on CD-ROM by CAB International (Anon., 2004). The CPC includes a Database on Seed-borne Diseases (DSD) and is updated annually (see also Section 7.2).

Quarantine for seed-transmitted viruses has been reviewed by Maury *et al.* (1998) and included in other information on quarantine, e.g. Kahn (1977, 1982, 1991), Foster and Hadidi (1998), Kahn and Mathur (1999) and Ebbels (2003).

References

Abdelmonem, A. (2003) Status report of the Capacity Enhancement Project between the Danish Government Institute of Seed Pathology for Developing Countries, Denmark, Ain Shams University, Cairo and Agricultural Research Centre, Giza, Egypt (unpublished).

Anon. (2004) *Crop Protection Compendium*, 2004 edn (CD-ROM). CAB International, Wallingford, UK.

Barker, I., Henry, C.M., Thomas, M.R. and Stratford, R. (1998) Potential benefits of the transgenic control of plant viruses in the United Kingdom. In: Foster, G.D. and Taylor, S.C. (eds) *Plant Virology Protocols: From Virus Isolation to Transgenic Resistance*. Humana Press, Totowa, New Jersey, pp. 557–566.

Borkhardt, B., Vongsasitorn, D. and Albrechtsen, S.E. (1994) Chemiluminescent detection of potato spindle tuber viroid in true potato seed using a digoxigenin labelled DNA probe. *Potato Research* 37, 249–255.

Bos, L. (1981) Wild plants in the ecology of virus diseases. In: Maramorosch, K. and Harris, K.F. (eds) *Plant Diseases and Vectors: Ecology and Epidemiology*. Academic Press, New York, pp. 1–33.

Bos, L. (1999) *Plant Viruses, Unique and Intriguing Pathogens – a Textbook of Plant Virology*. Backhuys Publishers, Leiden, The Netherlands.

Bottalico, G., Padula, M., Campanale, A., Sialer, M.F., Saccomanno, F. and Gallitelli, D. (2002) Seed transmission of *Artichoke Italian latent virus* and *Artichoke latent virus* in globe artichoke. *Journal of Plant Pathology* 84, 167.

Briddon, R. (2003) Tomato pseudo-curly top virus. *AAB Descriptions of Plant Viruses* No. 395. Available at: http://www.dpvweb.net

Brunt, A.A., Crabtree, K., Dallwitz, M.J., Gibbs, A.J. and Watson, L. (eds) (1996) *Viruses of Plants*. CAB International, Wallingford, UK.

Campbell, R.N. (1996) Fungal transmission of plant viruses. *Annual Review of Phytopathology* 34, 87–108.

Carroll, T.W., Hockett, E.A. and Zaske, S.K. (1983) Registration of mobet barley germplasm. *Crop Science* 23, 599–600.

Conti, M. and Masenga, V. (1977) Identification and prevalence of pepper viruses in northwest Italy. *Phytopathologische Zeitschrift* 90, 212–222.

da Graca, J.V. and van Vuuren, S.P. (2003) Viroids in Africa. In: Hadidi, A., Flores, R., Randles, J.W. and Semancik, J.S. (eds) *Viroids*. CSIRO Publishing, Collingwood, Victoria, Australia, and Science Publishers, Enfield, New Hampshire, pp. 290–292.

Diekmann, M. (1997) Pathogens in germplasm. In: *Crop Protection Compendium*, 2001 edn (CD-ROM). CAB International, Wallingford, UK.

Diener, T.O. and Raymer, W.B. (1971) Potato spindle tuber 'virus'. *AAB Descriptions of Plant Viruses* No. 66. Available at: http://www.dpvweb.net

Ebbels, D.L. (2003) *Principles of Plant Health and Quarantine*. CAB International, Wallingford, UK.

El-Hammady, M.H., Albrechtsen, S.E., Abdelmonem, A.M., El-Abbas, F.M.A., Rasmi, M.R. and Ghazalla, W. (2004) Interaction and frequencies of seed-transmitted faba bean (*Vicia faba* L.) viruses under natural conditions. *Arab Universities Journal of Agricultural Sciences* 12, 851–866.

Foster, G.D. and Taylor, S.C. (eds) (1998) *Plant Virology Protocols: From Virus Isolation to Transgenic Resistance*. Humana Press, Totowa, New Jersey.

Foster, J.A. and Hadidi, A. (1998) Exclusion of plant viruses. In: Hadidi, A., Khetarpal, R.K. and Koganezawa, H. (eds) *Plant Virus Disease Control*. APS Press, St Paul, Minnesota, pp. 208–229.

Frison, E.A., Bos, L., Hamilton, R.I., Mathur, S.B. and Taylor, J.D. (eds) (1990) *FAO/IBPGR Technical Guidelines for the Safe Movement of Legume Germplasm*. Food and Agriculture Organization of the United Nations, Rome/International Board for Plant Genetic Resources, Rome.

Giesler, L.J., Ghabrial, S.A., Hunt, T.E. and Hill, J.H. (2002) Bean pod mottle virus – a threat to US soybean production. *Plant Disease* 86, 1280–1289.

Gillaspie, A.G. Jr, Hopkins, M.S. and Pinnow, D.L. (1993) Relationship of cowpea seed-part infection and seed transmission of blackeye cowpea mosaic potyvirus in cowpea. *Plant Disease* 77, 875–877.

Gonsalves, D. and Garnsey, S.M. (1989) Cross-protection techniques for control of plant virus diseases in the tropics. *Plant Disease* 73, 592–597.

Green, S.K. and Kim, J.S. (1991) Characteristics and control of viruses infecting peppers: a literature review. *Technical Bulletin*, No. 18. Asian Vegetable Research and Development Centre Taipei, Taiwan, 60 pp.

Green, S.K., Hwang, L.L. and Kuo, Y.J. (1987) Epidemiology of tomato mosaic virus in Taiwan and identification of strains. *Journal of Plant Diseases and Protection* 94, 386–397.

Hadidi, A., Khetarpal, R.K. and Koganezawa, H. (eds) (1998) *Plant Virus Disease Control*. APS Press, St Paul, Minnesota.

Hull, R. (1998) Detection of risks associated with coat protein transgenics. In: Foster, G.D. and Taylor, S.C. (eds) *Plant Virology*

Protocols: From Virus Isolation to Transgenic Resistance. Humana Press, Totowa, New Jersey, pp. 547–553.

Hull, R. (2002) *Matthews' Plant Virology,* 4th edn. Academic Press, London.

Irwin, M.E. and Goodman, R.M. (1981) Ecology and control of soybean mosaic virus. In: Maramorosch, K. and Harris, K.F. (eds) *Plant Diseases and Vectors: Ecology and Epidemiology.* Academic Press, New York, pp. 182–220.

Jacobsen, B.J. (1997) Role of plant pathology in integrated pest management. *Annual Review of Phytopathology* 35, 373–391.

Jeyanandarajah, P. and Brunt, A.A. (1993) The natural occurrence, transmission, properties and possible affinities of cowpea mild mottle virus. *Journal of Phytopathology* 137, 148–156.

Jones, R.A.C. (2000) Determining 'threshold' levels for seed-borne virus infection in seed stocks. *Virus Research* 71, 171–183.

Jones, R.A.C. (2001) Developing integrated disease management strategies against non-persistently aphid-borne viruses: a model programme. *Integrated Pest Management Reviews* 6, 15–46.

Kahn, R.P. (1977) Plant quarantine: principles, methodology, and suggested approaches. In: Hewitt, W.B. and Chiarappa, L. (eds) *Plant Health and Quarantine in International Transfer of Genetic Resources.* CRC Press, Boca Raton, Florida, pp. 289–307.

Kahn, R.P. (1982) The host as a vector: exclusion as a control. In: Harris, K.F. and Maramorosch, K. (eds) *Pathogens, Vectors, and Plant Diseases: Approaches to Control.* Academic Press, New York, pp. 123–149.

Kahn, R.P. (1991) Exclusion as a plant disease control strategy. *Annual Review of Phytopathology* 29, 219–246.

Kahn, R.P. and Mathur, S.B. (eds) (1999) *Plant Protection and Quarantine.* APS Press, St Paul, Minnesota.

Kaniewski, W. and Lawson, C. (1998) Coat protein and replicase-mediated resistance to plant viruses. In: Hadidi, A., Khetarpal, R.K. and Koganezawa, H. (eds) *Plant Virus Disease Control.* APS Press, St Paul, Minnesota, pp. 65–78.

Khetarpal, R.K., Maisonneuve, B., Maury, Y., Chalhoub, B.A., Dinant, S., Lecoq, H. and Varma, A. (1998) Breeding for resistance to plant viruses. In: Hadidi, A., Khetarpal, R.K. and Koganezawa, H. (eds) *Plant Virus Disease Control.* APS Press, St Paul, Minnesota, pp. 14–32.

Lecoq, H. (1998) Control of plant virus diseases by cross protection. In: Hadidi, A., Khetarpal, R.K. and Koganezawa, H. (eds) *Plant Virus Disease Control.* APS Press, St Paul, Minnesota, pp. 33–40.

Lister, R.M. and Murant, A.F. (1967) Seed transmission of nematode-borne viruses. *Annals of Applied Biology* 59, 49–62.

Maddox, D.A. (1998) Implications of new technologies for seed health testing and the worldwide movement of seed. *Seed Science Research* 8, 277–284.

Makkouk, K.M., Erskine, W. and Attar, N. (1997) Detection of seed-borne viruses in lentil mother plants as compared to harvested seeds. *LENS Newsletter* 24, 49–51.

Martin, R.R. (1998) Alternative strategies for engineering virus resistance in plants. In: Hadidi, A., Khetarpal, R.K. and Koganezawa, H. (eds) *Plant Virus Disease Control.* APS Press, St Paul, Minnesota, pp. 121–128.

Matthews, R.E.F. (1991) *Plant Virology,* 3rd edn. Academic Press, San Diego, California.

Maury, Y., Duby, C. and Khetarpal, R.K. (1998) Seed certification for viruses. In: Hadidi, A., Khetarpal, R.K. and Koganezawa, H. (eds) *Plant Virus Disease Control.* APS Press, St Paul, Minnesota, pp. 237–248.

Myrta, A. and Savino, V. (1998) Study on the transmission of plum pox potyvirus through seeds. *Phytopathologia Mediterranea* 37, 41–44.

Naidu, R.A., Gowda, S., Satyanarayana, T., Boyko, V., Reddy, A.S., Dawson, W.O. and Reddy, D.V.R. (1998) Evidence that whitefly-transmitted cowpea mild mottle virus belongs to the genus *Carlavirus. Archives of Virology* 143, 769–780.

Noordam, D., Bijl, M., Overbeek, S.C. and Quiniones, S.S. (1965) Virussen uit *Campanula rapunculoides* en *Stellaria media* en hun relatie tot komkommermozaiek en tomaat-'aspermi'-virus. *Netherlands Journal of Plant Pathology* 71, 61 (Abstr.).

Palukaitis, P. and Zaitlin, M. (1997) Replicase-mediated resistance to plant virus disease. *Advances in Virus Research* 48, 349–376.

Ruesink, W.G. and Irwin, M.E. (1986) Soybean mosaic virus epidemiology: a model and some implications. In: McLean, G.D., Garrett, R.G. and Ruesink, W.G. (eds) *Plant Virus Epidemics – Monitoring, Modelling and Predicting Outbreaks.* Academic Press, Melbourne, Australia, pp. 295–313.

Saponari, M., Savino, V. and Martelli, G.P. (2002) Seed transmission in olive of two olive-infecting viruses. *Journal of Plant Pathology* 84, 167–168.

Shamloul, A.M., Hadidi, A., Zhu, S.F., Singh, R.P. and Sagredo, B. (1997) Sensitive detection of potato spindle tuber viroid using RT-PCR and identification of a viroid variant naturally infecting pepino plants. *Canadian Journal of Plant Pathology* 19, 89–96.

Singh, R.P., Ready, K.F.M. and Nie, X. (2003) Biology. In: Hadidi, A., Flores, R., Randles, J.W. and Semancik, J.S. (eds) *Viroids.* CSIRO Publishing, Collingwood, Victoria, Australia, and Science Publishers, Enfield, New Hampshire, pp. 30–48.

Skaf, J.S. and de Zoeten, G.A. (2000) Pea enation mosaic virus. *AAB Descriptions of Plant Viruses* No. 372. Available at: http://www.dpvweb.net

Slovakova, L., Davidova, I., Kudela, V. (2002) Annual herbs – possible reservoirs of sharka disease? In: Taborsky, V., Polak, J. and Lebeda, A. (eds) *Disease Resistance in Plant Pathology. Proceedings of the 6th Conference of European Foundation for Plant Pathology, Prague, Czech Republic, 2002. Plant Protection Science* 38 (Special 2), 364–366 (Abstr.).

Stace-Smith, R. and Hamilton, R.I. (1988) Inoculum thresholds of seedborne pathogens: viruses. *Phytopathology* 78, 875–880.

Teakle, D.S. (1988) The effect of environmental factors on fungus-transmitted viruses and their vectors. In: Cooper, J.I. and Asher, M.J.C. (eds) *Viruses with Fungal Vectors. Developments in Applied Biology* 2. *Proceedings of a Conference at the University of St Andrews, 1987.* Association of Applied Biologists, AAB Office, Wellesbourne, Warwick, UK, pp. 167–179.

Terrada, E., Kerschbaumer, R.J., Giunta, G., Galeffi, P., Himmler, G. and Cambra, M. (2000) Fully 'recombinant enzyme-linked immunosorbent assays' using genetically engineered single-chain antibody fusion proteins for detection of *Citrus tristeza virus. Phytopathology* 90, 1337–1344.

Thottappilly, G. (1992) Plant virus diseases of importance to African agriculture. *Journal of Phytopathology* 134, 265–288.

Timmerman-Vaughan, G.M. (1998) Releasing genetically engineered virus-resistant crops – risk assessment. In: Hadidi, A., Khetarpal, R.K. and Koganezawa, H. (eds) *Plant Virus Disease Control.* APS Press, St Paul, Minnesota, pp. 142–153.

Tomlinson, J.A. and Walker, V.M. (1973) Further studies on seed transmission in the ecology of some aphid-transmitted viruses. *Annals of Applied Biology* 73, 293–298 (Abstr.).

Walkey, D. (1991) *Applied Plant Virology.* Chapman & Hall, London.

Wylie, S., Wilson, C.R., Jones, R.A.C and Jones, M.G.K. (1993) A polymerase chain reaction assay for cucumber mosaic virus in lupin seeds. *Australian Journal of Agricultural Research* 44, 41–51.

4 Biological Assays

Biological assays are more time- and labour-consuming compared to laboratory assays, but still very important in the detection and diagnosis of plant viruses and viroids, and, not least, biological methods are the only means to propagate these pathogens. Infectivity assays are often needed to verify results of laboratory tests, and inoculation of differential hosts is an essential tool for virus identification at the species and strain levels. In seed-health testing, raising of seedlings from seeds under investigation gives the most reliable information on the presence of seed-to-plant-transmissible viral infection. In contrast to many other plant virus species, the vast majority of the seed-transmitted ones (exception: the cryptoviruses) are readily sap-transmissible and, therefore, well suited for infectivity assays. In fact, some seed-transmitted viruses can be detected from whole seeds by an infectivity test alone. Such tests were commonly used in the past for detection of, for example, lettuce mosaic in lettuce seeds and bean common mosaic in bean seeds and, needless to say, in the case of reagent deficits for laboratory assays and if growth facilities and labour are available, such methods are still valuable for seed assays. In addition, because the number of potential viruses seed-transmitted per host species is limited, their detection by biological assays is straightforward and qualitatively effective. An infectivity assay preceded by a growing-on test (Section 4.3) is also useful in testing for the absence of certain viruses in seed lots.

Viroids have no protein coat and are therefore serologically undetectable. They are, however, sap-transmissible and hence bioassays are one of the other options for their detection. For example, for infectivity tests for PSTVd that is seed-transmitted in true potato seed and tomato seed, see Diener and Raymer (1971), *AAB Descriptions of Plant Viruses* (No. 66) (http://www.dpvweb.net). Other methods are electrophoresis (Section 7.1) or molecular detection (Chapter 6).

A prerequisite for reliable and reproducible results of biological assays is the availability of adequate vector-proof growth facilities and skilled labour to produce indicator plants of optimal quality. Essential also is familiarity with symptomatology.

A disadvantage of bioassays compared to laboratory assays (serological or molecular techniques) is the higher risk of inoculum escape into the environment. It is therefore essential that tests are performed in insect-proof growth facilities and that plant waste is handled correctly (see Section 4.2.1 and the protocols).

4.1. Symptomatology

Viral diseases cause both external and internal symptoms in plants. External symptoms vary from mild to severe, but diseased plants will nearly always show reduced vitality and reduced growth or dwarfing, both in aerial parts and in roots (Fig. 4.1). Very often *colour variegation* occurs in the form of chlorosis, mosaic, mottling, rings, ringspots, line patterns, vein banding or vein chlorosis, etc. Such symptoms may sometimes be difficult to distinguish from physiological disorders caused, for example, by nutrient deficiency but often nutrient deficiency causes more uniform patterns, while virus symptoms usually have irregular patterns. Another common class of symptoms is *necrosis* (death of tissues), seen as browning and blackening, necrotic spots and phloem necrosis. Phloem necrosis may sometimes be systemic and lead to plant death, as results from BCMNV infection of certain bean cultivars. For full names of viruses, see Table 4.2. Typical symptoms also include different kinds of *malformation*, especially of leaves: leaf rolling, curling, narrowing, rugosity and wrinkling. Down folding of leaf margins or leaf halves is a common symptom in legumes affected by seed-transmitted viruses.

Fig. 4.1. Reduction in growth: bambara groundnut (*Voandzeia subterranea*) grown from a seed with seed-borne infection of a *Potyvirus* (probably PeMoV) (left) and healthy plant (right).

Some virus–host combinations are characterized by a 'shock' reaction followed by recovery, i.e. plant growth after infection is nearly or completely free of symptoms. This phenomenon is common, for example, among nematode-transmitted viruses, of which many are seed-transmitted (Lister and Murant, 1967; Bennett, 1969). Often the recovery phase continues in the offspring, i.e. seedlings from plants in the recovery phase tend to be symptomless or show faint symptoms only, whereas seedlings from mother plants with no recovery phase tend to show symptoms as severe as those of the mother plant (Bennett, 1969). Other factors are known to influence virus-symptom expression, e.g. symptoms caused by certain viruses are known to be masked by high temperatures.

An excellent source of pictures of external virus symptoms is *AAB Descriptions of Plant Viruses* (n.d.), accessible on http://www.dpvweb.net, which contains descriptions of over 400 viruses. Of these, about 55 have pictures in colour, and the remainder are in black and white. The collection is continuously being updated with new descriptions. A separate collection of about 50 virus-symptom pictures in colour is available on http://www.plbio.kvl.dk/~thluj3/.

Symptoms on seed

Apart from reduction in size, most seeds carrying viral infection appear normal. Only a few viruses induce characteristic symptoms on seeds of some hosts, such as brown or black mottling on soybean seeds caused by SMV, necroses on faba bean seeds due to BBSV (Fig. 4.2) or wrinkling of pea seed-coats associated with infection of PEBV. However, these symptoms are not correlated with presence of virus in the seeds, but reflect damage to the mother plant, and are thus only indicative of potential virus transmission in the seeds.

Fig. 4.2. Symptoms on seeds induced by seed-transmitted viruses. (A) Mottling of soybean seeds caused by SMV; (B) soybean seeds discoloured due to the 'soybean stunt' strain of CMV (note the ring and concentric ring pattern); (C) faba bean seeds with seed-coat necroses due to BBSV and normal seeds (upper row). Photo (B): courtesy Prof. R. Suseno, Bogor Agricultural University, Indonesia.

4.1.1. Indicator plant symptoms

A wide range of herbaceous plant indicator species and cultivars has been identified for virus diagnostic purposes, especially within the genera *Chenopodium* and *Nicotiana* and some *Cucumis*, *Phaseolus*, *Vigna* and *Vicia* spp. Indicator hosts may react either with local symptoms only or with symptoms of systemic infection (see below), or sometimes both, depending on the virus and indicator host.

Fig. 4.3. Indicator plant symptoms. (A) Chlorotic local lesions on *Chenopodium amaranticolor* inoculated with LMV; (B) diffuse local chlorotic spots on both cotyledons and a primary leaf of *Cucumis metuliferus* inoculated with *Squash mosaic virus* (SqMV, *Comovirus*); and (C) whitish-chlorotic ringspots, concentric rings and lines on *Nicotiana benthamiana* after inoculation with TRSV.

Species reacting locally to a virus are useful since symptoms appear quickly, i.e. from 2 to 7 days post inoculation (dpi), while symptoms on systemically reacting plants appear only after 7 to 14 dpi or more. On the other hand, local-reacting hosts may not be as sensitive as systemic hosts (Phatak, 1974). In host-range studies, host plants may sometimes respond with latency or faint symptoms to some viruses or strains, and must therefore be tested by back inoculation to specifically reacting hosts. In host-range studies, knowledge of both virus-susceptible plant species and non-hosts of viruses is useful.

Local and systemic reaction

As previously mentioned, locally reacting indicator species give a quick response. Such hosts are hypersensitive to the virus and react with local lesions only at the site of inoculation, i.e. roundish chlorotic or necrotic spots, which can vary in size (Fig. 4.3). The number of local lesions per leaf area unit is roughly proportional to the virion concentration in the inoculum, and such assays can therefore be used to approximate virion concentrations.

In systemically reacting hosts, usually no or only faint symptoms appear on the inoculated leaves, but new leaves emerging after the inoculation develop symptoms 7–14 dpi, most often in the form of colour variegation and leaf malformation. Some virus–indicator host combinations react both locally and systemically.

Indicator species, useful for diagnosis of major seed-transmitted viruses, are listed in Tables 4.1 and 4.2. Many more indicator species and non-hosts for a large number of viruses can be found in Brunt *et al.* (1996), *Viruses of Plants*, based on the Virus Identification Data Exchange (VIDE) Project, and also in the AAB Descriptions of Plant Viruses, mentioned above. However, only a few of the latter include non-hosts. Symptoms on indicator plants for each virus are described in both sources.

4.2. Facilities and Equipment for Biological Assays

The minimal facilities required for biological assays include: (i) an insect-proof greenhouse or growth room, preferably with temperature regulation and supplemental plant growth light; (ii) a tap water outlet in the room or in an adjoining room; and (iii) the availability of a vector- and pathogen-free growth medium (see Section 4.2.2).

4.2.1. Growth facilities

The optimal growth facility is a greenhouse in which temperatures of about 18°C at night (for tropical plants, like cowpea, not below 20°C) and 22–25°C in the daytime can be maintained. Such conditions are relatively easy to obtain in temperate climates, but difficult in warm climates. In the tropics and subtropics, a greenhouse must be equipped with an extensive cooling system, which can be rather costly to purchase and operate. In such climates, plants in greenhouses

Table 4.1. Important seed-borne viruses, their main hosts, reported seed transmission* and diagnostic assay species.**

Virus**	Main hosts	Seed transm. %	Assay hosts: Local	Assay hosts: Systemic	Assay hosts: Non-hosts
AMV	*Medicago sativa*	0.1–49	Ca, Cq, Vu	Ca, Cq, Pv	Cf, Ds
BSMV	*Hordeum vulgare*	≤ 100	Ca, Cq	Hv	–
BCMNV	*Phaseolus vulgaris*	+	Ca, Cq	Pv[a]	Cs, Nt
BCMV	*P. vulgaris*	0–83	Ca, Cq	Pv[b]	Cs, Nt
-BlCM	*Vigna unguiculata* sspp.	30	Ca	Vu[c]	Vu[g]
-PSt	*Arachis hypogaea*	25	Ca, Cq	Nb	Pvt
BYMV	Vegetables and legumes	0.1–14	Ca, Cq, Gg	Pv, Vf	Cs, Ds
BBSV	*Vicia faba, Lens culinaris*	1–22	Pv[d, e]	Pv[e]	Ca, Gg, Nt, Pv[f]
BBTMV	*V. faba*	1–17	–	Ps, Vf	Ca, Gg, Nt, Nc
CABMV	*Vigna unguiculata* sspp.	0–40	Ca	Vu[g]	Ds, Ng, Vf, Vu[c]
CPMV	*V. unguiculata* sspp.	1–5	Ca, Pv[f]	Vu	Ng, Cs, Bv
CPSMV	*V. unguiculata* sspp.	1–10	Ca, Pv	–	Ct
CGMMV	*Cucumis* spp., *Cucurbita* spp.	1–8	Ca	Cs	Ng, Ds, Ph
CMV	Vegetables and legumes	0–100	Ca, Cq, Vu	Nc, Ng, Nt	Ph, Ze
IPCV	*Arachis hypogaea*	6–14	Ca, Cq, Nb	Nb, Ng, Te, Ta	–
LMV	*Lactuca sativa*	3–15	Ca, Cq, Gg	Cq, Nb	Ds, Ng, Ph
PEBV	*Pisum sativum, V. faba*	1–37	Ca, Cq, Pv, Te	Pv, Te	Bv, Ms, Ng
PSbMV	*P. sativum, V. faba, Lens culinaris*	5–100	Ca, Cq	Ps, Vt	Cs, Ds, Ng
PCV	*A. hypogaea*	6–14	Ca, Cq, Nb	Nb, Ng, Te, Ta	–
PeMoV	*A. hypogaea*	0–20	Pv, (Ca)	Gm, Nc	Cs, Gg, Ds, Nt
PSV	*A. hypogaea*	0.1	Cq, Pv, Vu	Ds, Nt, Vu	Bn, (Gm), St
PMMoV	*Capsicum* spp.	22–29	Ntx, Ntw, Ng, Ca, Cq	Cc[h]	Cs, Gg, Le
PPV	*Prunus* spp.	+	Cf	Nb, Nc	Ca, Cs
PSTVd	*Solanum tuberosum, Lycopersicon esculentum*	0–100	–	Le[i]	–
PDV	*Prunus* spp.	+	Csp, Se	Cs	Ca, Cq, Gg
PNRSV	*Prunus* spp., *Rosa* spp., *Cucumis sativus*	≤ 70	Ct	Cs	Csp, Cm
SBMV	*P. vulgaris*	1–30	Pv[f]	Pv[f]	Ca, Cq, La
SCPMV	*V. unguiculata* sspp.	1–40	Vu[j]	–	Ca, Cq, La
SMV	*Glycine max*	0.1–30	Ct, Pvt[k]	–	Cs, Ph, Vf

Table 4.1.

SqMV	*Cucumis* spp., *Cucurbita* spp.	10–35	Cmet[l]	Cp, Cm	Ca, Ds, Ng
SCMoV	*Trifolium subterraneum*	3	Ps[m]	–	Ca, Cq, Cs, Nb
TMV	*L. esculentum, Capsicum* spp., *Nicotiana tabacum*	+	Cq, Ng, Ntx	Ntw	Ps, Ph, Ze
TSV	*G. max, P. vulgaris*	0–90	Ct, Pv[n]	–	Ds, La
TAV	*P. vulgaris*	19	Ca, Cq, Vu	–	–
ToMV	*L. esculentum, Capsicum* spp.	+	Ntx, Ntw	Nc	Cs, Pv
ULCV	*Vigna mungo, V. unguiculata*	6–18	–	Cs[o], Lc, Vm	Ca, Ds
ZYMV	*Cucumis* spp., *Cucurbita* spp.	0–5	Cp, Cl	Cs, Cm	Cq, Ca, Gg

*Sources: e.g. Brunt *et al.* (1996), Frison *et al.* (1990), *AAB Descriptions of Plant Viruses.*
**Full names in Table 4.2.
– No information.
[a]Cv. Widusa.
[b]Cvs Dubbele Witte, Stringless Green Refugee.
[c]Cv. Tvu 1582, Tvu 401.
[d]Cv. Prince.
[e]Cv. Tendergreen.
[f]Cv. Pinto (UI 114).
[g]Cvs TVu 2657, TVu 3433.
[h]Severe systemic necrosis.
[i]Cv. Rutgers.
[j]Cv. Clay.
[k]Detached leaves, 30°C.
[l]Some collections, e.g. PI 202681 or 292190.
[m]Cv. Greenfeast.
[n]Cv. Manteiga.
[o]Cv. National Pickling.

would also be quickly destroyed by heat during power failures. The use of screen houses would be better in these climates. Temperatures close to ambient can be maintained in a screen house consisting of, for example, a simple steel tube framework clad with insect-proof netting. A 32×32 mesh netting excludes aphids and also whiteflies, though less effectively (Rossel and Ferguson, 1979). Such houses, equipped with double doors and an entrance lock, are available as a do-it-yourself kit from Clovis Lande Associates Ltd, UK (http://www.clovis.co.uk/virology), which also supplies planting benches of stainless metal.

An alternative or supplemental solution in warm climates could be the establishment of growth rooms inside an existing building. In such rooms, temperature regulation by means of an air conditioner would be cheaper and safer than in a greenhouse, and optimal temperatures for virus infectivity assays could be obtained with a programmable thermostat. Further, growth chambers are easier to keep free of insect vectors and safer for work with exotic diseases than outdoor greenhouses. To optimize plant light conditions in a growth chamber, plants should be placed on shelves illuminated with several fluorescent tubes per shelf mounted inside reflectors and positioned as close as possible to the plants. Fluorescent tubes emitting 'cool white light' (standard daylight tubes) are superior to the alternative types of fluorescent tubes by emitting higher light intensity

Table 4.2.A. Abbreviations of indicator species in Table 4.1.

Abbreviation	Indicator species
Bv	*Beta vulgaris*
Bn	*Brassica napus*
Bp	*Brassica pekinensis*
Cc	*Capsicum chacoense*
Ca	*Chenopodium amaranticolor*
Cf	*Chenopodium foetidum*
Cq	*Chenopodium quinoa*
Cl	*Citrullus lunatus*
Cj	*Crotolaria juncea*
Csp	*Crotolaria spectabilis*
Cm	*Cucumis melo*
Cmet	*Cucumis metuliferus*
Cs	*Cucumis sativus*
Cma	*Cucurbita maxima*
Cp	*Cucurbita pepo*
Ct	*Cyamopsis tetragonoloba*
Ds	*Datura stramonium*
Gm	*Glycine max*
Gg	*Gomphrena globosa*
Hv	*Hordeum vulgare*
Lc	*Lagenaria cylindrica*
Lt	*Lavatera trimestris*
La	*Lupinus albus*
Le	*Lycopersicon esculentum*
Ms	*Medicago sativa*
Nb	*Nicotiana benthamiana*
Nc	*Nicotiana clevelandii*
Ng	*Nicotiana glutinosa*
Nt	*Nicotiana tabacum*
Nts	*N. tabacum* cv. 'Samsun NN'
Ntw	*N. tabacum* cv. 'White Burley'
Ntx	*N. tabacum* cv. 'Xanthi-*nc*'
Ph	*Petunia hybrida*
Pv	*Phaseolus vulgaris*
Pvt	*P. vulgaris* cv. 'Top Crop'

Table 4.2.A.

Ps	*Pisum sativum*
Se	*Sesbania exalta*
St	*Solanum tuberosum*
Te	*Tetragonia expansa*
Ta	*Triticum aestivum*
Vf	*Vicia faba*
Vm	*Vigna mungo*
Vu	*Vigna unguiculata*
Ze	*Zinnia elegans*

Table 4.2.B. Acronyms of viruses in Table 4.1.

Acronym	Virus	Genus[a]
AMV	*Alfalfa mosaic virus*	*Alfamovirus*
BSMV	*Barley stripe mosaic virus*	*Hordeivirus*
BCMNV	*Bean common mosaic necrosis virus*	*Potyvirus*
BCMV	*Bean common mosaic virus*	*Potyvirus*
-BlCM	Blackeye cowpea mosaic strain	
-PSt	Peanut stripe strain	
BYMV	*Bean yellow mosaic virus*	*Potyvirus*
BBSV	*Broad bean stain virus*	*Comovirus*
BBTMV	*Broad bean true mosaic virus*	*Comovirus*
CABMV	*Cowpea aphid-borne mosaic virus*	*Potyvirus*
CPMV	*Cowpea mosaic virus*	*Comovirus*
CPSMV	*Cowpea severe mosaic virus*	*Comovirus*
CGMMV	*Cucumber green mottle mosaic virus*	*Tobamovirus*
CMV	*Cucumber mosaic virus*	*Cucumovirus*
IPCV	*Indian peanut clump virus*	*Pecluvirus*
LMV	*Lettuce mosaic virus*	*Potyvirus*
PEBV	*Pea early-browning virus*	*Tobravirus*
PSbMV	*Pea seed-borne mosaic virus*	*Potyvirus*
PCV	*Peanut clump virus*	*Pecluvirus*
PeMoV	*Peanut mottle virus*	*Potyvirus*
PSV	*Peanut stunt virus*	*Cucumovirus*
PMMoV	*Pepper mild mottle virus*	*Tobamovirus*

Continued

Table 4.2.B. *Continued.*

PPV	*Plum pox virus*	*Potyvirus*
PSTVd	*Potato spindle tuber viroid*	*Pospiviroid*
PDV	*Prune dwarf virus*	*Ilarvirus*
PNRSV	*Prunus necrotic ringspot virus*	*Ilarvirus*
SBMV	*Southern bean mosaic virus*	*Sobemovirus*
SCPMV	*Southern cowpea mosaic virus*	*Sobemovirus*
SMV	*Soybean mosaic virus*	*Potyvirus*
SqMV	*Squash mosaic virus*	*Comovirus*
SCMoV	*Subterranean clover mottle virus*	*Sobemovirus*
TMV	*Tobacco mosaic virus*	*Tobamovirus*
TSV	*Tobacco streak virus*	*Ilarvirus*
TAV	*Tomato aspermy virus*	*Cucumovirus*
ToMV	*Tomato mosaic virus*	*Tobamovirus*
ULCV	Urdbean leaf crinkle virus	Unassigned
ZYMV	*Zucchini yellow mosaic virus*	*Potyvirus*

[a]Family affiliation, see Appendix 1.

(Vandre, 1996; Jauron, 1997). Moreover, standard daylight tubes are less expensive. The illumination system must include a programmable timer for selecting different photoperiod cycles. Using 16–18 h photoperiods should optimize light doses.

Whatever growth facility is used should have a water outlet/sink, washable floor and benches/shelves and a floor drain. In both the facilities and the work with bioassays, emphasis must be placed on: (i) excluding unwanted inoculum (especially viruliferous insects); and (ii) preventing escape of inoculum. At quarantine stations the latter is crucial. All openings in greenhouses must be secured with insect-proof netting, and a double door system providing an entrance lock is required. Plant debris and other waste from biological assays must be stored and transported in closed containers to the place where it can be destroyed by burning or sterilization. See also Prevention of contamination, Section 4.4.2.

4.2.2. Plant culture materials and equipment

Among commercially available growth media, potting soil that is pathogen- and vector-free is recommended. If this is not available, either an airy soil mix, consisting of peat soil, mineral soil and sand, or a mix of one-third mineral soil, one-third sand and one-third composted farmyard manure is usable after heat sterilization. Alternatively, a sterile growth medium like rock wool can be used. In warm climates, solarization may be sufficient treatment to eliminate vectors and

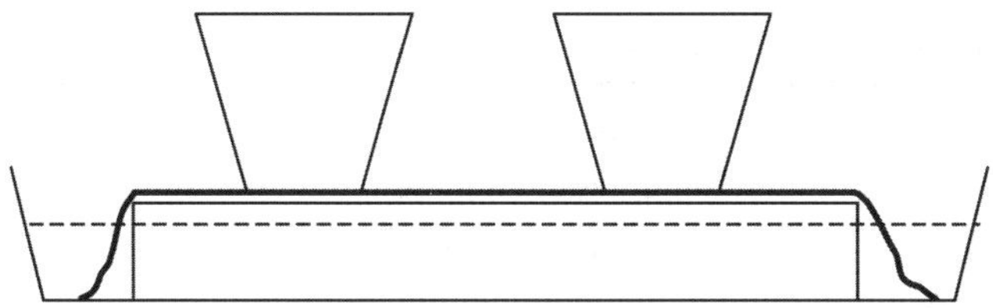

Fig. 4.4. Sketch of a system for water supply to pot plants. A capillary mat (heavy line), placed on a box or similar, placed upside down in the tray, supplies the plants in pots with perforated bottoms with the water they need. All four edges of the mat must be in contact with the water in the tray. The tray can contain water for several days and the plants do not suffer from water saturation of the soil.

pathogens from the soil mix. Information on soil solarization and links to relevant websites can be found at http://www.aces.edu/department/ipm/soils.htm. The soil mix should have a neutral pH and, if needed, be supplied with fertilizer to ensure normal growth.

Seeds and transplants are planted in pots (plastic or earthenware) with drain holes or in small trays with drains, placed in larger trays or on a clean plastic sheet. The large trays (without drains) can function as water reservoirs. However, watering must be adjusted so that the soil is water-soaked only for short periods. Automatic watering systems, being optimal for normal plant cultivation, could be problematic for assays for the presence of pathogens. A simple but efficient system using a capillary mat is shown in Fig. 4.4. Capillary mats are available from gardening supply companies at prices of about US\$1/m^2 (Denmark).

4.3. Growing-on Tests

A simple and often efficient way to determine whether seeds contain seed-transmitted viruses is to subject the seed sample to a growing-on test. Planting of seeds and growing of the progeny seedlings for 2–4 weeks under optimal conditions will in many cases (depending on the virus species and the host) reveal seed-transmitted virus infection by the appearance of symptoms. Further, and most importantly, only seed-to-plant-transmissible virus infections will be revealed (Section 2.1). As indicated, the availability of a vector-free growth facility and optimal germination and growth conditions are essential for reliable results. Optimal symptom development depends not least on sufficient light intensity; in growth rooms with artificial light, the daily light doses could be increased by using long photoperiods (16–18 h). Irrespective of light conditions, however, symptomless (latent) seedling infection or seedlings with only faint, hardly recognizable symptoms occur for a number of virus–host combinations (Section 4.1). Latent seedling infection is, for example, common among nematode-transmitted viruses (Lister and Murant, 1967), and only faint symptoms often occur with, for example, 'legume' strains of CMV in bean, soybean and cowpea.

In addition to that of revealing seed-to-plant-transmitted virus only, another advantage of growing-on tests is that the virus content is normally higher in

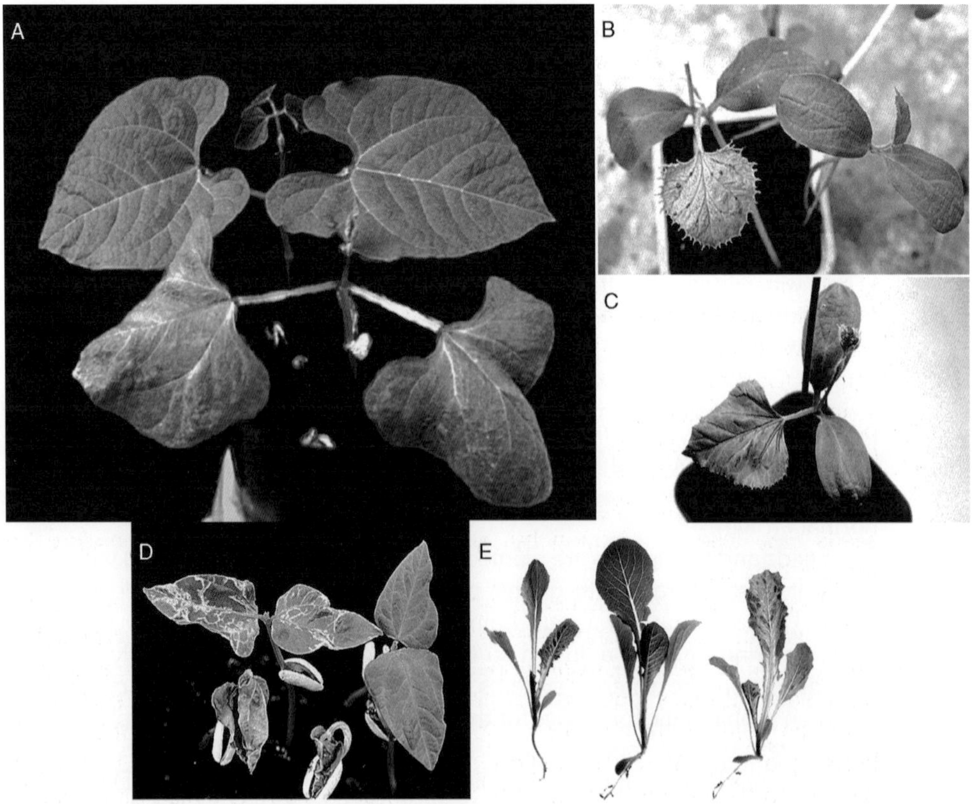

Fig. 4.5. Symptoms of seed-transmitted viruses on seedlings raised from infected seeds. (A) Malformed primary leaves of *Phaseolus vulgaris* caused by BCMV; note the characteristic abnormality of leaf bases; healthy seedling behind; (B) chlorosis and (C) vein banding on primary leaves of *Cucumis melo* due to SqMV; cotyledons are symptomless; (D) early vein chlorosis on *Vigna unguiculata* ssp. induced by a *Potyvirus* (CABMV or BCMV-BICM); healthy seedling on the right; (E) mottling and mosaic on *Lactuca sativus* caused by LMV; healthy seedling in the middle.

seedlings than in seeds and germinated seeds, and therefore easier to detect in a subsequent infectivity, serological or molecular assay. Higher virus concentration in progenies allows a larger number of samples to be pooled in group testing compared to whole-seed testing, thus reducing the cost of assays (Sample preparation, Section 5.2.2 and 7.2).

For establishing virus-free seed samples, e.g. for germplasm, the growth of seedlings can be continued (grow-out test) in containment until plant maturity and seeds can be harvested from plants proved to be virus-free (Section 1.4.2).

Visual inspection

Seedlings should be at least 2–3 weeks old, for scoring of symptom-bearing plants. In many virus–host combinations, only secondary leaves show symptoms, i.e. primary leaves and/or cotyledons remain symptomless. Early seedling symptoms caused by seed-transmitted viruses vary considerably in type, and range from faint to conspicuous. The most common symptoms are colour variegation and malformation; necrotic symptoms are less common. Examples of seedling symptoms are shown in Fig. 4.5.

4.3.1 Growing-on test, standard protocol

Following the International Seed Testing Association (ISTA) representative sampling procedure, the required number of seeds for testing is removed from the sample (Section 7.2). The number of seeds to be tested from a seed lot is governed by the maximum tolerable level of seed infection for the virus concerned (Section 7.2). Seeds to be tested for seed-transmitted viruses, especially for large-seeded species, should be treated with fungicides or surface-treated with 1% NaOCl for 1 min, followed by rinsing in water, before planting. Small-sized seeds may lose germination by NaOCl treatment and should therefore be fungicide-treated.

Treated seeds are planted in pots or small trays, as described in Section 4.2.2 above. Growth conditions should be conducive to optimal germination and growth. For seeds of tropical species, e.g. cowpea and groundnut, a temperature above 25°C is required for optimal germination. Alternatively, such seeds could be pre-germinated on moist paper in Petri dishes or similar at 25–30°C before planting. After germination, the temperature should be lowered a few degrees to facilitate normal growth; for tropical species the temperature should not be below 20°C.

EQUIPMENT AND MATERIALS.

a. Clean pots or trays with drain.

b. Clean trays without drain, or plastic sheets.

c. Pathogen- and vector-free soil mix, or a sterile growth medium (rock wool or similar).

d. Fertilizer, if not already added to the growth medium.

e. NaOCl, 1%, or fungicide (optional, but preferable).

f. Labels for pots or trays, and pencil or permanent marker.

g. Vector-free growth facility.

PROCEDURE.

1. Sample and count the required number of seeds. Sample according to the ISTA standard sampling procedure (see above).

2. Seed treatment: large-seeded species can be dipped, holding the seeds in a sieve or a piece of gauze, in 1% NaOCl for 1 min, followed by rinsing in water. NaOCl is not advisable for small seeds. Alternatively, treat the seeds with a fungicide. **Caution:** during seed treatment, wear protective gloves and mask and work in a fume cupboard.

3. Fill the pots or trays with growth medium and plant the seeds 3–6 cm apart, depending on seed size.

4. Label with date and seed sample number and place the planted seeds in a vector-free growth room or greenhouse for at least 2–3 weeks under optimal germination and growth conditions, as described above.
5. After emergence of seedlings, count the number of total emerged seedlings and of symptom-bearing seedlings, if any, after 2, 3, up to 4 weeks after sowing and/or harvest seedling material for testing by a biological, serological or molecular assay. On the basis of the number of virus-positive seedlings/total emerged seedlings, a seed infection percentage can be determined.
6. If desired, compare the number of symptom-bearing seedlings with the number of seedlings found virus-positive in the confirmative assay, in order to reinforce familiarity with seedling symptomatology.
7. After the growing-on test, do not place the remains (soil and seedlings) outdoors, especially if assaying for viruses that are quarantine objects. Instead, keep them in closed containers until infectious agents are destroyed by autoclaving or burning to prevent spreading to the environment (can happen quickly by insect vectors).

4.4. Infectivity Assays

The use of infectivity assays in testing for viruses is usually limited to: (i) inoculation of indicator plants for tentative virus identification; (ii) verification of serological or molecular test results; and (iii) virus strain determination using differential hosts. Among essential factors for optimal results are the choice and optimal quality of indicator host plants, correct inoculation techniques, optimal and vector-free growth conditions and familiarity with virus symptomatology (Section 4.1).

4.4.1. Production of indicator plants

The success of infectivity tests is highly dependent on the choice and strain of indicator species and the age and quality of the plants. Many species express maximum sensitivity only at a certain developmental stage. For example, *Chenopodium amaranticolor* typically reaches that stage with three well-developed secondary leaves, and *Nicotiana clevelandii* with five to six leaves. Primary leaves of bean (*Phaseolus vulgaris*) and cowpea (*Vigna unguiculata*) should be used for inoculation; the next-coming leaves are less susceptible. The very young primary leaves of bean and very young cotyledons of cucumber (*Cucumis sativus*) are resistant to several viruses. These plants do not reach their optimum susceptibility until they are about 10 days old, and become resistant after about 15 days (Horvath, 1993). The optimal developmental stage for a number of commonly used indicator species and their ages are shown in Table 4.3.

The susceptibility of test plants also depends on their genotype (variety) and physiological condition. Vigorously growing plants have a higher susceptibility than slow-growing, weak plants; therefore, optimal and reproducible growth conditions are essential for reliable and reproducible test results.

Table 4.3. Commonly used test plant species and cultivars, with their age, and stage of development (number and/or type of leaves) suitable for inoculation.[a]

Species/cultivar	Age (days)	Stage of development
Beta vulgaris	20–25	2
Brassica pekinensis	20–25	4–5
Capsicum annuum	35	3
Chenopodium amaranticolor	28	3 well-developed
Chenopodium quinoa	28	4 well-developed
Crotalaria juncea	8	Cotyledons
Cucumis sativus	10	Cotyledons
Cyamopsis tetragonoloba	10	Cotyledons + first undivided
Datura stramonium	22	2 well-developed
Glycine max	14	Cotyledons + 1
Gomphrena globosa	40–50	2-3 pairs
Helianthus annuus	16	1 pair
Lycopersicon esculentum	21–28	2–3
Nicotiana benthamiana	35	3 well-developed
Nicotiana clevelandii	35	5–6
Nicotiana glutinosa	35	2 well-developed
Nicotiana occidentalis-37B	30	5
Nicotiana rustica	30	1 well-developed
Nicotiana tabacum		
'Samsun NN'	28–30	2 well-developed
'White Burley'	28–30	2 well-developed
Petunia hybrida	29	3 well-developed
Phaseolus vulgaris	10	2 primary
Physalis floridana	14 (24)	2 (4)
Pisum sativum	13	2
Solanum melongena	30	2
Tetragonia expansa	28	8
Trifolium incarnatum	35	3–4
Vicia faba	14	1
Vigna unguiculata[b]	10	2 primary
Zinnia elegans	24	2 well-developed

[a]All plants are grown in a greenhouse with an average temperature of 20°C. All data are annual averages and may vary according to the season.
[b]For *V. unguiculata* an average temperature of 23°C is optimal; temperatures below 20°C should be avoided.
Table reproduced from Table 1 in Dijkstra and de Jager (1998) with permission from authors and the copyright holder © Springer-Verlag.

Small-seeded species like tobaccos and *Chenopodium* spp. are usually sown in large numbers in a pot or tray and transplanted from there into individual pots as soon as the seedlings reach a transplantable size, i.e. normally 1–2 weeks after sowing. Seedlings that are too old before transplanting produce poor test plants. Seeds of large-seeded species (legumes, etc.) are planted directly into individual pots.

Seeds of *N. clevelandii* usually do not germinate unless treated by presoaking overnight in water containing 10% Tween-20. *Nicotiana benthamiana* is a good alternative that can be used in many cases, is a readily germinating species and is susceptible to over 200 viruses belonging to 23 different genera (Bos, 1999). *Chenopodium amaranticolor* tends to flower at a very early development stage when grown under a 12/12 h light regime (such as in the tropics, as experienced by the author) and is then useless as an indicator. A photoperiod of 14–16 h can solve the problem.

Indicator plant seed is obtainable from different sources, such as plant pathological centres and botanical gardens, and can also be obtained from, for example, DSMZ – Plant Virus Collection, Germany (http://www.dsmz.de/nf-plvirus/). Seeds of most species are easy to propagate in the greenhouse. Some indicator plant species are known to be readily infected by seed-transmitted viruses (example: cowpea), which must be considered when propagating seed. It is advisable to have a gardener or a person with relevant training to be in charge of both the propagation of indicator plant seed and production of test plants.

4.4.2. Inoculation techniques and standard protocol

Preparation of inoculum

When preparing inoculum, infected plant parts (usually leaf pieces) are homogenized in water or buffer. A final dilution of 1/5–1/10 of the extract helps to dilute out the infection-inhibiting substances present in some plant species. This also applies for seed extracts. The majority of seed-borne viruses are rather stable and tolerate dilution in water; however, many workers routinely homogenize and dilute with a standard phosphate buffer at 0.01–0.1 M and pH 7–8. And, for inoculating labile viruses or viruses from seeds or dry leaf samples, a phosphate buffer with virus-protecting and anti-inhibitor additives is definitely recommended (see Inoculation: standard protocol below).

Inoculum from plants should be taken from young (but not the youngest) leaves showing clear symptoms, as these parts usually have the highest virus concentration. Seed extracts are best when made from seeds that have been soaked in water for some hours or overnight at 5°C, but dry seed flour can also be processed in this way. Homogenization of tissues together with 1–5 vols of water or buffer (seed flour, up to 10 vols or more) can be done with a mortar and pestle or, if leaves, in a sturdy plastic bag by thoroughly homogenizing against a solid support (table) the enclosed tissue and liquid from outside the bag with a pestle or a hand-held homogenizer (see protocol).

After further dilution of the slurry with water or buffer to a final dilution of 1/5–1/10 (dry seed flour, more diluted), the extract is ready for inoculation. If not inoculated immediately, the extract should be kept at 4°C or on ice, and for a short time only.

ABRASIVES. To infect plants with prepared inoculum, the virus must be actively introduced into plant cells. In mechanical inoculation of plant parts, usually the leaves, an abrasive, such as carborundum (silicon carbide) or Celite, is used to create micro-wounds in epidermis cells when rubbing the leaf surface, allowing virions in the inoculum suspension to enter into the cells. The abrasive carborundum powder (silicon carbide, 500–600 mesh) contains angular microparticles, small enough not to cause excessive damage to the plant, provided that only a small amount is present during the inoculation. Therefore, when dusting carborundum on to the leaves prior to inoculation, a thin, *hardly visible* uniform layer produces the best results. If Celite (diatomaceous earth) is used, this abrasive is usually added to the inoculum at final concentrations of 1–5% (w/v).

CONTROLS. In infectivity assays, one or more negative-control plants should be used, i.e. a plant that has been abraded with extract from a known healthy plant of the same species as the inoculum source (i.e. mock inoculum) and/or a plant inoculated with buffer and abrasive only. Symptoms caused by mechanical damage to negative-control plants can thus be compared to plants with virus symptoms. Positive controls, which are plants inoculated with known infected material (if the virus tested for is known), provide examples of specific virus symptoms.

Inoculation

Plants to be inoculated for infectivity tests must have fully turgid leaves for optimal results.

Application of inoculum to abrasive-dusted test plants can be done in several ways. Use of the forefinger, small foam blocks or glass spatulas to rub the leaf surface is common. Good results can also be obtained with disposable cotton swabs rubbed over the leaf surface. As previously mentioned, a thinly dusted carborundum layer produces the best result; excessive layers cause leaf damage. Apply the inoculum with long strokes from petiole to the tip, but do not rub more than once (or maximum twice) on the same leaf area (Fig. 4.6). Make sure that the entire leaf surface is inoculated. Immediately after inoculating the plant, rinse the leaves with tap water from a squeeze bottle or comparable. The rinsing and subsequent keeping of the inoculated plants in a humid environment for some hours reduce the risk of wilting of the inoculated leaves due to increased evaporation after the use of abrasive. After inoculation, avoid contact of plants to plants, healthy or inoculated with other viruses.

The susceptibility of test plants is usually increased if kept for 24–48 h in darkness and/or at elevated temperatures prior to inoculation. In contrast, after inoculation, plants should be kept under normal light and temperature conditions. In warm climates plants should be inoculated late in the day to avoid excessive post-inoculation temperatures.

Fig. 4.6. Mechanical inoculation.

PREVENTION OF CONTAMINATION. Although virus bioassay work does not require sterile conditions, certain hygienic precautions must be taken to avoid unwanted infections when inoculating and handling the plants. Some viruses remain infective outside their hosts for long periods. One of the most stable ones is TMV, which remains infective in dry tissue, e.g. smoking tobacco, for many years. A number of viruses can be easily transmitted by contact between plants or by touching plants with virus-contaminated hands or tools. The following precautions are important to prevent contamination (essentially as by Dijkstra and de Jager, 1998):

- Never smoke in the laboratory or greenhouse/growth room.
- Wash hands thoroughly with soap and water before and after inoculation. Use an alkaline soap.
- Clean the tables, equipment, etc. with soap and water or, preferably, with 10% trisodium phosphate, which inactivates viruses.
- The carborundum dust container may often be a source of contamination; wash hands very thoroughly after dusting the plants, prior to their inoculation.
- After washing hands, only the plants to be inoculated should be touched, and nothing else.
- Ensure that paper towels and inoculation tools, such as cotton swabs, are virus-free.
- Discard disposable materials (plastic bags, cotton swabs, etc.) after use.
- Never leave used equipment on benches.
- Plants inoculated with different virus suspensions should not be allowed to touch each other or non-inoculated plants.

- When inoculating different viruses on the same day, use a pair of disposable gloves for each new inoculation, in addition to the use of water and soap.
- Clean mortars and pestles thoroughly after use. An effective way is to scour first with an abrasive and then with water and soap (or 10% trisodium phosphate), followed by rinsing in tap water. Heat-treat in an incubator at 180°C for 30 min or 120°C for 2 h. Glassware is washed with water and soap, followed by rinsing in deionized water and dry heating at 120°C for 2 h.
- As a general rule, never touch with bare hands any plants in an infectivity test room, e.g. when inspecting for symptoms. Change gloves between each new virus–host to be examined.

Inoculation: standard protocol

The following are the steps of a general inoculation procedure. For details, see above. The inoculation buffer b below is recommended by the reagent and virus isolate supplier, DSMZ – Plant Virus Collection, Germany, for inoculum prepared from dried leaf samples (Section 4.5). The buffer is also recommended for inocula prepared from whole seeds, as well as for general use.

Always perform the infectivity test in the following sequence: negative control, sample and, if used, positive control.

EQUIPMENT AND MATERIALS.

a. Test plants for inoculation and for controls. The test plants must be at the correct developmental stage and with fully turgid leaves.

b. Inoculum source; if seed, it should preferably be presoaked in water for some hours or overnight. Dry seed flour may also be used.

c. Healthy material from the same species as the inoculum source, for inoculating negative controls.

d. Known infected material for positive control (optional). It may be in the form of fresh tissue from an infected plant or a purified virus suspension, kept at 4°C after adding an equal volume of glycerol. Infective, dehydrated plant material may also be used.

e. Carborundum powder, 500–600 mesh, in a sprinkler (e.g. a beaker covered with two to three layers of gauze). **Caution:** wear protective mask to avoid inhaling the dust.

f. Celite (diatomaceous earth): alternative to carborundum.

g. Mortars and pestles, pre-cooled, *or*

h. Polyethylene bags, sturdy type, approx. 7 × 10 cm. A pestle or a hand-homogenizer (available from, for example, Bioreba™; see Appendix 3) can be used to grind from outside the bags.

i. Cotton swabs, disposable, or disposable foam blocks, or glass spatulas.

j. Squeeze bottle with tap water.

k. Buffer a: phosphate buffer, 0.05 M, pH 7.5, *or*

l. Buffer b: inoculation buffer with additives ('DSMZ buffer', see above): 0.1 M Na_2HPO_4/KH_2PO_4 buffer, pH 7.5, containing 2% (w/v) polyvinylpyrrolidone and 0.2% (w/v) Na_2SO_3. (Phosphate buffers, see Appendix 2, p. 247.)

m. Disposable gloves.

n. Hand soap, alkaline.

o. Trisodium phosphate (Na_3PO_4), 10% solution (virus-inactivating cleaning agent).

p. Pot labels and pencil or permanent marker (not ballpoint pen).

q. Paper towels.

r. Gauze.

s. Insect-free greenhouse or growth room with an average temperature of 20–22°C.

PROCEDURE.

1. Wash hands with water and soap.

2. Arrange the plants to be inoculated. Use per experiment minimum three plants for test inoculum, and generally two plants for negative control (one non-inoculated and one mock-inoculated, i.e. using extract from healthy plant or seed material of same species as the inoculum source, and/or one inoculated with buffer and abrasive only). Also, include one or two plants for a positive control, if desired.

3. Label with date, inoculum specification and an inoculation number.

4. Dust the test plant leaves with a uniform, thin, i.e. *hardly visible*, layer of carborundum powder. Note that excess abrasive absorbs too much inoculum and may lead to serious injury of leaves. **Caution:** use protective mask when dusting. Alternatively, use Celite; see below.

5. Wash hands with water and soap and put on clean gloves when required (see precautions above).

6. Before preparing the inoculum, inoculate a negative-control plant as described below but with mock-inoculum (healthy extract) and/or one with buffer or water alone, using abrasive. Leave a non-inoculated plant as well.

7. Prepare the inoculum –

(a) From leaves: remove from the source plant one or more pieces (2–5 g) of younger leaves showing clear symptoms. Grind with 1–2 vols of ice-cold water or buffer, using mortar and pestle, or use a plastic bag, homogenizing from outside the bag with a pestle or a hand-homogenizer. Dilute the slurry in ice-cold water or buffer to give a final tissue concentration of 1/5–1/10.

(b) From seed: water-presoaked seeds are ground thoroughly by mortar and pestle in approx. 5 vols of cold buffer b and diluted further to make a thin paste. Instead of soaked seeds, dry, finely milled seed flour can also be used, but should be ground even more thoroughly with buffer to release virus particles.

(c) Filter by squeezing the suspension through several layers of gauze (especially needed for seed suspensions).

(d) If Celite is used instead of carborundum, add 1–5% (w/v) to the final seed or leaf suspension.

(e) Go immediately to step **8**. If not, keep the suspension on ice or at 4°C, but not for more than 2 h.

8. Dip a cotton swab into the suspension and apply the liquid on to the leaves (dusted, if carborundum is used) by gentle one-way strokes from petiole to tip while supporting the leaf from beneath. Inoculate the entire leaf surface, dipping the swab frequently in the liquid; rub only once or maximum twice on the same area. Other than cotton swabs, the gloved forefinger, a foam block or a glass spatula can be used for inoculation (Fig. 4.6).

For a positive control, if used, see **Note** below.

9. Immediately after plant inoculation, i.e. after maximum 2–5 min, rinse the leaves with tap water from the squeeze bottle. (The abrasive–inoculum mix is hard to remove, if dried up, and it can obscure symptom observation.)

10. Place the inoculated plants in an insect-free greenhouse or growth room with a 16 h photoperiod. Avoid excessively high temperatures, at least during the next 12 h. Keeping the inoculated plants in a humid environment for some hours reduces the chance for wilting (see Inoculation above).

11. Discard disposable materials, wash hands and clean the tools as described under Prevention of contamination above.

12. Inspect the plants daily. Local symptoms may occur after only 2 days and symptoms of systemic infection after 1–2 weeks. Record the symptoms.

13. It is advisable to enter into laboratory notes all data and observations made in each infectivity assay. By assigning a number to each inoculation, it is easy to trace back, for example, the origin of an isolate used and its history.

Note

If positive control(s) are used, then, *after* having inoculated with the test sample, prepare the known infected inoculum as described in step **7** and inoculate the control plant(s) as described in step **8**.

4.4.3. Seed as inoculum, protocols

In infected seed from some virus–host combinations, the content of infective virus is high enough to establish infection in susceptible indicator plants by simply inoculating with seed extracts. PEBV-infected pea seeds soaked overnight and ground in a mortar are an excellent source of PEBV, resulting in local lesions when inoculated onto *C. amaranticolor*, cucumber or *P. vulgaris* (Bos and van der Want, 1962, quoted from Bos, 1977, and Frison *et al.*, 1990). Wilson and Dean (1964, quoted from Phatak, 1974) used flour of bean seeds infected with BCMV as inoculum (Protocol I below), even after seed storage exceeding 4 years. A very sensitive bioassay for LMV in lettuce seed was described for routine use by Rohloff and Marrou (1981) (Protocol III below).

These methods have now generally been replaced by serological or molecular methods. However, for isolates kept as infected seeds, such methods can also be used to obtain and propagate virus for study or as positive control, etc., instead of obtaining it from seedlings raised from the seeds.

A method still in use is bioassay of tomato and pepper seed for infective tobamoviruses carried in testa and endosperm (Broadbent, 1965; Rast and Stijger, 1987); see Protocol II below, which also describes the elimination of these viruses from seed, etc.

Seeds to be tested must be representatively sampled according to the ISTA rules (Section 7.2). The number of seeds to be tested from a seed lot depends on the maximum tolerable level of seed infection of the virus concerned (Section 7.2).

The following three protocols are examples of infectivity assays, with seeds as inoculum.

I. Protocol for detection of Bean common mosaic virus *(BCMV) and* Bean common mosaic necrosis virus *(BCMNV) in seed*

As mentioned above, BCMV can be detected by using flour of infected bean seeds as inoculum. Quantz (1962) detected BCMV from bean seeds, pre-germinated for 3–4 days and inoculated onto *P. vulgaris* cv. 'Top Crop', which reacted with necrotic local lesions on detached leaves incubated for 3 days on moist filter paper. For the detached leaf method, see the protocol for SMV detection, Section 4.4.4. At DGISP, the method described below was used in class exercises. Extracts of bean seeds, soaked overnight, are inoculated on to systemically reacting cultivars of *P. vulgaris* for detection of BCMV and BCMNV.

EQUIPMENT AND MATERIALS. Equipment and materials are the same as those used in the standard protocol (Section 4.4.2, p. 65), except for the specific components:

a. Seeds of *P. vulgaris* to be assayed.
b. Seeds of *P. vulgaris* known to be healthy.
c. Infected plant or purified BCMV or BCMNV for positive control (optional).
d. Indicator plants of *P. vulgaris* cv. 'Dubbele Witte' or 'Stringless Green Refugee'. These cultivars react systemically to all strains of BCMV and BCMNV (Drijfhout, 1978). Use plants with two fully developed primary leaves (Table 4.3). Alternatively the local-reacting cv. 'Top Crop' can be used, but it may then require extracts from pre-germinated seeds as inoculum, see above and **Note 3** after procedure.
e. Mortars and pestles.
f. Petri dishes.

Not needed
Polyethylene bags, homogenizer and buffer a.

PROCEDURE.
Day 1
1. Take out a working sample from the seed lot according to the ISTA sampling procedure. The number of seeds to be tested depends on the purpose of testing (identification, or estimating seed infection rate; see Section 7.2).
2. Place the seeds in a container; add 2–3 vols of tap water and soak the seeds overnight.

Day 2
3. Prepare for inoculation as in steps **1–6** in the standard protocol, p. 66.

4. Drain off the water and divide the sample into groups of ten seeds (see **Notes 2** and **3** below) and grind each group individually with mortar and pestle together with 3–5 vols of *buffer b*. Dilute further with buffer to make a thin paste. Filter a part of the paste by squeezing through several layers of gauze, and collecting the filtrate in a clean Petri dish. Avoid cross-contamination between the groups.
5. Inoculate two or three bean plants per seed group as described in the standard protocol step **8**, taking care to avoid cross-contamination between the groups.
6. Positive control (optional): inoculate two plants with diluted known infected-plant sap or 10–50 times diluted purified BCMV or BCMNV.
7. Continue with steps **9–13** in the standard protocol. These test plants will after 1–2 weeks react systemically to BCMV or BCMNV if present in the seed. Typical symptoms are dark green/light green mosaic, malformed leaves and stunting.
8. By this test not only can presence of BCMV and BCMNV be determined, but also an estimate of the rate of seed infection by using the maximum likelihood estimate (MLE) formula (Section 7.2) may be possible. The reliability of the calculated MLE requires proof that no cross-contamination between groups has taken place. See also **Note 1**.

Notes

1. It should be noted that, although the two potyviruses BCMV and BCMNV are most likely to occur seed-transmitted in seed of *P. vulgaris*, relatively high seed infection rates in beans are also reported for other viruses, especially the two cucumoviruses CMV and *Tomato aspermy virus* (TAV), the sobemovirus SBMV and the ilarvirus TSV. These viruses may therefore also be detected in this test, but with symptoms differing from those of BCMV/BCMNV. However, in the case of bean seed samples suspected of carrying these other viruses or of having mixed infection, additional tests, involving a growing-on test with subsequent use of specific indicator plants, or serological or molecular tests, are required to determine the identity of the virus(es).

2. The indicated group size of ten seeds is arbitrary. The sensitivity of the test using this size has apparently been sufficient in experimental results at DGISP using a bean sample with a rather high infection level (unpublished results). However, the maximum group size allowing detection of one BCMV/BCMNV-infected seed out of *x* healthy seeds by this method has to be determined and can be done as follows: infected seeds from a bean sample are identified by testing individual half-seeds as described above, keeping track of their corresponding seed-halves, which are stored dry. When one or more infected seed-halves(s) are identified, the matching seed-halves are soaked, ground and diluted. A similarly diluted extract of known healthy seeds is then made. A two-fold dilution range of infected extract in healthy extract is made, and each dilution inoculated onto the susceptible bean hosts. The highest dilution still giving infection can then be used to determine the group size. However, to ensure detection of individual seeds having low virus titres, not the highest dilution but the second-highest or third-highest dilution should be used as the basis for determining the group size.

3. Alternatively, the hypersensitively reacting cv. 'Top Crop' can be used. This cultivar reacts with necrotic local lesions in only 3 days if the detached leaf method is used, but may not be as sensitive as the systemically reacting cultivars and therefore may require that the seeds to be tested have been pre-germinated on moist filter-paper for 3–4 days. The detached leaf protocol described for SMV in Section 4.4.4 can then be used with the source of inoculum as the only difference.

II. Tobamoviruses on tomato and pepper seed: elimination from seed and detection protocol

Some tobamoviruses are unique in their ability to retain infectivity in the seed-coat and fruit pulp remnants on dry seeds of, for example, tomato (*Lycopersicon esculentum*) and peppers (*Capsicum* spp.). Sometimes these viruses can also be located in the endosperm but at least TMV, ToMV and PMMoV have not been reported as being embryo-located in these species; however, the evidence speaks for itself: seedlings are not infected directly during seed germination. Virus concentration in the seed remnants is high enough, however, to invade the seedlings through wounds in the roots when seedlings are transplanted. If not transplanted but left undisturbed, no transmission from infected seeds occurs (Broadbent, 1965). When established even in a few plants, these contagious, sap-transmissible viruses may spread readily in the crop, especially tomato, through human contact with the plants, during pricking, pruning, etc.

ELIMINATION. In contrast to embryo-transmitted viruses, the infectivity of ToMV and TMV, the most common tobamoviruses in tomato and in pepper also PMMoV, can be eliminated from the seeds. Both externally and usually internally (endosperm) located tobamovirus can be inactivated by heat treatment of dry tomato seed for 2 days at 78°C (Green *et al.*, 1987). This treatment is more effective in eliminating internally located virus than a 20 or 30 min soak in 12.5% Na_3PO_4 and does not disturb germination.

For pepper seed, both Na_3PO_4 treatment and heat treatment effectively eliminate virus; the optimal soak treatment for pepper seed is 2 h in 100 g/l Na_3PO_4, and the optimal heat treatment is 76°C for 3 days (Rast and Stijger, 1987). However, in some pepper cultivars heat treatment may result in poor and delayed seed germination (Rast and Stijger, 1987). Reducing the seed moisture content to 3–4% by drying at 30–50°C prior to the heat treatment seems to be important (quoted in Green and Kim, 1991).

To minimize the spread of tobamoviruses while handling plants during pruning and harvesting, skimmed milk used as a dip for hands and tools has been shown to be effective, as demonstrated for TMV in sweet pepper (Rast, 1980, quoted from Green and Kim, 1991).

DETECTION. Presence of tobamoviruses in tomato and pepper seed can easily be detected by serological assay (e.g. by DISA, see Section 5.4) or a molecular test of whole seeds; but only a bioassay can reveal whether or not the detected virus is viable.

A simple and fast (3–5 days) test of seed for presence of infective tobamoviruses (these viruses remain infective for months or years in and on seeds) can be effected by inoculating susceptible indicator plants with an extract from seeds homogenized in buffer. The following procedure corresponds closely to the protocols used in a comparative test programme organized by the International Seed Health Initiative (ISHI; see Section 7.4). The protocols were accessible at http://www.worldseed.org; however, new revised protocols are expected to be published (as standard or generally recommended methods) on the Internet in 2005 (Dr R. Ranganathan, ISHI, 2004, personal communication). *ISHI states that Internet-published protocols may be updated at any time, and that the user must be aware that the most recent version of a method is being applied.*

EQUIPMENT AND MATERIALS. Equipment and materials are the same as those used in the standard protocol, p. 65, except for the specific components:

a. Test seed sample of tomato or pepper.

b. Healthy seed sample of tomato or pepper, for negative control.

c. Positive control: seed sample of tomato or pepper with known tobamovirus infection, *or* sap from a fresh plant or dried plant tissue, systemically infected with TMV, ToMV or PMMoV, *or* purified virions of one of the three viruses.

d. Laboratory mill or homogenizer.

e. Buffer a.

f. Indicator plants: *Nicotiana tabacum*, cv. 'Xanthi-*nc*', and *Nicotiana glutinosa* plants with five to seven fully turgid leaves. See also **Note 2** after procedure.

g. Cleaning solution: 10% Na_3PO_4 (w/v).

h. Laboratory soap, 2%, alkaline.

PROCEDURE.

1. Prepare for inoculation as in the standard protocol, steps **1–6**, p. 66, but use only one of each indicator plant per group of seeds. Negative control: inoculate one of each indicator species with buffer and abrasive only and one of each with healthy plant extract.

2. Inoculum: take out the required number of seeds (Section 7.2) from the working sample. Sample according to the ISTA rules; see also **Note 1** below.

3. Divide the sample into groups, each of 100 seeds (approx. 0.3 g) and grind each group individually in a laboratory mill or in a homogenizer. If ground in a homogenizer, grind with 20 ml of buffer. If ground in a dry mill, add the 20 ml of buffer to the flour after grinding, and mix well.

4. Clean the grinder carefully between seed *samples* as follows:

(a) Rinse seed grinding parts with distilled water.

(b) Rinse three times (three different beakers) with *cleaning solution.*

(c) Rinse three times (three different beakers) with *2% alkaline laboratory soap.*

(d) Rinse three times (three different beakers) with distilled water.

5. Clean the grinder between *groups* from the same sample, as follows: Rinse the seed grinding parts by activating the grinder in three changes of *cleaning solution* followed by two or three changes of distilled water.
6. Leave the extract for about 30 min at 4°C to allow seed parts to settle, or filter the extracts through a wad of cotton wool. Inoculate with either the supernatant or filtrate of each seed group well-developed leaves of one plant of each indicator species following the procedure in the standard protocol, steps **8–9**. Use at least 100 µl per leaf and make sure that the entire surface is inoculated and well wetted.
7. Inoculate the positive control plants with similarly made seed extracts or infective, diluted plant sap. If plant sap or purified virus, use 10^{-2}, 10^{-3} and 10^{-4} dilutions on different leaves on the plants.
8. Place the inoculated plants in an insect-free greenhouse or growth room at 22°C ± 3°C (should not exceed 28°C) and 12–16 h photoperiod. Note that insufficient light reduces test sensitivity.
9. Discard disposable materials, wash hands and clean the tools thoroughly as described under Section 4.4.2, Prevention of contamination.
10. Symptom recording: examine the plants for local lesions occurring as necrotic spots with a dark brown halo after 2–5 days in case of infection (Fig. 4.7). Record the results (standard protocol, step **13**). If only one or two local lesions occur per leaf, the presence of virus should be confirmed by re-inoculation. This can be done by cutting out the suspect lesion, crushing it in a minute amount of buffer and inoculating two leaves of the two different detector-plant species.

Notes
1. The total number of seeds to be tested from a seed lot depends on the maximum acceptable percentage of infested seeds (Section 7.2). Groups of 100 seeds should allow detection of one infested seed per group.
2. By using two differently reacting cultivars of *N. tabacum*, a partial virus differentiation is possible (Table 4.4). If local lesions are present on 'Xanthi-*nc*' but absent on 'White Burley' plants, leave the latter for 2 weeks after inoculation and examine them for systemic infection caused by TMV (seen as dark green/ light green mosaic and leaf malformation).

By further including tomato (Table 4.4), a full differentiation of virus species can be made. Jacobi *et al.* (1998) used a molecular assay to differentiate TMV and ToMV. Identification of TMV, ToMV and PMMoV strains requires a number of other differential hosts (Green *et al.*, 1987; Green and Kim, 1991), a serological assay (using monoclonal antibodies) or a molecular assay (Letschert *et al.*, 2002).

III. Testing of lettuce seed for Lettuce mosaic virus *(LMV), protocol*
The third example of methods for direct seed testing is assaying of lettuce (*Lactuca sativa*) for LMV, a seed bioassay of unusually high sensitivity, where results correspond closely to the actual rate of transmissible infection. This method is described by Rohloff and Marrou (1981) in Working Sheet No. 9 (2nd edn), ISTA *Handbook on Seed Health Testing* (Anon., 1981). It was once

Fig. 4.7. Necrotic local lesions on a leaf of *Nicotiana tabacum* cv. 'Xanthi-*nc*' 3 days after inoculation with extract of ToMV-infested tomato seeds.

Table 4.4. Reactions of indicator hosts to three tobamoviruses.

	Virus		
Host	TMV	ToMV	PMMoV
Nicotiana tabacum			
cv. 'Xanthi-*nc*'	NLL	NLL	NLL
cv. 'White Burley'	M	NLL	NLL
Tomato	M	M	IS
(Pepper[a]	M	M	M)

NLL, necrotic local lesions; M, mosaic due to systemic infection; IS, insusceptible.
[a]Not needed as indicator in this test.

used as a routine method for indexing lettuce seed for LMV but, since then, routine methods based on ELISA by Falk and Purcifull (1983) and van Vuurde and Maat (1983) have been described. Van der Vlugt *et al.* (1997) have since described an immunocapture-reverse-transcription polymerase chain reaction (IC-RT-PCR) assay, shown to be 1000 times more sensitive than ELISA for detection of LMV in lettuce seedlings.

Conditions required for reliable results from the following bioassay are: (i) the use of well-developed, vigorously growing *Chenopodium quinoa* plants of optimal development and full turgor; and (ii) strict adherence to each procedural step.

EQUIPMENT AND MATERIALS. Equipment and materials are the same as those used in the standard protocol, p. 65, except for the specific components:

a. Test sample of lettuce seed.

b. Lettuce seed, known to be healthy, for negative control (may be omitted).

c. Sap from LMV-infected plants or purified LMV for positive control.

d. Mortars and pestles.

e. Instead of buffer a or b, prepare an 'LMV buffer': Na_2HPO_4 containing 0.2% sodium diethyldithiocarbamate (Na-DIECA) and 0.5% $NaHSO_3$, adjusted to pH 7.0. **Caution:** $NaHSO_3$ (sodium hydrogen sulphite) is a strong poison; protect eyes and skin against contact; the chemical develops toxic gas in contact with acids. Na-DIECA is also harmful.

f. Activated charcoal.

g. Polyethylene bags, approx. 20 × 40 cm.

h. Pipette (Pasteur or bubble pipette with ml graduation).

i. *Chenopodium quinoa* test plants: Seeds are sown in soil, and the seedlings transplanted at the cotyledon or two-leaf stage into soil in pots with an upper diameter of about 10 cm. Maintain plants at 18–20°C and at 16 h photoperiod. If grown in continuous light, plants should have a 24 h period of darkness before inoculation. Plants are ready for use when four to six leaves have developed. Plants with flowers should not be used.

j. Wooden stakes, 30–40 cm long.

PROCEDURE.

1. Prepare for inoculation as in the standard protocol, steps **1**–**4**, p. 66.

2. Negative controls: inoculate one plant with '*LMV buffer*' containing a pinch of *charcoal* and leave one plant non-inoculated. Further, if healthy seed is available, one mock-inoculated plant may be included.

Wash hands.

3. From a working sample, take out groups of 300 seeds each, using the ISTA sampling method. The total number of seeds to be tested depends on the maximum tolerable percentage of infected seed; see Section 7.2.

4. Place each group in a mortar, add 3 ml of '*LMV buffer*' and grind the seeds thoroughly; see also **Note** below. Then add a pinch of *charcoal* and mix.

5. Squeeze the slurry through layers of gauze and inoculate three plants with each homogenate (group) as described in steps **8–9**, standard protocol. Make sure not to cross-contaminate between groups.
6. Positive control: it is advisable to inoculate two plants with either diluted extract from a known-infected plant or purified LMV diluted 10–50 times.
7. Place three stakes in each pot. Cover each plant with a polyethylene bag, held free of (not touching) the plant by the stakes. Place the plant in a vector-free greenhouse or growth room at 25°C and 16 h photoperiod.
8. Continue with steps **11–13** in the standard protocol.
Remove the plastic bags after 24 h.
Inoculated leaves of plants may show chlorotic, local lesions after approx. 6 days, but only systemic infection is conclusive for a positive reaction. Systemic infection appears 14–21 days after inoculation and is observable in the upper leaves as yellowish mosaic and spots and leaf distortion.
9. On the basis of the number of positively reacting groups, the rate of seed infection in the sample can be estimated using the MLE formula (Section 7.2).

Note
Each group may also be ground separately in a dry laboratory mill, followed by adding of the buffer. The slurry must then be ground thoroughly by mortar and pestle before adding charcoal, and inoculated. The laboratory mill must be carefully cleaned between seed groups.

4.4.4. Plants and seedlings as inoculum, protocol

Infectivity assays, using plants or seedlings as inoculum, will in most cases be for verifying results of serological or molecular assays, as a part of identification and characterization of new or unusual viruses/viroids, or for distinguishing viruses in mixed infections. However, in seed-health testing, it is possible to perform tests for several virus–host combinations by bioassay only. Of course, such tests, not involving laboratory tests, take a longer time and require sufficient greenhouse space and labour.

A quick bioassay for SMV in soybean seedlings or plants is possible by inoculating detached leaves of hypersensitively reacting bean cultivars, e.g. cv. 'Top Crop'; see protocol below. Similar tests are described for other seed-transmitted viruses, such as those quoted for legume germplasm in Frison *et al.* (1990). See also Section 4.4.3, Protocol I, **Note 3**, p. 70. Diagnosis by infectivity assays is possible for a number of viruses; however, it usually requires differential indicator hosts for reliable identification (Section 4.4.5).

Detection of Soybean mosaic virus *by the detached leaf method protocol*
SMV can be detected by inoculation of extract from soybean seedlings onto primary leaves of young plants of *P. vulgaris* cv. 'Top Crop', followed by incubation of the detached leaves at 30°C under constant light for 3–5 days (Milbrath and Soong, 1976). In the detached leaf method, first locally reacting leaves of a hypersensitively reacting host are inoculated, then detaching of the

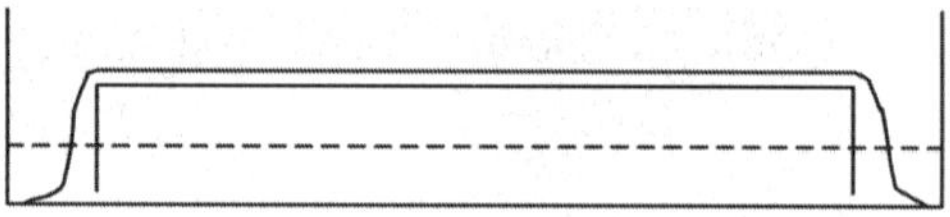

Fig. 4.8. Container for incubating detached leaves. The box, placed upside down inside the larger box, is covered with a sheet of filter paper kept moist by the water reservoir in the bottom.

leaves followed by their incubation on moist filter paper in transparent containers under artificial light.

EQUIPMENT AND MATERIALS. Equipment and materials are the same as those used in the standard protocol, p. 65, apart from the specific components:

a. Soybean seedlings, 2–3 weeks old, to be tested (growing-on test).

b. Soybean seedlings or plants, known to be healthy, for negative control.

c. SMV-infected plant(s) or purified SMV for positive control (advisable if not familiar with SMV symptoms on detached 'Top Crop' leaves).

d. Plants of *P. vulgaris* cv. 'Top Crop' with fully developed primary leaves, about 10 days old, but before emergence of trifoliate leaves.

e. Petri dishes, polystyrene, 9 cm in diam. *or*

f. Boxes of polystyrene or similar with a transparent lid, size order: 20–25 cm × 30 cm × 8 cm.

g. Boxes, e.g. polystyrene or expanded polystyrene (Styrofoam), etc., slightly smaller than those in (f), e.g. 17–22 cm × 25 cm × 5 cm, and waterproof (see below and Fig. 4.8).

h. Filter paper, circular for Petri dishes or sheets for boxes.

i. Inoculation buffer a or b.

j. Incubator at 30°C with light for incubation or an illuminated shelf in a room at 30°C. Standard daylight tubes with intensity of 1600–2100 lux, at 15–30 cm distance from leaves.

PROCEDURE.

1. Soybean seedlings raised in a growing-on test (Section 4.3.1, p. 59), 2–3 weeks old, with developed primary leaves, are inspected for symptom-bearing seedlings. Typical signs of SMV infection are mottling and downward curling of primary leaves. For the number of seeds required for determination of seed infection percentages, see Section 7.2.

2. Prepare for incubation of detached leaves:

(a) Place three to four filter-papers moistened with tap water in each Petri dish, *or*

(b) Place the small boxes in the large ones, cover with a sheet of filter-paper as shown in Fig. 4.8. Add tap water.

3. Prepare for inoculation, steps **1–4** in the standard protocol, p. 66. Use at least one bean plant (two leaves) per inoculum source.

4. Controls: inoculate two carborundum-dusted bean plants, one with buffer and one with known healthy soybean extract.

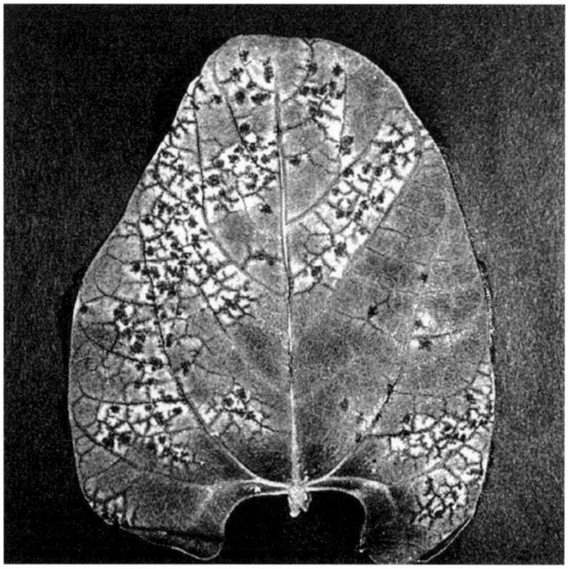

Fig. 4.9. Necrotic spots and veins on a detached leaf of *Phaseolus vulgaris* cv. 'Top Crop' 5 days after inoculation with SMV.

5. Continue with steps **6–9** in the standard protocol; prepare inoculum as described in step **7a** (see also **Note 1** below). Avoid cross-contamination between inocula.
6. Positive control (advisable): inoculate one or two plants with diluted, known infected plant extract or 10–50 times diluted purified SMV.
7. Wash hands with water and soap.
8. Cut off the inoculated bean leaves leaving approx. 1 cm of petiole on the leaf. Start with the negative controls. Place the leaves on the moist filter paper in Petri dishes or boxes, such that the petiole end is in contact with the filter paper.
9. Label with date and inoculation number, cover the containers with their lids, and place them under constant light at 30°C for 3–5 days (see **Note 2** below). Make sure that the filter paper remains wet all the time (easiest when using boxes).
10. Discard disposable material, wash hands and clean the tools.
11. Inspect the detached leaves daily. Symptoms due to SMV in the form of small necrotic, local lesions and vein necroses may appear after only 48 h. Chlorotic areas develop later (Fig. 4.9).
12. Record the result (step **13**, standard protocol).

Notes

1. This assay may be used primarily as a test for confirming the identity of the agent causing seedling symptoms, i.e. testing only symptom-bearing seedlings. It could also be used to provide a direct estimate of the actual per cent SMV seed infection in the sample. In the case of doubtful seedling symptoms and as a way to potentially reduce the number of required inoculations, all emerged seedlings

could be tested. In this case groups of *x* seedlings should be extracted for inoculation. The highest number of seedlings per group still allowing detection of one infected seedling out of *x* seedlings, has to be determined prior to the assay, as follows. Make two similarly diluted extracts of known SMV-infected and known healthy seedling material, respectively. Make a twofold dilution range of the infected extract in healthy extract and inoculate a set of 'Top Crop' plants with the dilutions. The highest dilution or second-highest dilution still resulting in conspicuous symptoms can be used as a basis for setting the group size. (Milbrath and Soong (1976) found that local lesions were produced with up to 1 : 256 dilution in *buffer*). After finalizing an assay, an estimate of the actual rate of seed infection can then be made on the basis of positive groups, using the MLE formula (Section 7.2).
2. According to Milbrath and Soong (1976), 30°C is optimal. At 25°C fewer local lesions developed, and at 35°C no local lesions developed.

4.4.5. Differential hosts

Infectivity tests using differential hosts are important means of identifying species or strains of viruses and viroids. Differential hosts are also indispensable for separating mixed infections, which occur frequently in nature (Bos, 1999; Hull, 2002).

For example, in distinguishing the two potyviruses *Cowpea aphid-borne mosaic virus* (CABMV, *Potyvirus*) and the blackeye cowpea mosaic strain of BCMV (BCMV-BlCM), a set of four cultivars of cowpea (*V. unguiculata*) can be used. These cultivars are TVu 1582, TVu 401 (BCMV-BlCM-specific), TVu 2657 and TVu 3433 (CABMV-specific) (Taiwo *et al.*, 1982; Dijkstra *et al.*, 1987). According to Huguenot *et al.* (1993), a number of isolates could be differentiated into BCMV-BlCM and CABMV by only two of the differentials, TVu 1582 and TVu 2657. This set of differentials was also used to identify a virus found seed-transmitted in a Vietnamese *V. unguiculata* ssp. as BCMV-BlCM (Hao *et al.*, 2003).

A set of differential cultivars of *P. vulgaris* for classification of ten strains of BCMV and BCMNV was identified by Drijfhout (1978) and can be found in the *AAB Descriptions of Plant Viruses* No. 337 (Morales and Bos, 1988), accessible on http://www.dpvweb.net. Three of the strains, NL3, NL5 and NL8, have since been classified as strains of BCMNV. Sets of differentials that can distinguish tobamoviruses affecting tomato and peppers (Green and Kim, 1991; Section 4.4.3) and strains of ToMV (Green *et al.*, 1987) have also been identified.

Indicator plant species are listed in Tables 4.1 and 4.2, including some non-host species, usable for full or partial identification of important seed-transmitted viruses at the species level.

4.5. Maintenance of Isolates

For virus or viroid assays in the laboratory and greenhouse, the availability of a collection of plant and seed material with known, well-identified infection for use as positive controls is essential. Maintenance of isolates in plants as stock cultures is one possibility, but this has the disadvantages that isolates may become contaminated with other viruses, or change over time with repeated transfers during maintenance, and in that process stock cultures occupy valuable space in the greenhouse. An example of reported changes of isolates is loss of aphid transmissibility after even a few mechanical transfers, e.g., PEMV-1, (R. Hampton, USA, 2004, personal communication). A better way to maintain isolates is dehydration and storing over a desiccant. The majority of isolates of seed-transmitted viruses stored in dry plant tissue at 4°C retain both serological reactivity and infectivity for years (see protocol below).

Another and obvious way to store seed-transmitted virus isolates is in seeds. Viruses carried in the embryo of seeds often persist as long as the seeds remain viable, which may be 10–20 years or more if kept at low temperature and low humidity.

Other methods of maintenance are storing of infected leaves at – 20°C, or lower, in sealed polyethylene bags, by which infectivity may sometimes be retained for up to a year. Better than freezing at – 20°C is preservation in liquid nitrogen. For the latter method, freshly harvested infected leaves are ground in a chilled mortar with a 1 : 1 ratio of 0.08 M TRIS-HCl buffer, pH 8, containing 8% (w/v) glucose and 6% (w/v) Na-glutaminate. The grinding should be done at low temperature, e.g. in a cold room. After squeezing the pulp through three to four layers of gauze, vials are filled with 2.5 ml of the sap and placed in liquid N_2 (Dijkstra and de Jager, 1998). According to the same authors, freeze-drying is less preferable, as some viruses when so treated seem to rapidly lose infectivity.

In general, plants known to contain inhibitors of infection should be avoided as hosts of viruses to be preserved. Plants such as *Nicotiana* spp. and *P. vulgaris* have not caused virus inhibition.

Important aspects of virus preservation and virus collections in general are discussed by Bos (1999), and detailed protocols for different storing methods are described by Dijkstra and de Jager (1998).

Maintenance of viroids in plants, vegetative propagules or seed is an option (G. Adams, University of Hamburg, Germany, 2004, personal communication). According to the same source, some viroids can be stored and retain infectivity in desiccated plant tissues, but an ideal means for their preservation is cloned DNA. Detectable levels of PSTVd in true potato seeds remained even after 21 years' storage at room temperature (Singh *et al.*, 1991, quoted from Singh *et al.*, 2003), indicating that viroid longevity may exceed that of viable seed (Singh *et al.*, 2003).

Short-term storage

Virus-infected leaves may be stored for a few days between damp filter-paper pieces in sealed polyethylene bags kept at 4°C. If samples cannot be refrigerated, e.g. during travel, leaves can be placed between filter-paper pieces saturated in 50% glycerol and sealed in polyethylene bags. Such samples may retain infectivity for up to 10 days (Dijkstra and de Jager, 1998).

4.5.1. Long-term storage of dehydrated samples, protocol

Dehydration of infected tissue and storing it over $CaCl_2$ or $CaSO_4$ at 4°C (McKinney, 1953) seem equal to or better than most other long-term storing methods for a large number of viruses (Dijkstra and de Jager, 1998; Bos, 1999). According to the same authors, infectivity is retained from 3 to 20 years for numerous virus species, and in rare cases up to even 40 years. We obtained similar results by this simple method at DGISP. Dehydrated infected tissues might also be a well-suited form of reference material to be used in comparative testing for viruses (Section 7.4.1).

EQUIPMENT AND MATERIALS. All equipment and materials must be virus-free; strict precautions against contamination by other viruses must be taken during the work.

a. Scissors.
b. Cheesecloth or filter-paper.
c. Petri dishes, 10 cm or, better, 15 cm in diameter.
d. Cotton wool.
e. Adhesive tape.
f. Test tubes, glass, airtight-sealable by rubber stopper or screw lid.
g. Refrigerator.
h. Calcium chloride ($CaCl_2$), anhydrous, granulated (3–5 mm grains); see **Note 1** after procedure.
i. Calcium sulphate ($CaSO_4$), anhydrous, granulated (optional).
j. Silica gel, commercial ('Blue-gel').

PROCEDURE.

1. Freshly harvested symptomatic leaves from young plants are cut finely (2 mm strips) and placed in an even layer on a circular piece of cheesecloth or filter-paper over the granulated desiccant $CaCl_2$ in a Petri dish. Use at least 5 times more $CaCl_2$ than tissue (w/w). Avoid any direct contact between plant tissue and $CaCl_2$.

2. Place the other half of a Petri dish on top and tape the two halves together airtight (easiest by combining two Petri dish lids or two bottoms) and store at 4°C until the tissue is completely dry (crispy). This usually takes at least a week. The low temperature helps to minimize virus-degrading activities in the tissue during the drying.

3. Transfer the dry material to a sealable glass test tube, containing a little granulated $CaCl_2$ (about one-fifth of the tube volume) separated from the tissue with a wad of cotton wool. Addition of a few grains of Blue-gel to the desiccant can serve as indicator for moisture absorption by changing from blue to pink. Instead of $CaCl_2$, $CaSO_4$ can be used as an even better desiccant. Make sure that the tube is airtight-sealed.

4. Enclose relevant information (virus, host, date, registration number, etc.) and store the samples at approx. 4°C (see **Note 2**). For safety, leave a label (indicated on a tiny slip of paper) inside the tube as well.

5. Alternatively, dehydration can be achieved in 1 day or less over $CaCl_2$ under vacuum in a desiccator at room temperature or, preferably, at 4°C (see **Note 3**).

6. For long-distance shipping, samples should be dehydrated. Fresh samples, even if glycerol-preserved, are very often useless at receipt. For shipping dehydrated samples, use small vials with a few small granules of $CaCl_2$ in the bottom of each vial under a piece of cotton, and seal airtight.

Notes

1. A granulated form of $CaCl_2$ should be used as desiccant, as the powder form absorbs poorly. The ideal grain size is 3–5 mm. Instead of $CaCl_2$, silica gel can be used. It absorbs moisture rapidly, but has a lower desiccant capacity than $CaCl_2$, and a larger proportion of silica gel/tissue must therefore be used (about 50/1, v/v). Silica gel can be reused after regeneration at 120–150°C. Temperatures above 175°C ruin the gel. Heating to 250°C overnight can regenerate $CaCl_2$. At the same time, the high temperature degrades any remaining virus from prior use.

2. As the number of samples collected often tends to increase more than anticipated, it is essential to have a consistent labelling and numbering system and systematic registration.

3. Dehydration over $CaCl_2$ under vacuum for 8 h or less and then storage at – 30°C were found most efficient for preservation of infected tissues; after more than 20 years, > 90% of all viruses survived so stored (R. Hampton, USA, 2004, personal communication).

Reactivation of virus in dehydrated samples

Grinding the dry tissue pieces gently into a coarse powder with a clean spatula or similar before taking out any portions of the sample for reactivation will ensure a better homogeneity of the sample and thus a uniform concentration of virus in each subsample taken out.

For bioassays, grind the dried material thoroughly with a cold mortar and pestle together with approx. 1.9 ml of *buffer b* per 0.1 g of tissue (buffer, see Section 4.4.2, Inoculation: standard protocol, p. 65). This mixture corresponds approximately to fresh plant sap. Dilute the suspension further approx. 1 : 5 in buffer b before use for inoculation.

For laboratory assays, grind the material in a similar way, using the assay buffer as diluent. Filter. Remember that dried material is up to 20 times concentrated compared to fresh plant sap.

Literature on bioassays

General aspects of biological assays are described in, for example, Matthews (1993), Bos (1999) and Hull (2002), and there are excellent protocols in Dijkstra and de Jager (1998).

References

AAB Descriptions of Plant Viruses (n.d.) Nos 1–409. Available at: http://www.dpvweb.net

Anon. (1981) *Handbook on Seed Health Testing, Working Sheets.* International Seed Testing Association, Zurich, Switzerland.

Bennett, C.W. (1969) Seed transmission of plant viruses. *Advances in Virus Research* 14, 221–261.

Bos, L. (1977) Seed-borne viruses. In: Hewitt, W.B. and Chiarappa, L. (eds) *Plant Health and Quarantine in International Transfer of Genetic Resources*. CRC Press, Boca Raton, Florida, pp. 39–69.

Bos, L. (1999) *Plant Viruses, Unique and Intriguing Pathogens – a Textbook of Plant Virology*. Backhuys Publishers, Leiden, The Netherlands.

Bos, L. and van der Want, J.P.H. (1962) Early browning of pea, a disease caused by a soil- and seed-borne virus. *Tijdschrift over Planteziekten* 68, 368–390.

Broadbent, L.H. (1965) The epidemiology of tomato mosaic, XI. Seed transmission of TMV. *Annals of Applied Biology* 56, 177–205.

Brunt, A.A., Crabtree, K., Dallwitz, M.J., Gibbs, A.J. and Watson, L. (eds) (1996) *Viruses of Plants*. CAB International, Wallingford, UK.

Diener, T.O. and Raymer, W.B. (1971) Potato spindle tuber 'virus'. *AAB Descriptions of Plant Viruses* No. 66. Available at: http://www.dpvweb.net

Dijkstra, J. and de Jager, C.P. (1998) *Practical Plant Virology: Protocols and Exercises*. Springer-Verlag, Berlin, Heidelberg, New York.

Dijkstra, J., Bos, L., Bouwmeester, H.J., Hadiastono, T. and Lohuis, H. (1987) Identification of blackeye cowpea mosaic from germplasm of yard-long bean and from soybean, and the relationships between blackeye cowpea mosaic virus and cowpea aphid-borne mosaic virus. *Netherlands Journal of Plant Pathology* 93, 115–133.

Drijfhout, E. (1978) Genetic interaction between *Phaseolus vulgaris* and bean common mosaic virus with implications for strain identification and breeding for resistance. *Agricultural Research Reports Wageningen* 872, 98pp.

Falk, B.W. and Purcifull, D.E. (1983) Development and application of an enzyme-linked immunosorbent assay (ELISA) test to index lettuce seeds for lettuce mosaic virus in Florida. *Plant Disease* 67, 413–416.

Frison, E.A., Bos, L., Hamilton, R.I., Mathur, S.B. and Taylor, J.D. (eds) (1990) *FAO/IBPGR Technical Guidelines for the Safe Movement of Legume Germplasm*. Food and Agriculture Organization of the United Nations, Rome/International Board for Plant Genetic Resources, Rome.

Green, S.K. and Kim, J.S. (1991) *Characteristics and Control of Viruses Infecting Peppers: a Literature Review*. Asian Vegetable Research and Development Centre. Technical Bulletin No. 18, Taipei, Taiwan.

Green, S.K., Hwang, L.L. and Kuo, Y.J. (1987) Epidemiology of tomato mosaic virus in Taiwan and identification of strains. *Journal of Plant Diseases and Protection* 94, 386–397.

Hao, N.B., Albrechtsen, S.E. and Nicolaisen, M. (2003) Detection and identification of the blackeye cowpea mosaic strain of *Bean common mosaic virus* in seeds of *Vigna unguiculata* sspp. from North Vietnam. *Australasian Plant Pathology* 32, 505–509.

Horvath, J. (1993) Host plants in diagnosis. In: Matthews, R.E.F. (ed.) *Diagnosis of Plant Virus Diseases*. CRC Press, Boca Raton, Florida, pp.15–48.

Huguenot, C., Furneaux, M.T., Thottappilly, G., Rossel, H.W. and Hamilton, R.I. (1993) Evidence that cowpea aphid-borne mosaic and blackeye cowpea mosaic viruses are two different potyviruses. *Journal of General Virology* 74, 335–340.

Hull, R. (2002) *Matthews' Plant Virology*, 4th edn. Academic Press, London.

Jacobi, V., Bachand, G.D., Hamelin, R.C. and Castello, J.D. (1998) Development of a multiplex immunocapture RT-PCR assay for detection and differentiation of tomato and tobacco mosaic tobamoviruses. *Journal of Virological Methods* 74, 167–178.

Jauron, R. (1997) Indoor lighting for houseplants. Available at: http://ipm.iastate.edu/ipm/hortnews/1997/1-7-1997/hplantlight.html

Letschert, B., Adam, G., Lesemann, D.-E., Willingmann, P. and Heinze, C. (2002) Detection and differentiation of serologically cross-reacting tobamoviruses of economical importance by RT-PCR and RT-PCR-RFLP. *Journal of Virological Methods* 106, 1–10.

Lister, R.M. and Murant, A.F. (1967) Seed transmission of nematode-borne viruses. *Annals of Applied Biology* 59, 49–62.

Matthews, R.E.F. (ed.) (1993) *Diagnosis of Plant Virus Diseases.* CRC Press, Boca Raton, Florida.

McKinney, H.H. (1953) Plant virus type culture collections. *Annals of the New York Academy of Sciences* 56, 615–620.

Milbrath, G.M. and Soong, M.M. (1976) A local lesion assay for soybean mosaic virus using *Phaseolus vulgaris* L. cv. Top Crop. *Phytopathologische Zeitschrift* 87, 255–259.

Morales, F.J. and Bos, L. (1988) Bean common mosaic virus. *AAB Descriptions of Plant Viruses* No. 337. Available at: http://www.dpvweb.net

Phatak, H.C. (1974) Seed-borne viruses – identification and diagnosis in seed health testing. *Seed Science and Technology* 2, 3–155.

Quantz, L. (1962) Zum Nachweis des Gewöhnlichen Bohnenmosaikvirus im Bohnensamen mit Hilfe des Schalentests. *Nachrichtenblatt des Deutsche Pflanzenschutzdienstes* 14, 49–54.

Rast, A.T.B. (1980) Tobacco mosaic virus in sweet pepper. In: *Annual Report* 1980. Glasshouse Crops Research and Experiment Station, Naaldwijk, The Netherlands, pp. 92–93.

Rast, A.T.B. and Stijger, C.C.M.M. (1987) Disinfection of pepper seed infected with different strains of capsicum mosaic virus by trisodium phosphate and dry heat treatment. *Plant Pathology* 36, 583–588.

Rohloff, I. and Marrou, J. (1981) Lettuce, lettuce mosaic, *Lactuca sativa*, Lettuce mosaic virus. Working Sheet No. 9, 2nd edn. In: *ISTA Handbook on Seed Health Testing.* International Seed Testing Association, Zurich, Switzerland, 4 pp.

Rossel, H.W. and Ferguson, J.M. (1979) A new and economical screenhouse for virus research in tropical areas. *Plant Protection Bulletin* 27, 74–76.

Singh, R.P., Boucher, A. and Wang, R.G. (1991) Detection, distribution and long-term persistence of potato spindle tuber viroid in true potato seed from Heilongijang, China. *American Potato Journal* 68, 65–74.

Singh, R.P., Ready, K.F.M. and Nie, X. (2003) Biology. In: Hadidi, A., Flores, R., Randles, J.W. and Semancik, J.S. (eds) *Viroids.* CSIRO Publishing, Collingwood, Victoria, Australia, and Science Publishers, Enfield, New Hampshire, pp. 30–48.

Taiwo, M.A., Gonsalves, D., Provvidenti, R. and Thurston, H.D. (1982) Partial characterization and grouping of isolates of blackeye cowpea mosaic and cowpea aphidborne mosaic viruses. *Phytopathology* 72, 590–596.

Van der Vlugt, R.A.A., Berendsen, M. and Koenradt, H. (1997) Immunocapture reverse transcriptase PCR for the detection of lettuce mosaic virus. In: Hutchins, J.D. and Reeves, J.C. (eds) *Seed Health Testing: Progress towards the 21st Century. Proceedings of the 2nd Symposium of the International Seed Testing Association Plant Disease Committee, Cambridge, UK, 1996.* CAB International, Wallingford, UK, pp. 185–191.

Vandre, W. (1996) Fluorescent lights for plant growth. Available at: http://www.uaf.edu/coop-ext/publications/freepubs/HGA-00432.html

Van Vuurde, J.W.L. and Maat, D.Z. (1983) Routine application of ELISA for the detection of lettuce mosaic virus in lettuce seeds. *Seed Science and Technology* 11, 505–513.

Wilson, V.E. and Dean, L.L. (1964) Flour of infected bean as a source of virus. *Phytopathology* 54, 489.

5 Serological Testing Methods

Serological detection, i.e. the use of a specific reaction between an antibody and its corresponding antigen is one of the most important methods for diagnosis and identification of virus diseases, particularly for those attacking plants. Immunizing a warm-blooded animal with an agent called an *antigen* (*anti*body *gen*erator) or *immunogen* such as virus particles stimulates the animal to produce proteins called *antibodies* in its blood serum. When formed, these antibodies react specifically with the antigen that stimulated their production, and the animal is said to have acquired immunity against this *homologous* antigen. Immunity means that a subsequent invasion of the animal by the same agent is met by antibodies already present in the serum and ready to bind to the agent and efficiently neutralize it. This *in vivo* reaction can also be accomplished *in vitro*, providing the means for serological detection: specific recognition and binding between an antibody and antigen.

The protein subunits of the outer coat of virus particles (virions) are ideal antigens for eliciting formation of antibodies in suitable animals. Viroids, on the other hand, are agents lacking a protein coat and therefore unable to induce production of antibodies. Viroids, consequently, are serologically undetectable.

Serological detection of viruses has been used since the 1930s, but the more recent development of enzyme labelling and monoclonal antibodies two to three decades ago represents important improvements of its sensitivity and specificity.

This chapter will concentrate on methods suited for routine testing, i.e. enzyme-based immunoassays (EIAs). Some of the other serological techniques are briefly dealt with in Section 5.5.

5.1. Antigens and Antibodies

Antigens

Most plant viruses are excellent antigens, eliciting good immune responses in animals and inducing production of specific antibodies. For production of *polyclonal* antibodies (see below), virus suspensions should be as free of host constituents as possible to minimize formation of antibodies specific to the plant hosts. It is also important, in producing virus-specific antibodies, that the virus antigen is intact when injected into the animal. If viral coat proteins (*antigenic determinants* or *epitopes*) are structurally compromised, antibodies generated may not recognize the intact form of the virions, or only react weakly with them, when used for virus detection.

Purification methods differ and are highly dependent on both the virus and the host in which it is propagated. Purification protocols for individual viruses can be found in *AAB Descriptions of Plant Viruses* (n.d.) (http://www.dpvweb.net) and (references to protocols) in *Viruses of Plants* (Brunt *et al.*, 1996). Excellent purification principles and protocols are also described by Brakke (1990), Stace-Smith and Martin (1993) and Hull (2002). Detailed stepwise protocols are described in Foster and Taylor (1998) for members of 20 virus genera, and in Dijkstra and de Jager (1998) for 16 viruses, representing 15 genera.

Generalized steps for the purification of most viruses are as follows:

1. Isolation of the virus, determination of its identity and absence of mixed infection.
2. Multiplication in a suitable host.
3. Maceration of tissue in the presence of a suitable buffer with additives to prevent virus inactivation, followed by filtration.
4. Clarification by one of various means, such as suspension in organic solvents followed by low-speed centrifugation.
5. Concentration of the virus from the clarified suspension by means of: (a) ultracentrifugation (at 80,000–200,000 *g*); or (b) precipitation by polyethylene glycol and low-speed centrifugation.
6. Resuspension of pellet followed by clarifying by low-speed centrifugation.
7. Repetition of steps **5** and **6** once or twice if necessary. (Alternation between **5a** and **6** is called differential centrifugation.)
8. Further separation of the virions from host constituents by density gradient centrifugation using sucrose or caesium salts.
9. Removal of sucrose or salt by suspending the virus band in buffer and ultracentrifuging (settling the virus and resuspending) or by dialysis.
10. Assessing the virus concentration by ultraviolet (UV)-absorption measurement and its relative purity by electron microscopy.
11. A purified preparation should be used as soon as possible. For short-term storage it can be frozen, lyophilized or kept liquid below 0°C by addition of an equal volume of glycerol. Most purified viruses are degraded by long-term storage.

A method for purifying plant viruses without ultracentrifugation and gradient centrifugation, but by ultrafiltration (through polyethersulphone membranes), has been successfully used by Michel *et al.* (2004). The method was effective for spherical virions, and might be applicable for other virion types also.

Regardless of method, buffers with virus-protecting additives must be used throughout the purification procedure, as well as the temperature of the material being kept below 4°C. Yields of purified virus vary considerably: from 1 g/kg of tissue (*Tobacco mosaic virus*) to 1 mg/kg (luteoviruses).

As an alternative to purified virions, synthetic viral coat protein produced in genetically altered bacteria can be used as immunogens (e.g. Van Regenmortel *et al.*, 1988). Thereby problems of contamination with plant host proteins are circumvented.

Antibodies

The antibody proteins, also called *immunoglobulins* (Igs) are secreted by lymphocytic cells in the immune system. Each cell or cell clone produces antibodies of only *one specificity*. Among the different classes of Ig molecules, the Y-shaped *IgG* is the most frequent. The two arms of the IgG molecule have structures capable of recognizing and binding to its corresponding antigen epitope much the same way as a key fits in a lock. When a suitable immunogen (= antigen) is brought into the blood system, the lymphocytes multiply and produce IgG, after a series of events through which IgG develops specific recognition/binding ability (Fig. 5.1). The details and principles of this process can be found in immunological textbooks and in concentrated form in Mernaugh *et al.* (1990). The IgG molecule consists of two identical heavy chains and two identical light chains held together with disulphide bridges (Fig. 5.1). The outer parts of the arms are called the variable regions (V) and contain the sites that bind specifically to the antigen. The rest of the molecule has an amino acid sequence that is the same for different antibodies of the same type and is called C (constant).

The IgG molecule can be fragmented by proteases into two fragments, Fab and Fc. The Fab part (fragment antigen binding) is the antigen-specific fragment, while one of the functions of the Fc (fragment crystalline with the constant amino acid sequence) is the binding of IgG to cell membranes. Another important

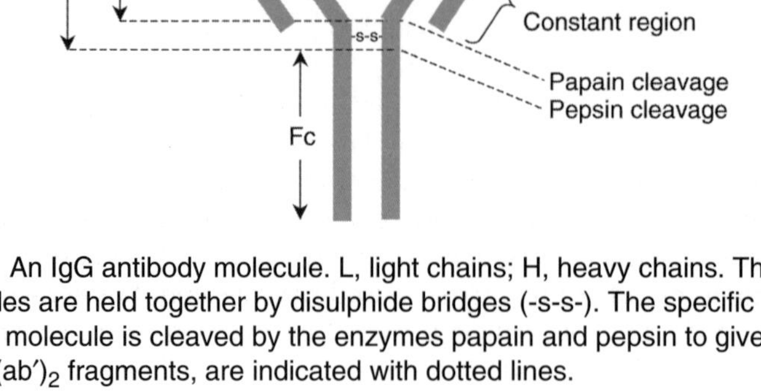

Fig. 5.1. An IgG antibody molecule. L, light chains; H, heavy chains. The four polypeptides are held together by disulphide bridges (-s-s-). The specific sites, where the molecule is cleaved by the enzymes papain and pepsin to give the Fab, Fc, and F(ab′)$_2$ fragments, are indicated with dotted lines.

feature of the Fc of most antibodies is its ability to bind to protein A with very high affinity. Protein A is a molecule isolated from the cell wall of the bacterium *Staphylococcus aureus*. The specific binding of protein A to Fc is used in a number of serological procedures.

An IgG can also be used as an immunogen itself. Thus IgG from, for example, a rabbit injected into other animals, for example goats or pigs, induces production of *anti-rabbit IgG* in that animal. If this anti-antibody is directed against the Fc part of rabbit IgG, antibody is obtained that is able to react with all possible rabbit antibodies. Such antibodies are commonly used in serological procedures (Section 5.2). Depending on the protease used (papain or pepsin), either the Fab or the so-called $F(ab')_2$ fragment can be cleaved from the Fc fragment (Fig. 5.1). Preparation of antibody fragments is described in detail by Ball *et al.* (1990) and Van Regenmortel and Dubs (1993). Especially the $F(ab')_2$ fragment is used in some versions of enzyme-linked immunoassays (Section 5.2).

5.1.1. Polyclonal antibodies

Specific antibodies obtained from the blood of an animal are called polyclonal because they are a mixture of slightly differing antibodies capable of binding to many different epitopes on the corresponding antigen. In principle, all animals can be used to produce antibodies, but typically only a few species are used. For production of polyclonal antibodies (Pabs) rabbits are most commonly used because they are easy to handle and respond well to plant virus antigens. The general steps in antibody production are as follows:

1. Purification of antigen (see Antigens above).
2. Bleeding of the animal before immunization to obtain control serum.
3. Injection of antigen into the animal (two or three animals per antigen).
4. Bleeding to obtain antiserum (after at least 2 weeks).
5. Titre determination.
6. Booster injection of the animal with antigen 4 weeks after the first one.
7. Alternating bleeding and booster injection, typically every fortnight, normally up to a total of ten bleedings.
8. Evaluation of the antiserum.

Comments on the individual steps of the general procedure above:

Re **3**: The antigen is mixed with Freund's adjuvant, a mineral oil, which emulsifies the antigen suspension. Thereby the antigen is slowly released into the blood (depot effect). If an adjuvant mixed with killed mycobacteria – Freund's complete adjuvant – is used for the first injection, an inflammatory reaction occurs and thereby a stronger immunological response is obtained. Other adjuvant systems (Quil-A® and Ribi®) also encourage good antibody titres and have fewer side-effects than the Freund's formulations (Brattey and Burns, 1998). Commonly 1 to 10 mg of purified virus is used per injection, but antisera of adequate titre (antibody content) can be obtained by injecting animals with as little as 50 to 100 μg of virus. Immunization with large doses does not result in proportionally higher antibody levels; instead, contaminants may then reach levels at which an immune response results (Van Regenmortel and Dubs, 1993).

Injections are normally given either subcutaneously or intramuscularly (in rabbits).

Re **4**: Up to 50 ml blood can be obtained per bleeding per rabbit, which yields 20–25 ml of antiserum. The blood is allowed to clot by leaving it for an hour at room temperature, heating to 37°C for 30 min, followed by cooling for 4–12 h at 4°C, after which the serum is decanted from the clot and centrifuged at low speed. After addition of a preservative, e.g. an equal volume of glycerol, the antiserum can be stored frozen for years without losing activity. Store in small aliquots to avoid repeated freezing/thawing. In glycerol it even keeps intact for years at 4°C.

Re **5**: Titre determination means to assess the concentration of specific antibodies. This can be done by testing the antiserum at increasing dilutions against its homologous antigen, using one of the common serological methods. It must also be tested against healthy plant sap to check for cross-reaction with plant proteins. The titre is the highest dilution at which the antibody discernibly reacts with the antigen. However, this depends on the type of assay used. For example, in a precipitation assay (Section 5.5) the titre for a particular antiserum may be 1 : 500, whereas the same antiserum may have a titre of 1 : 10,000 in an enzyme-linked immunoassay. Therefore, the indication of a titre must also include the assay conditions.

Re **6** and **7**: After the first injection, a relatively moderate immune response is obtained, whereas a much higher yield of specific antibodies usually occurs after the second and following injections (booster injections). This is because the antibody-producing cells now 'remember' the antigen and are therefore able to produce larger amounts of antibodies. The antiserum titre is determined after each bleeding. The titre gradually declines after each injection, prompting a booster injection.

Re **8**: After a series of injections and bleedings, the quality of the antiserum, i.e. its titre and specificity, is tested.

There are many variants of the above-described procedure regarding doses, sites and schedules of injection. Detailed procedures for the immunization and handling of experimental animals are described by, for example, Van Regenmortel (1982), Ball *et al.* (1990) and Hau and Van Hoosier (2002).

5.1.2. Monoclonal antibodies

An antigenic protein molecule has several or many different epitopes on its surface. Such epitopes are recognized by the surface receptor of various lymphocytic B-cells (IgG-producing cells). This means that for any given protein antigen there may be many different B-lymphocytes with slightly differing binding sites that can be stimulated by the antigen. The resulting antiserum is *polyclonal*, which indicates that it contains many different antibodies with specificity to the same antigen, but each arising from a different clone of B-lymphocytes (Hull, 2002).

B-lymphocytes cannot be cultured *in vitro*. This obstacle was circumvented by Köhler and Milstein (1975), who took B-lymphocytes from an immunized mouse and fused these *in vitro* with an 'immortal' mouse myeloma cell line. Each of such *hybridomas* produces only a single kind of antibody. A hybridoma producing the desired antibody can be selected and multiplied, or cloned, thus providing a vehicle for production of a *monoclonal antibody* (Mab) for a practically unlimited period of time. Depending on specified properties of antibody-secreting hybridomas, antibodies of very narrow specificity (recognizing a specific virus strain) or broad reactivity (recognizing all strains of a virus or all viruses of a genus) can be obtained. Similarly, it is possible to select antibodies with very high binding affinity to a specific epitope (Matthews, 1991).

Production of Mabs

Although the principles of Mab production are relatively simple, a large number of often critical steps are required. Mab production involves immunogen preparation and immunization followed by fusion, selection and screening, cloning and growth, and maintenance and cryopreservation of the hybridomas obtained. Knowledge of and experience with basic cell culture techniques are required for correct preparation of media, maintenance and preservation of cell lines and sterile culturing. Also required is a rational, sensitive and fast serological assay, as hundreds or thousands of hybridomas need to be screened for each project (Jordan, 1990). A brief outline of the steps is given below (essentially according to Andersen *et al.*, 1991):

1. The preparation of immunogen and immunization of the mouse (or rat) is largely as for the production of Pabs, except that the purity of the immunogen is less critical, because a single clone – and thereby a suitable antibody – is selected later in the process.
2. When the animal's blood has been tested and found to be antibody-responsive, the animal is killed and the spleen taken out. The antibody-producing spleen cells are suspended, washed and mixed with mouse myeloma cells. Hybridomas of the two kinds of cells are formed by adding polyethylene glycol to the mixture.
3. To sort out the hybridoma cells formed, the mixture is multiplied in titre plates containing a selective medium. Surviving cells are tested for antiserum production by assaying the culture fluid. This is done by a sensitive serological assay, often an enzyme-linked immunosorbent assay (ELISA) (Section 5.2). Positive cultures are then cloned by diluting and seeding them in titre plates with growth medium. The degree of dilution is adjusted so as to give (statistically) one progenitor cell per well.
4. Cell lines both having good growth abilities and producing antibodies with the desired properties are selected. Positive cell lines are diluted and seeded as single cells and then cultured and tested for antibody production. Dilution, seeding and testing are repeated until monoclonality of the culture is secured.

5. When a cell line that produces antibodies of the desired quality is finally selected, the cell line is multiplied, either by *in vitro* culturing in a CO_2 incubator or *in vivo* by injecting it intraperitoneally in mice followed by incubation and harvesting. By the *in vitro* method, fluid with a Mab concentration of 5–50 μg/ml can be harvested, while the yield is considerably higher by the *in vivo* method (up to 10 mg/ml). The product prepared *in vitro* can, however, be purified and concentrated by a simple procedure.
6. When an adequate amount of the Mab has been harvested, the cell line can be stored below –135°C or in liquid nitrogen until next use.

Several reviews are available on Mabs and their production. Detailed procedures for the production of Mabs to plant viruses are excellently described by Sander and Dietzgen (1984) and Jordan (1990).

One of the main advantages of Mabs is that producing antibodies of unchanged quality and specificity is virtually unlimited, as long as the hybridoma clones are kept viable. When the supply of a polyclonal antiserum is exhausted, it is nearly impossible to prepare a new antiserum with the same specificity and quality due to the fact that virus preparations are variable and animals react to injections differently. Moreover, by use of Mabs, standardized determination of virus strains is possible. For instance, if Mab reagents were distributed worldwide to all interested laboratories, uniform results for virus-strain identification or seed certification could be assured (Matthews, 1991; Van Regenmortel and Dubs, 1993; Hull, 2002).

The limitations of Mabs are the high initial demands of time, labour and cost, and that they may be assay-specific, i.e. they may react only in the assay environment in which they were developed. Some Mabs may not be able to precipitate viral antigens, or may be sensitive to changes in assay conditions.

Nevertheless, since their first production in 1975, Mabs against a large number of plant viruses have been prepared. Thus, Pabs are steadily being replaced by Mabs in commercially available diagnostic kits.

An eventual alternative to conventional antibody-production methods may be production of recombinant antibodies from large expression libraries. Ways to produce antibodies have been developed using DNA technology to clone and express fragments of antibody genes in bacterial systems, thus avoiding animal immunization and culture of hybridoma cells (Torrance, 1995; see also Section 5.2.7).

5.2. Enzyme-linked Immunosorbent Assay (ELISA)

Enzyme-linked immunosorbent assay (ELISA), introduced in plant virology mainly by Clark and Adams (1977), has become the most widely used serological method for routine detection of plant viruses. Nucleic acid-based (NAB) detection methods, especially the highly sensitive polymerase chain reaction (PCR), are in use in routine testing at some places, and intense research is going on worldwide to improve and simplify these methods, enabling a more general use for mass-testing purposes (Chapter 6).

Despite detection sensitivity and specificity less than that of the NAB methods, ELISA and other enzyme-based immunoassays (EIAs) have a great potential for use in routine seed-health testing for viruses. One of the characteristics of seed-transmitted viruses is their relatively high concentration in their hosts compared to many non-seed-transmitted viruses, such as members of the *Luteoviridae* and *Geminiviridae* families. Most seed-transmitted viruses, therefore, are readily detectable by EIAs. Tissue blotting immunoassay (TBIA) has shown promising results, even for luteovirus detection (Section 5.4). Furthermore, EIAs are low-cost and simple techniques, requiring less equipment, skilled manpower and laboratory facilities compared to the NAB methods. However, the high sensitivity of the NAB methods, especially PCR, usually allows detection of one infected seed or plant in a larger amount of pooled samples than EIAs.

5.2.1. General principles of ELISA

As its full name indicates, ELISA involves use of antibodies linked, or conjugated, with a suitable *enzyme*. The fact that even a small amount of an enzyme effectively enhances degradation or change of a relatively large amount of a specific compound, called the *enzyme substrate*, is utilized in ELISA to produce a colour reaction indicating the presence of a viral antigen. The method is therefore far more sensitive than earlier serological techniques. To be useful in ELISA, an enzyme should fulfil a number of criteria. It should: (i) have a high turnover; (ii) be highly stable under storage and assay conditions; (iii) be easy to conjugate; (iv) require no noxious substrate; and (in plant virus detection) (v) should not interfere with or be activated by compounds in the host plant material (Porstmann and Kiessig, 1992).

The enzyme best fulfilling these demands and used most widely in plant virus ELISA is alkaline phosphatase (AP), which dephosphorylates the colourless substrate *p*-nitrophenyl phosphate (*p*NPP), producing the bright yellow *p*-nitrophenol (*p*NP).

The ELISA reaction takes place on a solid plastic surface, to which proteins in suspension bind readily. Protein molecules like antibodies and the protein coat of virus particles will therefore adhere to the surface when suspended in an appropriate buffer and allowed to remain (i.e. incubate) in contact with the surface for a prescribed period of time. The product most widely used is a 96-well polystyrene microtitre plate. In the case of a 'positive' reaction at least three layers of reactants are gradually built up on the wall surface of the wells (classically): virus-specific antibody + antigen (virus particles) + virus-specific antibody conjugated with enzyme (Table 5.1). When the substrate is finally added to the well, an enzyme-catalysed colour development takes place in the substrate solution. In the case of a 'negative' reaction, i.e. when the test material is free of virus, no binding of antibody–enzyme conjugate occurs, thus no (or only very weak) colour will appear. The colour intensity (when using the AP/*p*NPP system) is usually proportional to the concentration of antigen present in the well, but also depends on other factors. The colour intensity can be estimated visually or, more conveniently and with greater accuracy, be measured by a spectrophotometric device, an *ELISA reader*, whose printout also serves as documentation.

LABORATORY HYGIENE AND PRECAUTIONS. ELISA of plant viruses does not need to be carried out aseptically, but, like other sensitive detection methods, ELISA is vulnerable to contamination of different kinds and when there is inappropriate handling, such as insufficient washing, poor hygiene and deviations from the procedure prescribed for the ELISA type in question. All glassware to be used for buffers and reagents must be thoroughly washed with the use of a detergent and rinsed in distilled or deionized water.

Normal laboratory safety precautions should be taken, including when handling the few dangerous chemicals used as additives to ELISA buffers (see Appendix 2).

ELISA formats

Several variants of ELISA have been developed. They can be divided into two main types: direct and indirect ELISA. In the *direct* format, the detecting virus-specific antibody is labelled with an enzyme (Table 5.1A). In the *indirect* format, the detecting virus-specific antibody is not labelled with enzyme, but is (most often) detected by an 'anti-antibody' labelled with enzyme. This second antibody is produced against the IgG of the animal in which the virus-specific antibody was produced. The virus-specific antiserum could, for instance, be produced in rabbits and the anti-rabbit serum in goats. In such a procedure the last serological step is thus an animal-to-animal reaction (Table 5.1B, C).

DIRECT ELISA. The double-antibody sandwich (DAS)-ELISA is a direct ELISA and was the first ELISA procedure developed for plant virus detection. It is still widely used, but requires preparation of an antibody–enzyme conjugate for each virus to be detected (Table 5.1A).

Table 5.1. Overview of the steps of three commonly used ELISA formats.*

	Reactant no.[a]				
ELISA format	1	2	3	4	5
A. DAS	Antibody	Antigen	Virus-specific IgG[b]–enzyme conjugate	Substrate	
B. Indirect AgF (PTA, DAC, OS, ACP)	Antigen	Antibody[b, c]	Anti-rabbit IgG[d]–enzyme conjugate	Substrate	
C. TAS	Antibody[c]	Antigen	Virus-specific monoclonal antibody[b]	Anti-mouse (-rat) IgG[d]–enzyme conjugate	Substrate

*See also Figs 5.2, 5.4, 5.6, 5.7 and 5.8.
[a]Wash between each step. Before step 2, a blocking step may be included.
[b]Detecting or primary antibody.
[c]In these cases produced in rabbit.
[d]This IgG is also called the secondary antibody.

INDIRECT ELISA. The main advantage of indirect ELISA is that a single enzyme conjugate (for example, enzyme-labelled goat anti-rabbit immunoglobulin) can be used for detecting any virus. Such conjugates are commercially available (e.g. Sigma™, DakoCytomation™), as also are other generally applicable conjugates, such as protein A–enzyme and streptavidin–enzyme (Section 5.2.7).

Indirect ELISA can be performed in two ways. In the first, the antigen is the first reactant and the ELISA format therefore named 'plate-trapped antigen' (PTA), 'direct antigen coating' (DAC), 'antigen-coated plate' (ACP) or 'open sandwich' (OS) ELISA. However, the name 'antigen-first ELISA' or *indirect AgF-ELISA* is thought to be more appropriate and is used hereafter (Table 5.1B). In the second way, virus-specific antibody is added first, followed by antigen and two distinct antibodies. The procedure is thus designated triple-antibody sandwich ELISA or TAS-ELISA (Table 5.1C). This procedure requires a trapping antibody, which is not recognized by the antibody–enzyme conjugate. In the example shown, the trapping antibody is produced in rabbit and the detecting monoclonal antibody in mouse (or rat), and the conjugate (Table 5.1C, step 4) is an anti-mouse (or rat) antibody labelled with enzyme. Instead of using antibodies produced in different animals, the same antiserum can be used both as coating and detecting antibody, provided that:

- the detecting antibody is labelled with biotin and a streptavidin–enzyme conjugate is used, or
- the coating antibody is used after digestion with pepsin, leaving only the $F(ab')_2$ fragments. A protein A–enzyme conjugate will then only recognize the intact detecting antibody.

Use of biotin–streptavidin and protein A is further described in Section 5.2.7.

Koenig and Paul (1982) have extensively studied the characteristics of indirect ELISA compared to direct ELISA. Use of indirect ELISA enables the detection of a broader range of strains and serologically related viruses than is possible by DAS-ELISA. The broadest range is detected by the antigen-first (non-pre-coated) type of indirect ELISA, and its sensitivity is often higher than that of DAS-ELISA. The method, however, is highly susceptible to interference by crude plant sap compared to direct ELISA (Koenig and Paul, 1982; Lommel *et al.*, 1982). This problem can, however, be considerably lessened by cross-absorption of the detecting antiserum with healthy plant sap (see p. 95). Also, the use of more diluted plant or seed antigen extract (1/50–1/100), rather than the 1/20 dilution often used in DAS forms, enhances detection of infected samples (Lommel *et al.*, 1982; Mowat and Dawson, 1987; Suzuki *et al.*, 1990; James and Mukerji, 1996; see also Section 5.2.8).

5.2.2. General components and steps of ELISA

Below are described the successive steps, reactants, other ingredients and techniques which are common to all types of ELISA. Specific guidance for selected ELISA variants is given in each of their protocols following this section.

Loading and incubation

Each of the reactants, adequately diluted in a suitable buffer, is added to the well, and each reactant is allowed to incubate, usually 1 to 4 h. The loading is most conveniently done by an adjustable micropipette with changeable tips. For reactants other than the antigens, a multiplex pipette with eight or 12 tips can be used to increase loading efficiency. The amount to be loaded in each well may vary from 50 to 200 μl. The volume of reactants once chosen must be maintained throughout the assay. Use of *two wells per test sample* is highly recommended for the sake of test reliability.

Incubation provides time for each sequential reaction to occur. To minimize evaporation during incubation the plate must be either covered with its lid or wrapped in thin household plastic foil (cling film) before being placed in an incubator at 37°C for 1 to 4 h, a temperature that is optimal for serological reactions. Alternatively, incubation can take place at room temperature or in a refrigerator at 5°C, but it will then take longer, e.g. overnight at 5°C. When using an incubator, the plate should not be placed directly on metal, but on an insulating plate (cork, etc.). Polystyrene (see Microtitre plates, this section) is a poor heat conductor, and in direct contact with hot metal the outer rows of wells may be warmer than the other wells, leading to inconsistent results. Gentle shaking or vibration during incubation can be used to reduce incubation periods. ELISA incubators with a built-in vibrator device are commercially available.

Washing

Thorough and careful washing between each step is imperative for optimal ELISA results. Washing, if done properly, removes remnants and unbound substances from the well, but does not remove protein bound to the wall (see below). Washing might be done with water only, but for efficient removal of non-specific compounds each washing should preferably be done with a buffer containing a detergent. The most commonly used detergent is Tween 20, added to retard binding of extraneous materials (Converse and Martin, 1990).

After incubation with a reactant, the plate is emptied by turning it upside down with a flinging movement, pouring the contents into a sink. While inverted, the plate's upper face is wiped by pressing it onto a clean tissue or piece of paper before it is turned upward and rinsed once by filling the wells with washing buffer, which is then poured out. The washing is repeated at least three times, allowing the liquid to remain in the wells for 3–5 min per wash. If a phosphate buffer is used for washing, the washing process must be concluded with deionized water, to remove traces of phosphate before adding the substrate. Traces of phosphate have an inhibitory effect on AP activity (Andersen *et al.*, 1991). The washing fluid is poured into the wells with a wash bottle having a standard nozzle or an eight-channel head (e.g. Plate Washer Bottle®, Agdia Inc., USA), providing a forceful stream of fluid. Automatic washing devices of different constructions are available, giving the advantage of providing a reproducible washing procedure in each assay. However, careful manual washing is adequate. Before addition of the next reactant, the remains of the washing liquid or bubbles are carefully removed from the wells by pounding the microtitre plate upside down against layers of clean cloth or paper.

Blocking

Protein (antibodies or antigen) bound after the first step may leave some of the well surface empty. A *blocking agent* is required to saturate these areas, as otherwise a passive binding of the next protein reactant may take place, resulting in non-specific reactions. In other words, the blocking secures that the reactants bind specifically to each other only. The blocking agent usually recommended (e.g. Zimmermann and Van Regenmortel, 1989) is albumin, typically in the form of skimmed milk powder. If a *blocking step* is included, the blocking substance suspended in buffer is added to all the wells and incubated for ½ to 1 h at room temperature or for 1 h at 37°C, as recommended for skimmed milk (1%) by Zimmermann and Van Regenmortel (1989). A very fast blocking step can be obtained with polyvinyl alcohol (PVA), requiring only 1 min of incubation, as reported by Miranda *et al.* (1993) for use in immunoblots (Section 5.3), but also recommended for ELISA (Copeland, 1998).

Subsequent reactants, except the substrate, must also contain a blocking agent (albumin) in order to maintain the blocking effect throughout the serological process. In most ELISA procedures, no separate blocking step is included, but the blocking agent is added to the buffer used for each of the reactants following the first. It should be remembered, however, that:

- the first reactant must not contain any blocking agent, and
- a blocking agent must not be added to the substrate (see Substrate below).

Antibodies

For the DAS- and TAS-ELISA procedures (Table 5.1), the first step is coating with the antibody. To maximize the amount of antibody bound to the solid surface and thereby optimize sensitivity of the ELISA, crude antiserum cannot be used, because of its content of competing, non-reactive proteins. Therefore, the immunoglobulins (IgG) containing antibodies specific to the virus being detected must be separated from the serum. Isolation of virus-specific IgG is usually accomplished by precipitating the IgG from serum with a salt, normally ammonium sulphate. After the precipitation, the salt is removed from the IgG by several rounds of dialysis against an appropriate buffer (Section 5.2.3).

The detecting (primary) antibodies in indirect AgF- and TAS-ELISA (Table 5.1) need no fractionation. A crude antiserum can be used directly because it is applied as the second and third reactant, respectively. Consequently, its virus-specific immunoglobulin cannot bind non-specifically to the solid surface, nor can its other proteins be bound, as the surface is already occupied by the first reactant (and blocking agent), i.e. the IgG can bind only to the homologous antigen, if present.

Cross-absorption

Immunoglobulins of polyclonal antisera produced against plant viruses include varying amounts of antibodies specific to plant host proteins. This is due to the difficulty of removing plant proteins entirely during purification of the virus to be used for immunization of the animal (rabbit). Consequently, anti-plant antibodies may also be produced. Monoclonal antibodies do not usually react with plant proteins. Polyclonal antibodies used in ELISA for a test of crude plant or seed material typically react with such proteins, giving rise to non-specific reactions with healthy samples, also called 'background noise' or 'background'. It is, however, possible to reduce greatly the amount of plant protein-specific antibodies from the antiserum by simply mixing and incubating it with a suspension of extract from healthy plant or seed tissues of the same species as being tested. The non-specific antibodies will then bind to the non-virus proteins, i.e. be inactivated, after which the mixture is added to the microtitre plate. Cross-absorption is normally never needed with Mabs, or with Pabs having high specificity to their homologous viruses. The type of ELISA used also plays a role in test specificity. For some reason, indirect AgF-ELISA gives higher background reaction than DAS or TAS procedures (Koenig and Paul, 1982) and use of cross-absorption is therefore generally recommended for indirect AgF-ELISA except when monoclonal antibodies are used.

Antigen

See Sample preparation (this section).

Antibody–enzyme conjugate

The last serological reactant to be used in ELISA is the antibody–enzyme conjugate. Coupling of the fractionated immunoglobulins to the enzyme results in a complex capable of both reacting serologically and functioning enzymatically to degrade a substrate. Should the complex be bound in the well, a signal appears in the subsequent substrate step in the form of a coloured product. The enzyme can be conjugated either to virus-specific IgG for use in DAS-ELISA or to IgG of anti-antiserum (anti-rabbit, anti-mouse, etc.) for general use in indirect ELISAs.

Specific conjugates can be made in the laboratory (Section 5.2.3) or be obtained from suppliers of phytodiagnostic reagents (Appendix 3). Anti-rabbit and anti-mouse conjugates are also commercially available from, for example, Sigma-Aldrich™ or DakoCytomation™.

As mentioned previously in this section, alkaline phosphatase (AP) is by far the most used enzyme in detection of plant viruses, as it fulfils a number of requirements. AP type VII is from bovine intestinal mucosa (Sigma P5521 or similar), stabilized with 100 μM $ZnCl_2$ and 1 mM $MgCl_2$. Type III (Sigma) from *Escherichia coli* may also be used. AP from *E. coli*, however, has a lower activity than bovine AP, and its optimal pH is 8 (Andersen *et al.*, 1991), while that of bovine AP is 9.8. Therefore, if *E. coli* AP is used, TRIS-HCl is recommended as substrate buffer, whereas diethanolamine buffer is recommended for bovine AP.

Glutaraldehyde (at a final concentration of 0.06%) is used to link IgG to the enzyme. A procedure of conjugation is described under the DAS-ELISA protocols. More details on enzymes and the principles of conjugation are given by Converse and Martin (1990) and Kurstak (1986), and are also reviewed by Abraham (1995). Among other enzyme/substrate systems suitable for ELISA is horseradish peroxidase/tetra-methyl benzidine dihydrochloride (HRP/TMB), which is briefly dealt with in Section 5.2.7.

Unlike AP and HRP, the enzyme penicillinase (β-lactamase) is not so widely used, although it should have the potential of becoming more popular due to its easy availability, inexpensive substrate (penicillin) and clear-cut colour reaction. Penicillinase (PNC) can be conjugated by glutaraldehyde to IgG in the same way as AP. The enzyme hydrolyses penicillin into penicillinoic acid, and the slight decrease in pH due to the acid production causes a colour change in different types of liquid pH indicators used in the mixture with the substrate solution (Yolken *et al.*, 1984; see also protocol in Section 5.2.6).

Substrate

AP-ELISA: The substrate for the most common enzyme, AP, is *p*-nitrophenyl phosphate (*p*NPP), which is colourless but produces the yellow *p*-nitrophenol (*p*NP) in the presence of the enzyme. The substrate is available as pills (e.g. Sigma N9389, containing 5 mg *p*NPP per pill). It must be kept dark and cold, preferably below 0°C. Contact with the skin must be strictly avoided due to the skin's excretion of AP. Equipment for substrate preparation must also be free of fingerprints. Excess substrate solution can be used as quality control in the substrate step; it should remain water-clear throughout this period. If desired, the enzymatic reaction can be stopped by addition of 25 μl 3 M NaOH per well (with the use of 100 μl reactant volume), followed by mixing on a suitable shaker.

PNC-ELISA: Among the pH indicators to be mixed with the penicillin substrate, bromothymol blue was found to be most suitable by Sudarshana and Reddy (1989) and Abraham and Albrechtsen (2001). With this indicator, the production of penicillinoic acid changes the colour from blue at pH 7.2 to yellow at lower pHs, depending on the amount of acid produced. The distinct colour change of the substrate–indicator mixture is very convenient for visual determination of positive or negative reactions and does not necessitate the use of an ELISA reader (Premier *et al.*, 1985; Sudarshana and Reddy, 1989). Due to the reduced costs and fewer demands for equipment, PNC-ELISA should be advantageous for use in less equipped laboratories (Section 5.2.6).

On substrates for HRP, see Section 5.2.7.

Buffers

Throughout the ELISA procedure the test reagents and the substrate are suspended in their respective appropriate buffers at dilutions ranging from around 10^{-3} to 10^{-6}.

DAS-ELISA AND TAS-ELISA. A sodium carbonate buffer of high pH (9.6) is used for efficient coating with virus-specific antibodies.

For the antigen, use is made of a phosphate-buffered saline buffer (PBS) at 0.01–0.05 M, pH 7.2–7.4, containing Tween-20, polyvinylpyrrolidone (PVP, MW 10,000–40,000) and a blocking agent. Tween-20 is added to retard non-serological attachment to the solid surface after the trapping antibody or antigen has been added, and PVP is added to bind polyphenols that might otherwise modify viral antigenic structures (Converse and Martin, 1990). The blocking agent is added for the reasons described above. Further, chemicals like 2-mercaptoethanol or sodium diethyldithiocarbamate (Na-DIECA) may be added for protection of the virus. Also sodium sulphite can be used for this purpose (Adgen ELISA Protocols, 2000).

The buffer for antibody and conjugate is usually 0.01–0.05 M PBS, pH 7.2–7.4, containing 0.05% Tween-20 and 0.2% blocking agent.

For preservation of the carbonate and phosphate buffers, sodium azide (NaN_3) is added to a final concentration of 0.02%.

The buffer used for the last reactant, the substrate, must provide optimal conditions for the enzyme/substrate reaction. The most common enzyme, alkaline phosphatase, type VII (see above), functions optimally in a diethanolamine buffer, pH 9.8.

INDIRECT AGF-ELISA. Sodium carbonate buffer, pH 9.6 is most commonly used as extraction and coating buffer for the antigen. However, 0.01–0.05 M PBS, pH 7.2–7.4, functions very well as extraction and antigen buffer in our laboratory, though rarely reported for this purpose. Chemicals such as 0.01 M Na-DIECA (Hobbs *et al.*, 1987) may be added to both buffers to protect the antigen.

For the next two reactants, the crude, unfractionated antiserum and the conjugate, the same buffer as in DAS- and TAS-ELISA is used. In the substrate step also, the same buffer as in DAS- and TAS-ELISA is used.

Note: Na-DIECA has a limited shelf life when added to buffers.

Caution: The additives mentioned above are harmful, especially Na-DIECA and 2-mercaptoethanol. NaN_3 is highly toxic. See also Appendix 2.

Microtitre plates

Microtitre plates in an internationally standardized 96-well format are available in a variety of brands. Polystyrene plates are by far the most commonly used. Each well holds about 300 μl and has a flat or round bottom. Those with a round bottom are the easiest to wash, but unfortunately the reading of these is not reliable in a conventional ELISA reader (Andersen *et al.*, 1991). It is important that the plates have a high and uniform protein-binding capacity. These properties differ somewhat from brand to brand and may also depend on price. It is also of importance that the plates have a high and uniform transparency to secure a correct spectrophotometric reading. So-called ELISA strips or mini-plates with only two rows of eight wells each are available and are more economical in cases where only a few samples are to be tested. One or more strips can be mounted in a frame of the same size as a 96-well plate for easy handling and reading.

During the ELISA procedure the upper surface should only be touched wearing gloves. Human skin excretes AP, and even traces of AP on the well edges due to fingerprints may lead to unspecific reactions. Labelling of plates or other marking must be made only on the plate sides and not on the upper surface, as it may disturb the reading.

Reuse of plates requires their complete regeneration by special washing procedures, of which a number have been described (Bar-Joseph *et al.*, 1979; Manandhar *et al.*, 1996; and others). Some of the methods, however, result in reduction of the plates' protein-binding property, but a method employing clothes washing powder allowed reuse of the same plate up to four times without disturbing the protein binding (Manandhar *et al.*, 1996; protocol, Section 5.2.8).

Recording of results

The intensity of colour, revealing presence of virus (positive reaction) or absence (negative reaction), may be assessed visually by use of a uniform light source as background when comparing with a predetermined colour scale. For *p*-nitrophenol (*p*NP) the human eye can determine colour at an intensity corresponding to approximately 0.15 optical density (OD) at A_{405} (see below) (Converse and Martin, 1990). A small light box is useful for such readings or to decide when the colour is sufficiently developed for recording by an ELISA reader. Visual evaluation can be difficult, especially for *p*NP, going from colourless to yellow. With other enzyme/substrate systems, such as penicillinase/penicillin, in which the colour of the substrate shifts from, for example, blue to yellow, visual evaluation is much easier (Section 5.2.6).

For evaluation of results where *p*NP is used, an automatic ELISA reader is highly recommended. An ELISA reader is a spectrophotometer designed for measuring the colour intensity in the plate wells. It prints out the absorbance (*A*) values at a given wavelength, e.g. at 405 nm as required for the yellow colour of *p*NP (A_{405}). In each plate three types of control wells are required:

- One or more so-called zero wells or 'blanks' in the plate as a reference value for the apparatus. These wells are kept empty until the substrate step, when they are loaded with substrate *buffer* only.
- Four to six negative controls, i.e. wells loaded with material from known healthy tissue.
- One or more positive controls, i.e. wells loaded with material containing the virus concerned.

The readers can be manually operated or have automatic reading of the plate. After reading, the values are printed out, and a predetermined cut-off value for infected/healthy samples may also be indicated. When the reader is connected to a computer, the results can be stored electronically and further analysed. Several types of readers with varying degrees of automation and facilities are on the market at prices that have dropped considerably in recent years. Thus the price difference between the least costly automatic readers and the manual ones is now almost negligible.

Like other electronic instruments, ELISA readers are sensitive to excessive power fluctuations and high humidity. These facts should be considered when installing such instruments, especially in tropical developing countries where unstable power supplies and/or high air humidity may occur and cause serious damage to the electronic circuits unless adequate precautions are taken.

Two or three readings may be taken at intervals during the colour development, and the optimal printout used to evaluate the result. Absorbances at A_{405} are usually most accurate near OD = 1.0 (Converse and Martin, 1990). This fact should be taken into account when choosing among the readings made. On the other hand, one should also make sure that the substrate incubation has been sufficient or has not run too long. A good indicator for a sufficient incubation period is when, in a series of, for example, 15 min readings, the ratios, each calculated from (mean of positive control values)/(mean of negative control values), become equal or start to decrease. In routine tests for the same virus conducted under identical conditions, setting a standard incubation time should be feasible. These matters and how to set a threshold value to distinguish between healthy and infected samples are treated in Section 5.2.9.

Sample preparation

Sample preparation for ELISA comprises the collection of material in the form of individual seeds, seedlings or plants, or parts thereof, and the making of extracts from these materials. The outlines for representative sampling of seeds for testing are described in Section 7.2.1. Sample preparation is the most time-consuming part of the ELISA work. See also Routine seed-health testing, Section 3.2.2.

SEEDS. In the majority of cases where a transmissible virus occurs in seeds, virus is present in the embryo, endosperm and testa. Only the virus located in the embryo will remain infective and able to infect the progeny, while the testa- and endosperm-located virus have lost infectivity during seed maturation (exceptions: some tobamoviruses). Such surface-located virus is, however, *still serologically detectable* (Section 2.1). Often, an infected seed lot will, apart from seeds with embryo infection, also contain a high proportion of seeds carrying detectable virus in the testa and endosperm only. Whole-seed assays of such seed lots by a serological or molecular test will thus give rise to a high number of false positives, indicating a rate of seed transmission much higher than the actual rate. For example, decortication of soaked seeds of pea seed lots infected with PSbMV and testing the separated individual seed-coats by ELISA revealed that up to 50% of the seed-coats contained the virus. A subsequent test of germling tissue made after germination of the naked seeds (embryos) showed only 2–3% seed-to-plant-transmitted infection, i.e. only 2–3% of the embryos carried the virus (the author, unpublished data). (For abbreviated virus names, see full names in Table 4.2.)

This example demonstrates what is a common experience among seed virologists that a test of whole seeds in general leads to unreliable results regarding determination of the rate of transmissible virus (Johansen *et al.*, 1994; and others). When testing whole seeds of some crops belonging to *Leguminosae*, it is possible to overcome this problem by soaking the seeds in water for some hours prior to decortication and rinsing off any remaining virus outside the embryo before assaying them (Maury *et al.*, 1998). Direct testing of crop seeds of other plant families, e.g. barley for BSMV, requires the embryo to be separated not only from the seed-coat but also from the endosperm (Hamilton, 1965; Lister *et al.*, 1981).

In large-scale routine testing, mechanical decortication of soaked seeds may give complications (Maury *et al.*, 1998), not least for non-legume seeds. However, in some cases it is possible to test whole seeds without decortication and determine the actual rate of transmissible virus due to the absence of detectable viral antigen in the testae of non-embryo-infected seeds: the peanut stripe strain of BCMV (BCMV-PSt) and PeMoV in groundnut (Bharathan *et al.*, 1984; Zettler *et al.*, 1993), LMV in lettuce (Falk and Purcifull, 1983), SMV in soybean (Maury *et al.*, 1985) and CMV in lupin (R.A.C. Jones, University of Western Australia, 2004, personal communication).

For direct testing of seeds of certain species, indirect AgF-ELISA seems less efficient than DAS- and TAS-ELISA, probably because binding of seed tissue components blocks a sufficient amount of virus particles that would be bound in the wells. Increasing the dilution of seed extracts may, however, alleviate this problem.

NON-DESTRUCTIVE TESTING OF SEEDS. The possibility of testing a small part of a seed for virus while leaving the rest viable would be useful for small seed lots, such as germplasm, in order to select seeds without seed-to-plant-transmissible virus. This approach, however, assumes: (i) a uniform distribution of detectable virus in both the seed's embryo axis and the cotyledons; (ii) that there is complete correlation between virus absence in the cotyledons and virus absence in the progeny; (iii) that the testae of seeds without transmissible virus do not carry virus; and (iv) that the seed germination is not affected by such testing.

So far, these requirements have only been fulfilled for lettuce infected with LMV (Falk and Purcifull, 1983), groundnut with PeMoV (Bharathan *et al.*, 1984) and BCMV-PSt (Demski and Warwick, 1986; Xu *et al.*, 1991) and soybean with SMV (Higley *et al.*, 1993, 1994). For another seed-transmitted virus of groundnut, PCV, Reddy *et al.* (1998) found that, whether the embryos were infected or not, seed-coats of all seeds harvested from infected groundnut plants contained detectable virus. Therefore, to assess the level of seed transmission in a direct test or non-destructive test of groundnut seeds for PCV, removal of the seed-coat is required.

A non-destructive testing of seeds is virtually impossible for small seeds like lettuce seeds. However, this approach seems feasible for the screening of groundnut seeds for PeMoV and BCMV-PSt and soybean seeds for SMV. For cleaning groundnut germplasm stocks, a slice of cotyledon tissue can be separated from the part of the seed distal to the embryo axis, avoiding seed-to-seed contamination and keeping track of individual seeds. After testing of the sliced-off seed parts by ELISA, seeds found to be virus-free can be released to users (Bharathan *et al.*, 1984; Zettler *et al.*, 1993). A similar technique has been described for cleaning soybean stocks, where 3 mm cores of seed tissue were excised and assayed for SMV by ELISA (Higley *et al.*, 1993, 1994); however, apparently the germination of cored seeds was affected to some extent.

Takeuchi *et al.* (1999) described a method to test non-destructively for tobamoviruses on pepper seeds by a serological technique (Section 5.4).

GERMINATED SEEDS AND SEEDLINGS. Testing of progeny rather than seeds provides obvious advantages. Only transmissible virus is detected, seeds need no initial treatment (except occasionally fungicide treatment) and the antigen titre is generally higher in progeny than in seeds. For example, when testing soybean for SMV by ELISA, Lister (1978) found four times higher absorbances in leaves than in seed extracts. In addition, growing-on tests (see Chapter 4) provide the opportunity to disclose possible symptom developments in the seedlings prior to ELISA.

A disadvantage as compared to direct testing of seeds is the time it takes to raise seedlings (2–3 weeks for most vegetable and field crops).

When sampling seedling tissue for ELISA or any other test, care must be taken to obtain representative samples, as the virus can be irregularly distributed within the plant. Representative sampling is of special importance when latent seedling infection occurs. Latency or weak symptom expression in seedlings raised from infected seeds of various host species is characteristic for some virus genera, such as *Cucumovirus* and those that are nematode-transmitted.

GROUP TESTING. The possibility of combining seeds or seedlings into groups of a certain size and testing each group, followed by a statistical estimate of the infection rate, would reduce workload, cost and time consumption. Detection of one LMV-infected lettuce seed in groups of 100 seeds or more by DAS-ELISA has been reported (Falk and Purcifull, 1983; Van Vuurde and Maat, 1983), but this is an exception, and most other virus–host combinations would require far smaller group sizes for effective testing. Thus, in his laboratory for large-scale testing of seeds for viruses, Dr G. Mink (Washington State University, USA, 1987, personal communication), routinely used group sizes of only two seeds/seedlings. One of the factors implicated by direct testing of seeds is the great variation of virus concentrations existing among embryos of different virus–host combinations (Maury *et al.*, 1998). Although virus concentrations are generally higher in the progeny, such a variation also occurs in seedlings. The two most important factors for determining group size are the virus–host combination in question and the sensitivity of the assay. Thus, Mowat and Dawson (1987) consider the sensitivity of indirect AgF-ELISA to be too low for pooled samples compared to DAS- and TAS-ELISA because of the increased competition for binding sites produced by non-viral components of any non-infected samples. Irrespective of test method, determination of group sizes, allowing reliable detection of one infected seed in *x* healthy seeds, can only be done on the basis of a trial, including tests of increasing dilutions of known infected material in healthy material. Such a check must be done once prior to routine testing, not only for each new virus–host combination but also when introducing new detection reagents into screening procedures (for details see Section 5.2.8).

PREPARATION TECHNIQUES. Whether whole soaked seeds, seed flour, embryos or seedlings are to be tested, the tissue must be thoroughly homogenized groupwise or individually in the buffer recommended for the ELISA type to be used. For detection of CMV in lupin seeds, assay of dry seed flour is possible (see Nucleic acid extraction protocols, Section 6.2.3) due to the absence of detectable virus in the seed-coat (R.A.C. Jones, University of Western Australia, 2004, personal communication). For detection of PSbMV in pea seeds, Torok and Randles (2001) described a technique to remove the seed-coats prior to milling (Section 6.2.2).

Careful homogenization is important for maximum release of virus particles and must be carried out in a reproducible manner from sample to sample and from assay to assay. Tissue grinding is normally done in the presence of buffer (see also Lommel *et al.* (1982) under Increasing the sensitivity, Section 5.2.8), at least for labile viruses, for which the buffer should further contain virus-protecting additives, and can be done manually by mortar and pestle. A more rational way is to place tissue and a little buffer in a small, strong polyethylene bag, e.g. 7 cm × 10 cm, place the bag on the table and grind the material from outside the bag with a pestle or, more efficiently, with a hand homogenizer (see Equipment in Section 5.2.3). Machines such as Meku™, which disrupts the tissue between rollers with simultaneous addition of buffer, or Bioreba™, which grinds the tissue while inside plastic bags, greatly increase the speed of sample preparation. A rack for photographic slides is well suited as a holder for the bags before and after hand or machine grinding. Plastic bags with a built-in filter prevent clogging of the pipette with plant material during pipetting (available from, for example, Adgen Diagnostics™ or Bioreba™).

Especially when homogenizing by hand, the initial amount of buffer added should be relatively small, 1 : 5 or 1 : 10 (w/v) of tissue:buffer, as larger amounts will impede the grinding. After grinding, the suspensions are further diluted as recommended.

Preparation of the correct *control samples* is important. Positive control samples must contain the virus verified to be identical to the virus/virus strain being tested for. A number of negative control samples (four to six per plate) must also be made, each representing different individual seeds or seedlings, which should match the test samples with respect to cultivar, age, growing conditions and tissue type (Sutula *et al*., 1986; *see* also Section 5.2.9).

When performing large-scale testing, storing of large amounts of samples prior to testing may be needed. According to Van Regenmortel and Dubs (1993), leaf samples are best stored at –20°C after homogenization in 1% K_2HPO_4, 0.1% Na_2SO_3 (1 g leaf in 1 ml buffer). This method, however, may be adequate for DAS-ELISA but not for indirect antigen-first ELISA according to Lommel *et al*. (1982), who found that extraction in PBS results in lower absorbances than extraction in distilled H_2O (see also Section 5.2.8). Sampling of leaf parts may also be done directly in the grinding bags mentioned above, which, after immediate sealing to avoid evaporation, can be stored at 4°C for a day or so until processing.

Throughout the sample preparation, care must be taken to avoid cross-contamination between samples. To further minimize this risk, always prepare negative control samples first, then test extracts and finally the positive control samples.

5.2.3. Double-antibody sandwich ELISA (DAS-ELISA), protocols

There are several variants of DAS-ELISA. Described here is the classical version, as reported by Clark and Adams (1977), involving three layers of serological reactants (Table 5.1, Fig. 5.2). The method requires use of:

1. The antibody-containing IgG fraction, henceforth called IgG, separated from crude antiserum specific to the virus being tested for.

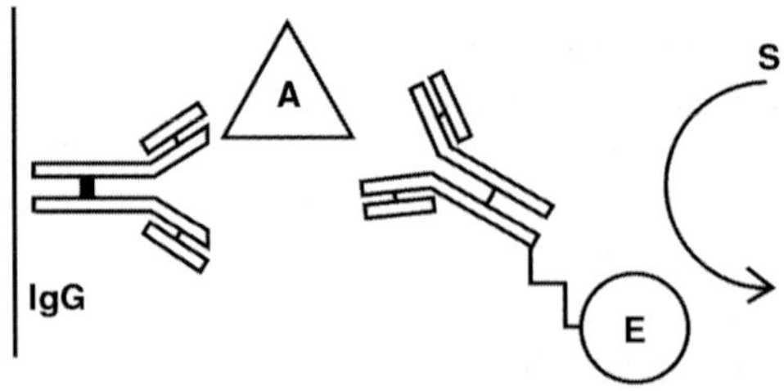

Fig. 5.2. DAS-ELISA. Specific Pabs (IgG) are bound to the solid surface. Antigen (A) in the sample binds to the coating antibody. Specific Pab (IgG)/AP conjugate is then added and binds to antigen if present. E, enzyme (AP); S, substrate. Mabs may be used instead of Pabs. Courtesy of ADGEN Phytodiagnostics, Neogen Europe Ltd, UK.

2. Antibody–enzyme conjugate, consisting of virus-specific IgG linked to an enzyme. The most commonly used enzyme, alkaline phosphatase (AP), will be dealt with here.

Unless ready-made ELISA kits are used, these two components have to be prepared in the laboratory before starting the detection of the virus. The shelf life of the preparations with an adequate preservative added and correct storage conditions is, however, reasonably long.

The specificity of DAS-ELISA is higher than that of indirect ELISA. On the other hand, with DAS-ELISA there is a risk of not detecting serologically divergent isolates (Koenig, 1981; see also Section 5.2.8).

IgG fractionation

Fractionation of IgG from the crude antiserum can be done in different ways. The most commonly used method is a partial purification, in which IgG is precipitated by a high concentration of a salt, leaving the other serum proteins in solution. The IgG is subsequently released from the resuspended precipitate by dialysis. A fractionation method (McKinney and Parkinson, 1987) involving use of octanoic acid to precipitate non-IgG protein prior to salt precipitation of IgG resulted in a high yield of IgG according to Abraham (1995). Methods for further purification of the IgG involve chromatographic procedures, such as ion exchange or protein A affinity chromatography following the salt fractionation. A reasonable yield and a high degree of purity can, however, be obtained by the following method (Ball *et al.*, 1990), producing up to 5–10 mg IgG per ml crude antiserum, which is sufficient for coating antibody and conjugate for several hundreds of DAS-ELISA test wells.

EQUIPMENT AND MATERIALS.

a. Glass beaker, 10 ml.

b. Glass beakers or Erlenmeyer flasks, 500 ml.

c. Test tubes and centrifuge tubes.

d. Dialysis tubing (44104 Visking® 8/32, 6 mm diameter, Serva™ or Slide-A-Lyser®, Pierce™, or similar) (see **Note 1** after procedure).

e. Stirring apparatus, magnetic, with rods.

f. Low-speed centrifuge.

g. UV spectrophotometer.

h. Saturated ammonium sulphate: add $(NH_4)_2SO_4$ to 25 ml deionized water under stirring until crystals remain at the bottom at room temperature – approx. 800 g/l. Adjust pH to 7.0–7.2 with ammonia (NH_3) or HCl and store at 5°C. **Caution:** handle the strong acid, HCl, with care – see Appendix 2, p. 247.

i. Phosphate buffered saline (PBS) *without* addition of NaN_3 (0.01 M phosphate buffer, pH 7.4, containing 0.15 M NaCl) – see Appendix 2, p. 247.

j. 0.2 µm nitrocellulose filter.

k. Sodium azide (NaN_3), stock solution – see Appendix 2, p. 247. **Caution:** highly toxic.

PROCEDURE.

1. Add in a small glass beaker the following per ml of crude antiserum: 1 ml of deionized water and 1.3 ml cold, *saturated ammonium sulphate*, pH 7.0–7.2 drop-wise with slow stirring (final ammonium sulphate concentration of 40%).

2. Mix for 30 min and keep at 4°C overnight.

3. Centrifuge at 8000 *g* for 10 min and collect precipitated pellet.

4. Re-suspend precipitate gently in 1 ml *PBS*.

5. Add 1 ml water and 1.02 ml *saturated ammonium sulphate* per ml of re-suspended precipitate (to a final ammonium sulphate conc. of 33%).

6. Mix for 30 min and centrifuge as above.

7. Re-suspend the pellet in *PBS* at one-half the original volume of antiserum.

8. Transfer the solution to a dialysis tubing presoaked in PBS, and dialyse the preparation against several changes of PBS over 24–36 h at 4°C (use 300–500 ml fresh PBS per change, and dialyse under slow stirring of the PBS).

9. Filter through the 0.2 µm filter and assess the IgG concentration by UV-spectrophotometry. At A_{278} (0.1%, 1 cm), the optical pathway is 1 cm, and a 0.1% solution (1 mg/ml) has an absorbance of 1.35 ± 10% at 278 nm (Tijssen, 1985). Thus, for example, for a 0.2% (2 mg/ml) solution the A_{278} will be 2.7 and so on. *Note*: spectrophotometric measurement must be done before adding sodium azide as this may skew the result due to its high UV absorbance.

10. Adjust to 1 mg IgG/ml with PBS. Indicate name and concentration on the tube. Sodium azide at 0.02% may be added before long-term storing at –20°C or short-term storage at 4°C; however, see **Note 2**.

Notes

1. Brattey and Burns (1998) describe the use of centrifugation in 'Concentrator tubes' (Centriprep-30®, Amicon™) to free the IgG from salt instead of using dialysis tubing.

2. Do not add sodium azide to the IgG to be used for enzyme conjugation (Dijkstra and de Jager, 1998).

Antibody–enzyme conjugation

Linking of the enzyme AP to IgG is done by mixing the two components in the presence of glutaraldehyde. This chemical provides an effective link, still allowing the reactive parts of both AP and IgG to function, provided that the conjugation is correctly carried out. It is important to mix the ingredients in the optimal proportions and to use the correct buffers and conditions. The one-step glutaraldehyde procedure below, based on Avrameas (1969), is adopted from Converse and Martin (1990) with slight modifications.

EQUIPMENT AND MATERIALS.

a. Virus-specific IgG without NaN_3.
b. Test tube, hydrophobic.
c. Dialysis tubing (see materials for IgG fractionation above).
d. Glass beakers or Erlenmeyer flasks, 500 ml.
e. Stirring apparatus, magnetic, with rods.
f. Alkaline phosphatase (AP), type VII, from bovine intestinal mucosa, stabilized with 100 μM $ZnCl_2$ and 1 mM $MgCl_2$/ml (Sigma P5521 or similar).
g. PBS (0.01 M, pH 7.4) without NaN_3 – see Appendix 2, p. 247.
h. Glutaraldehyde: a good quality, e.g. 25% aqueous, electron microscopy grade, in ampoules (Merck 112179 or similar), is recommended. **Caution:** toxic, handle with care.
i. TRIS-HCl buffer, 0.05 M, pH 8.0 (dissolve 6.05 g TRIS base in 800 ml deionized water. Adjust to pH 8.0 with HCl and bring the volume to 1000 ml with deionized water). **Caution:** HCl is a strong acid; handle with care – see Appendix 2, p. 247.
j. $MgCl_2$.
k. Bovine serum albumin (BSA), Sigma A7638 or similar.
l. Sodium azide (NaN_3), 2% stock solution – see Appendix 2, p. 247. **Caution:** highly toxic.

PROCEDURE.

1. Dilute IgG to 1 mg/ml in *PBS* in a hydrophobic test tube. Add 5 mg AP, *type VII*, per ml of IgG solution.
2. Transfer the solution to a dialysis tubing presoaked in *PBS*, and dialyse overnight at 4°C in three changes of PBS, using a total of 1500 ml. Dialyse as described for IgG.
3. Replace the PBS with PBS containing *0.06% glutaraldehyde* (1 ml of 25% glutaraldehyde in 420 ml PBS) and dialyse for 1–2 h at room temperature under slow stirring.
4. Dialyse overnight at 4°C in three changes of *PBS*, total 1500 ml as before.
5. Replace PBS with *0.05 M TRIS-HCl buffer, pH 8.0*, containing *1 mM* $MgCl_2$, and dialyse overnight (or 16 h) at 4°C, with two changes, using a total of 1000 ml.
6. Add *BSA* to 1% (w/v) and NaN_3 to 0.02% (w/v) to the conjugate and store at 4°C in the dark in a glass vial. Do not freeze.
7. Determine optimal concentration for ELISA by a chequerboard test. A series of conjugate dilutions is evaluated against a series of trapping globulin dilutions using known virus (homologous antigen) and healthy plant preparations of the same species as the virus host (see Fig. 5.3 and the DAS-ELISA procedure below).

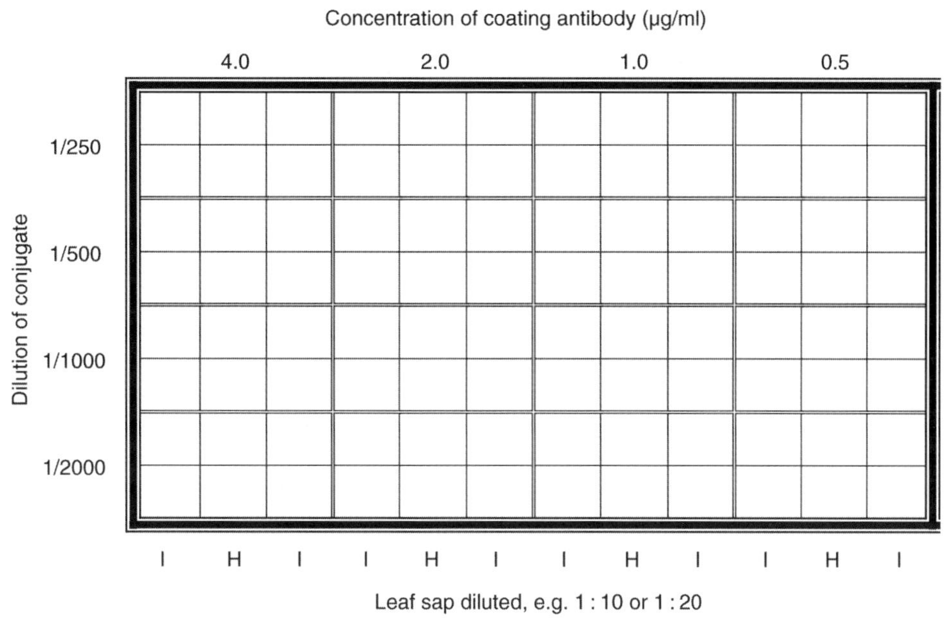

Fig. 5.3. Example of a chequerboard arrangement of coating and conjugate concentrations on an ELISA test plate to optimize concentrations for ELISA systems (see DAS-ELISA procedure). Duplicate wells are used for each treatment. I, infected (use, for example, 1 : 10 and 1 : 20); H, healthy (use 1 : 10).

Double-antibody sandwich (DAS) ELISA

The following procedure is adopted with slight modifications from Converse and Martin (1990) and is based on the method of Clark and Adams (1977).

EQUIPMENT AND MATERIALS.

Equipment

a. Adjustable pipettes, single- and multi-channel types.
b. Facility for incubation at 37°C.
c. Wash bottle or automatic washing device.
d. Glassware, thoroughly cleaned.
e. Hand homogenizer (e.g. Bioreba™).
f. Automatic homogenizer (optional).
g. Test tube rack(s).
h. A rack made for photographic slides is used for holding samples in polyethylene bags.
i. ELISA reader (recommended), measuring absorbances at 405 nm wavelength (A_{405}) and with printing facility.

Materials

j. Microtitre plates with or without lids.
k. Tips for adjustable pipettes.
l. Household cling film (in absence of plate lids).

m. Polyethylene bags, sturdy type, approx. 7 cm × 10 cm, and test tubes (size 3–4 ml, glass or disposable) for preparation of test samples (see also p. 103). Plastic bags with a built-in filter, preventing clogging of the pipette with plant material, are available from, for example, Adgen Diagnostics™ or Bioreba™.

n. Test samples.

o. Controls: known *healthy* material of the same species, cultivar, age and tissue type as that being tested, and known *infected* material.

p. Known healthy material the same as that used for control if cross-absorption is to be carried out.

Reagents, solutions and buffers

q. Virus-specific IgG for coating/trapping.

r. Blocking agent: skimmed milk powder, bovine serum albumin (BSA) or ovalbumin; see also buffers below. For use in a separate blocking step, polyvinyl alcohol (PVA, e.g. Sigma P8136) is an option.

s. Conjugate: virus-specific IgG conjugated to AP.

t. Substrate: *p*-nitrophenyl phosphate (*p*NPP), e.g. Sigma N9389 (5 mg pills).

u. Coating buffer: 0.05 M carbonate buffer, pH 9.6 – see Appendix 2, p. 247. After storing, check pH before use.

v. Sodium azide (NaN_3), 2% stock solution – see Appendix 2, p. 247. **Caution:** highly toxic.

w. PBS – see Appendix 2, p. 247.

x. Washing buffer: PBS-Tween (add 0.5 ml Tween-20 per litre of PBS; use PBS *without* addition of NaN_3).

y. Polyvinylpyrrolidone (PVP), MW 10,000–40,000.

z. Sodium sulphite (anhydrous).

aa. Extraction buffer – see Appendix 2, p. 247. *This buffer is used in DAS- and TAS-ELISA only.*

bb. Conjugate buffer: PBS containing NaN_3, 0.5 ml Tween-20 and 2 g blocking agent (BSA, ovalbumin, or skimmed milk powder) per litre.

cc. Substrate buffer: diethanolamine, 10% (v/v), pH 9.8 – see Appendix 2, p. 247. After storing, check pH and adjust, if necessary, before use.

dd. NaOH, 3N.

PROCEDURE (see also flow schedule Table 5.2). Before using the ELISA in routine testing, the optimal dilutions of IgG and conjugate for the virus to be detected should be determined as shown in Fig. 5.3.

In the case of any uncertainty about parts of the following, consult Section 5.2.2. For trouble shooting, see p. 123.

Preparation

1. Fill in the loading pattern in a test schedule for each microtitre plate. Indicate both antigens (serial number, etc.) and detecting antibody in their respective, predetermined dilutions. Test samples are to be loaded in at least duplicate wells. Leave in each plate: wells for blanking (zero references), negative controls (four or more) and positive controls (two).

Table 5.2. Flow schedule for DAS-ELISA.

Step (step no. in procedure)		Buffer	Time	Temperature
a. Coating with IgG	(**1**–**4**)	Carbonate	2–4 h or overnight	37°C or 4°C, respectively
b. Washing[a]	(**5**–**6**)	Washing	≥ 3 × 3 min	
c. (Blocking)	(**7**)	Washing	30 min	Ambient or 37°C
d. Antigen	(**8**–**10**)	Extraction	2–4 h or overnight	37°C or 4°C, respectively
e. (Cross-absorption)	(**11**)	Conjugate	45–60 min	37°C
f. Washing	(**12**)	Washing	≥ 3 × 3 min	
g. Conjugate[b]	(**13**)	Conjugate	2–4 h or overnight	37°C or 4°C, respectively
h. Washing	(**14**)	Wash → H_2O	≥ 4 × 3 min	
i. *p*NPP	(**15**–**17**)	Substrate	≥ 10–60 min	Ambient
j. Recording	(**18**–**20**)			

[a]After step b, a blocking step may be included. No washing after the blocking step.
[b]Conjugate may be cross-absorbed before use.

2. Label the plate on the side with a permanent marker for identification (virus being tested for, antibody used, etc.). Do not put any marks on the top as it may interfere with the plate reading at the end of the procedure. Avoid throughout the procedure any fingerprints on the plate upper surface. Human skin excretes AP, and fingerprints may, therefore, give rise to unspecific colour development.

Coating with IgG

3. Add trapping antibodies appropriately diluted (often 1 μg IgG/ml) with *coating buffer*, using 100 μl per well. Use the same reagent volume throughout the procedure except in the blocking step, if included. Leave the blank wells empty.

Coating buffer can be stored for later use but must be pH-checked just before use.

4. Cover the plate with a lid, or wrap the plate in thin household cling film. Incubate the plate for 2–4 h at 37°C or overnight at 4°C.

Go to step 8, while the plates incubate.

Washing

5. Empty wells by machine or hand. If by hand, turn the plate upside down and at the same time empty the plate into a sink with a flinging movement. While inverted, wipe the top surface of the plate dry (press onto a clean tissue).

6. Rinse the wells once with *washing buffer*, followed by at least three thorough washes, soaking 3 min each time (fill the wells, let soak and empty plate).

Note: after this and each subsequent washing procedure, remove carefully the remains of washing liquid and bubbles from the wells by repeated pounding of the plate upside down on a stack of dry paper towels. If using aspiration, take care that the floor of the wells and the underside of the plates are not scratched, as this may interfere with the ELISA reading.

After coating, sealed plates can be stored at 4°C.

For details, see Section 5.2.2.

Blocking

7. Optional step: Before adding the next reactant a blocking step may be included (Section 5.2.2). In that case, add 150 µl per well of a blocking agent such as dry skimmed milk powder at 1–2% (w/v) in *washing buffer* and incubate for 30 min at room temperature or 37°C to help block any remaining uncovered well surfaces. Alternatively, block for 1 min with PVA at 1 µg/ml in washing buffer. After incubation, empty plate and blot the surface.

Do not wash.

Antigen

8. Calculate the volume needed of each of the antigens for the number of wells to be loaded. Prepare antigen samples in the following sequence: negative controls, test samples and positive controls as described in Sample preparation, pp. 102–103. *Extraction buffer* is used for grinding and for making a final dilution of 1/10 or 1/20 in the test tubes.

For easy plate loading, arrange the test tubes in the test tube holder according to the pattern for loading of the antigens.

9. Use the adjustable pipette to load 100 µl of the diluted antigens into the wells according to the test schedule. Load in the following sequence: healthy controls, test samples and positive controls. Leave the blank wells empty.

In case of tissue debris clogging the pipette tips, filter the suspensions, cut a piece of the pipette tips, or consider using bags with built-in filter (see Materials).

10. Cover and incubate 2–4 h at 37°C or overnight at 4°C.

(*Cross-absorption*)

11. Depending on the quality of the detecting antibody, a cross-absorption step (see p. 95) may be included here:

(a) Tissue from healthy plants or seeds of the same species as is being tested is ground up and diluted in *conjugate buffer* at 1 : 20–1 : 100 (w/v). Make about 18 ml per plate.

(b) Filter the extract through several layers of gauze, and

(c) use the *filtrate* to dilute the conjugate (usually 1/1000 or 1/2000) instead of pure conjugate buffer in step **13** below. About 12 ml is required per plate. Mix in a glass beaker or similar.

(d) Stir thoroughly and let the mixture incubate under cover for 45–60 min at 37°C or overnight at 4°C.

Conjugate

12. Empty and wash the plate as in steps **5–6** above, taking special care not to cross-contaminate wells with antigens from adjoining wells when emptying and washing the plate. Make sure that wells are absolutely free from plant or seed debris.

13. Add *either* conjugate appropriately diluted (usually 1/1000–1/2000) in *conjugate buffer or* the cross-absorption mixture made in step **11**. Use 100 µl/well. Leave blank wells empty.

In the case of stable viruses incubate for 2–4 h at 37°C; for unstable viruses incubate overnight at 4°C. Most seed-transmitted viruses are rather stable.

Substrate

14. Empty the plate and wash as in steps **5–6** above, but finish with one to two washings in deionized H_2O (to remove phosphate remains, which have a significant inhibitory effect on the subsequent AP/substrate reaction).

15. Prepare the substrate immediately before use by dissolving *p*NPP pills in *substrate buffer* at 0.25–1 mg/ml. About 12 ml is required per plate. Use gloves to avoid fingerprints on tools and, not least, on the pills.

16. Load the wells with 100 µl per well, except the blank wells, which are loaded with substrate *buffer* only.

If an ELISA reader is to be used, switch it on so it can warm up.

17. Incubate the plate at room temperature until yellow colour appears (at least in the positive control wells), generally 10–60 min. To speed up the reaction, incubation at 37°C, e.g. for 15 min, may be included.

Recording

18. When colouring starts to appear, evaluate visually using a uniform light source (e.g. a light box), or read the plate on an ELISA reader at 405 nm wavelength (A_{405}). Prior to reading, wipe the top surface of the plates to prevent substrate solution from contaminating the photocells in the reader. Repeat the reading one to three times during the next 1 to 6 h (see also Sections 5.2.2 and 5.2.9).

19. When the incubation is completed, and if desired, stop the reaction by adding 25 µl of 3 M NaOH per well, followed by thorough mixing on a suitable shaker.

20. Interpretation of the result is discussed in Section 5.2.9.

5.2.4. Indirect antigen-first (AgF) ELISA, protocol

The principles of the indirect ELISA procedures are described in Sections 5.2.1 and 5.2.2.

The variant dealt with here, the indirect AgF-ELISA (Fig. 5.4), has the advantages of: (i) requiring no fractionation of IgG – whole (crude) antiserum can be used directly; and (ii) using no virus-specific conjugate, but a generally applicable conjugate.

In the following procedure polyclonal antibodies and anti-rabbit IgG/AP conjugate are used. Cross-absorption of the detecting virus-specific antibody as standard is normally included in this method (see p. 95). The procedure is adopted from Hobbs *et al.* (1987), and has functioned well in virus identification and routine testing at our institute. The entire test may be completed within 1 day (however, note *step **10***).

EQUIPMENT AND MATERIALS. The *equipment* is the same as listed for the DAS-ELISA protocol, p. 107.

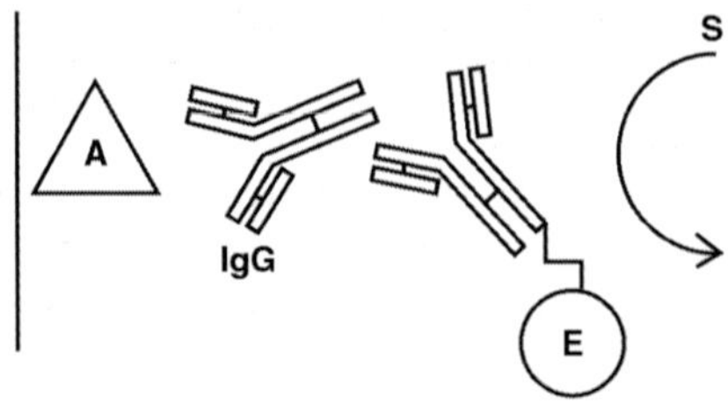

Fig. 5.4. The principles of indirect AgF-ELISA. A, antigen; IgG, virus-specific antibodies in crude antiserum; E, enzyme (AP), which is conjugated to anti-rabbit IgG produced in goat or pig; S, substrate. Courtesy of ADGEN Phytodiagnostics, Neogen Europe Ltd, UK.

The *materials* and *buffers* are as listed under DAS-ELISA, except for the following:

Use of	Instead of
✓ Coating buffer/extraction buffer = carbonate buffer with addition of 0.01 M Na DIECA*	Coating buffer without additions
✓ Unfractionated (crude) antiserum = detecting antibody (optimal dilution, see step **5** and Fig. 5.5 below)	Coating IgG
✓ Conjugate = anti-rabbit IgG/AP (DakoCytomation™ D-306 or similar)	Conjugate = virus-specific IgG/AP

Note: The *serum/conjugate buffer* used in this procedure is the same as the conjugate buffer used in DAS-ELISA.
*Others, e.g. James and Mukerji (1996), use 2% PVP instead of DIECA (because DIECA is hazardous). DIECA is to be added just before use, as it is unstable in solution.

PROCEDURE (see also flow schedule, Table 5.3). Before using the ELISA in routine testing, the optimal dilutions of IgG and conjugate for the virus to be detected should be determined, as shown in Fig. 5.5. Use the flow schedule (Table 5.3) to plan the work (e.g. avoid overnight incubation in step **10**).

Preparation
1. Prepare for ELISA in the same way as described in *DAS-ELISA, steps* ***1*** *and* ***2***, pp. 108–109. Remember: two wells per test sample.

Antigen
2. Calculate the volume needed of each antigen for the number of wells to be loaded.

Prepare antigen samples in the following sequence: negative controls, test samples and positive controls, as described under Sample preparation, pp. 102–103. Use *coating/extraction buffer* (carbonate buffer) for grinding and for making a final dilution of 1/50 or 1/100.

Table 5.3. Flow schedule for indirect AgF-ELISA.

Step (step no. in procedure)		Buffer	Time	Temperature
a. Coating with antigen	(**1–4**)	Extraction/coating (carbonate)	1–2 h or overnight	37°C or 4°C, respectively
b. Cross-absorption[a]	(**5**)	Serum/conjugate	≥ 45 min or overnight	37°C or 4°C, respectively
c. Washing[b]	(**6–7**)	Washing	≥ 3 × 3 min	
d. Specific antiserum	(**9–10**)	Serum/conjugate	1–3 h, not overnight[c]	37°C
e. Washing	(**11**)	Washing	≥ 3 × 3 min	
f. Conjugate	(**12–14**)	Serum/conjugate	1–2 h or overnight	37°C or 4°C, respectively
g. Washing	(see DAS **14**)	Washing → H_2O	≥ 4 × 3 min	
h. Substrate	(see DAS **15–17**)	Substrate	≥ 10–60 min	Ambient
i. Recording	(see DAS **18–20**)			

[a]Cross-absorption may also be initiated prior to ELISA.
[b]After step c, a blocking step may be included.
[c]If cross-absorption is used.

For easy plate loading, arrange the test tubes in the test tube holder according to the pattern for loading of the antigens.

3. Use the adjustable pipette to load 100 μl of the diluted antigens into the wells according to the test schedule. Load in the following sequence: healthy controls, test antigens and positive controls.

Load with the same reagent volume (100 μl) throughout the procedure except in the blocking step, if included.

Leave the blank wells empty.

In the case of tissue debris clogging the pipette tips, filter the suspensions or cut a piece off the pipette tips (for plastic bags with built-in filter, see DAS-ELISA).

4. Cover with lid or wrap in thin household cling film and incubate the plate for 1–2 h at 37°C or overnight at 4°C.

Short-term storage of covered plates at 4°C after incubation and washing (step **6**) before continuation of test has been reported (Hobbs *et al.*, 1987).

Cross-absorption

5. Cross-absorption of the detecting antibody is generally recommended in this type of ELISA (see p. 95) and can be carried out while incubating antigen – or before start of test.

(a) Tissue from healthy plants or seeds of the same species as that being tested is ground up in *serum/conjugate buffer*, final dilution *c.* 1 : 20 (w/v). Make about 18 ml per plate.

(b) Filter the extract through several layers of gauze, and

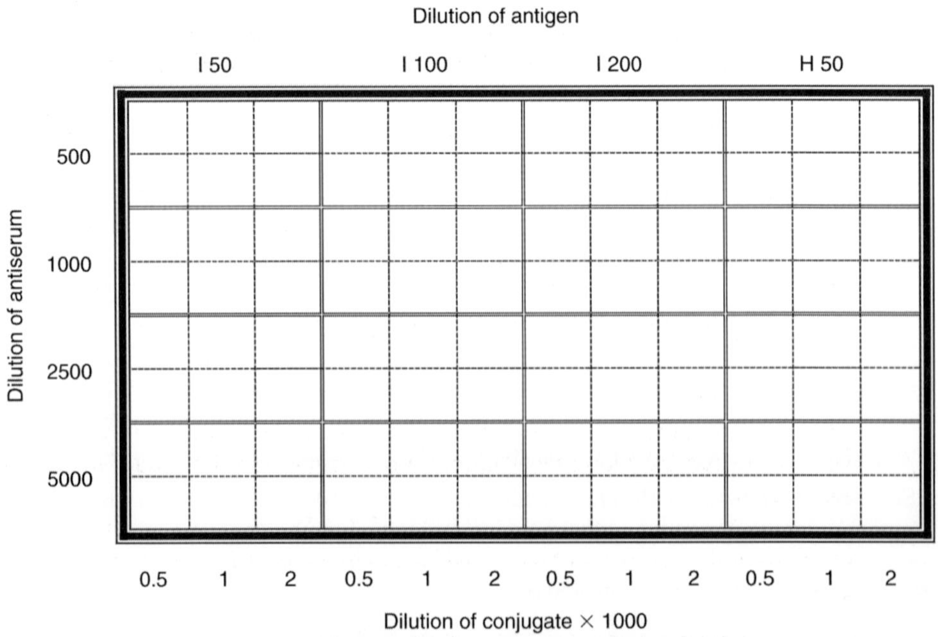

Fig. 5.5. Example of a chequerboard arrangement with examples of buffer dilutions (reciprocal figures) of antigen, antiserum and conjugate in an ELISA test plate to optimize concentrations for an indirect AgF-ELISA system (see procedure). High-quality antisera may also work well at dilutions over 5000 ×. Duplicate wells are used for each treatment. I, infected; H, healthy.

(c) use the *filtrate* to dilute the crude antiserum instead of pure serum/conjugate buffer. About 12 ml of the mixture is needed per plate. Mix in a glass beaker or similar.

Optimal dilution of antiserum is to be assessed before start of test; it is often in the range of 1/500–1/2000 but dilutions of 1/5000 or more may also be optimal (see chequerboard test (Fig. 5.5)).

(d) Stir thoroughly and let the mixture incubate under cover for 45 min or more at 37°C while continuing steps **6–8** below (alternatively, use overnight incubation at 4°C).

Washing and blocking

6. Empty the antigen solution from the wells by machine or hand. By hand, turn the plate upside down and at the same time empty the plate into a sink with a flinging movement. While inverted, wipe the top surface of the plate dry (press on to a clean tissue).

7. Rinse the wells once with *washing buffer*, followed by at least three thorough washes, soaking 3 min each time (fill the wells, let soak and empty plate).

Note: after this and each subsequent washing procedure, remove carefully the remains of washing liquid and bubbles from the wells by repeated pounding of the plate upside down on a stack of dry paper towels. If using aspiration, take care that the floor of the wells and the underside of the plates are not scratched as this may interfere with the ELISA reading.

Make sure that wells are absolutely free of plant or seed debris.

For details, see Section 5.2.2.

8. Optional step: Before adding the next reactant a blocking step may be included (see Section 5.2.2). In that case, add *150 μl* per well of a blocking agent such as dry skimmed milk powder at 1–2% (w/v) in *washing buffer* and incubate for 30 min at room temperature to help block any remaining uncovered well surfaces. Alternatively, block for *1 min* with PVA at 1 μg/ml in washing buffer. After incubation, empty plate and blot the surface.

Do not wash.

Specific antiserum

9. Load the diluted and cross-absorbed virus-specific crude antiserum, 100 μl per well. Leave the blank wells empty. For cross-absorption and dilution, see step **5** above.

10. Incubate under cover for 1–3 h at 37°C. Do not incubate with cross-absorbed antisera overnight as this will result in increased background.

Conjugate

11. Empty the plate and wash as in steps **6–7** above.

12. Dilute anti-rabbit IgG/AP conjugate in *serum/conjugate buffer* into a pre-determined concentration according to a preliminarily undertaken chequerboard test (Fig. 5.5), usually 1/1000–1/2000.

13. Load each well with 100 μl. Leave blank wells empty.

14. Incubate for 1–2 h at 37°C or overnight at 4°C.

Wash, substrate and recording

15. Continue with wash, substrate incubation and recording as described in DAS-ELISA, steps **14** to **20**.

Optimizing of indirect AgF-ELISA

In order to obtain maximum sensitivity of ELISA, i.e. the maximum difference between A_{405} values of healthy and infected antigen, it is highly recommended to determine the optimal dilutions of reagents for each new virus–host combination and when introducing new reagents. Figure 5.5 shows the set-up for such a check, which may reveal that even high reagent dilutions can result in maximum sensitivity (see also Increasing the sensitivity under Section 5.2.8). Importantly, determination of optimal dilutions may also lead to lower consumption of costly reagents. For trouble shooting, see p. 123.

5.2.5. Triple-antibody sandwich ELISA (TAS-ELISA), protocol

TAS-ELISA is an indirect ELISA format where the antigen is trapped in wells coated with a virus-specific antibody, for instance a polyclonal antibody (Pab). The next reagent could be a monoclonal antibody (Mab) as detecting antibody, followed by a Mab-recognizing anti-mouse (or rat) IgG/enzyme conjugate and substrate (Fig. 5.6; see also Sections 5.2.1 and 5.2.2).

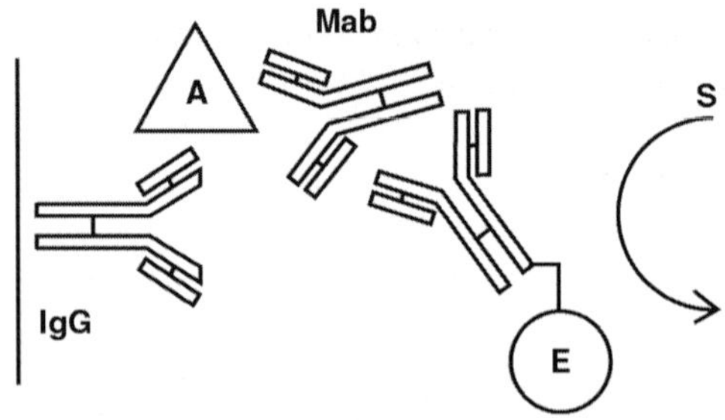

Fig. 5.6. TAS-ELISA. Antigen (A) binds to a coating Pab. A virus-specific Mab binds to the antigen, and trapped Mab is identified by anti-mouse or anti-rat IgG/enzyme. E, enzyme; S, substrate. Courtesy of ADGEN Phytodiagnostics, Neogen Europe Ltd, UK.

ELISA using Mabs is mainly employed in cases where high sensitivity or strain specificity is required. Several variants of TAS-ELISA exist (Van Regenmortel and Dubs, 1993) and some of them are briefly described under Section 5.2.7. In the following procedure, primarily as described by *Adgen ELISA Protocols* (2000), the plate is coated with a virus-specific Pab IgG, and the trapped antigen is detected with a Mab raised in mice. Unlike Adgen's procedure a separate blocking step is included here, as recommended by Huguenot *et al.* (1993) and Hamilton and Huguenot (1995).

EQUIPMENT AND MATERIALS. The equipment and materials for this version of TAS-ELISA are the same as for the DAS-ELISA described earlier (p. 107), except for the following:

Use of	Instead of
✓ Virus-specific monoclonal antibody fluid prepared from mouse ✓ Anti-mouse IgG/AP conjugate, available from e.g. DakoCytomation™	Virus-specific antibody/AP conjugate

Note: The *serum/conjugate buffer* used in this procedure is the same as the conjugate buffer used in DAS-ELISA.

PROCEDURE (see also flow schedule, Table 5.4). Before using the test for routine purposes, optimal reagent dilutions should be determined as described for DAS- and AgF-ELISA.

Preparation, IgG coating, washing, blocking and antigen

1–10. Perform these steps as described for DAS-ELISA, p. 108, with the following exceptions:

Step **7**, *blocking* is required in this protocol: Add *150 μl* per well of blocking solution: 1% BSA or skimmed milk powder in *washing buffer* to block any unoccupied sites of the solid surface. Alternatively, use polyvinyl alcohol (PVA) at 1 μg/ml – note: *1 min* blocking time (see Section 5.3.2).

Table 5.4. Flow schedule for TAS-ELISA.

Step (step no. in procedure)		Buffer	Time	Temperature
a. IgG coating	(see DAS **1–4**)	Carbonate	2–4 h or overnight	37°C or 4°C, respectively
b. Washing	(see DAS **5–6**)	Washing	≥ 3 × 3 min	
c. Blocking[a]	(see DAS **7**)	Serum/conjug.	30 min	37°C
d. Antigen	(see DAS **8–10**)	Extraction	4 h or overnight[b]	37°C or 4°C, respectively
e. Washing	(**11**)	Washing	≥ 3 × 3 min	
f. Detecting antibody	(**12–13**)	Serum/conjug.	2 h or overnight	37°C or 4°C, respectively
g. Washing	(**14**)	Washing	≥ 3 × 3 min	
h. Conjugate[c]	(**15–16**)	Serum/conjug.	2 h or overnight	37°C or 4°C, respectively
i. Washing	(see DAS **14**)	Wash → H_2O	≥ 4 × 3 min	
j. Substrate	(see DAS **15–17**)	Substrate	≥ 10–60 min	Ambient
k. Recording	(see DAS **18–20**)			

[a]Do not wash after the blocking step.
[b]Adgen Diagnostics recommends 16 h at 4°C.
[c]Anti-mouse (or rat) IgG–AP conjugate.

Step ***10****, incubation* (with antigen): Use incubation for 4 h at 37°C or overnight at 4°C (Adgen Diagnostics recommends overnight or at least 16 h at 4°C).

Detecting antibody

11. Empty the plate and wash in *washing buffer* as before.

12 Load 100 µl per well of the monoclonal antibody fluid at an appropriate dilution in *serum/conjugate buffer* (usually 1/1000 or more). Leave the blank wells empty.

13. Incubate under cover for 2 h at 37°C (or overnight at 4°C).

Conjugate

14. Empty the plate and wash in *washing buffer* as before.

15. Dilute anti-mouse IgG/AP conjugate into a predetermined concentration in *serum/conjugate buffer*, often 1/1000–1/2000, and load each well with 100 µl. Leave blank wells empty.

16. Incubate for 2 h at 37°C (or overnight at 4°C).

Washing, substrate and recording

17. Continue with washing, substrate incubation and recording as described in DAS-ELISA, steps **14** to **20** . For trouble shooting, see p. 123.

5.2.6. Penicillinase-ELISA, protocol

Penicillinase (PNC), also called β-lactamase, which is capable of hydrolysing penicillin into penicilloic acid, can be used as the enzyme in ELISA. A coloured pH indicator added to the penicillin substrate can detect the penicilloic acid produced in the presence of PNC. Among pH indicators, bromothymol blue dye (BTB) has been found the most suitable (Sudarshana and Reddy, 1989; Abraham and Albrechtsen, 2001) and is used in the following procedure. At pH 7.2 this pH indicator shows a distinct blue colour, which changes to yellow at lower pH values. The distinct colour shift makes it easy to determine positive and negative reactions visually, thus obviating the need for an ELISA reader.

Sudarshana and Reddy (1989) found the sensitivity of the PNC system comparable to those of the AP and HRP systems. The substrate, penicillin, is readily available, also in developing countries, and at substantially lower cost than *p*NPP. Thus, the cost of enzyme and substrate components of PNC-ELISA was found to be less than a seventh of those of AP-ELISA. Additionally, PNC when used with pH indicator (but not with starch–iodine complex, also reported as an indicator) was suitable for quantitative virus estimation.

This enzyme and its substrate should be usable in all ELISA formats. The only differences from the ELISA procedures previously described are the conjugation of PNC to IgG (although fairly similar) and the preparation and use of the substrate/BTB mixture; therefore, only these procedures are described in the following.

EQUIPMENT AND MATERIALS.

Conjugation

a. Test tube, hydrophobic, 2–3 ml.

b. Eppendorf tube, 1.5 ml.

c. Dialysis tubing, see IgG fractionation in Section 5.2.3, p. 104.

d. Glass beakers or Erlenmeyer flasks, 500 ml.

e. Stirring apparatus, magnetic, with rods.

f. IgG, virus-specific, without NaN_3 (for DAS-ELISA), or

g. IgG, anti-rabbit, or anti-mouse, without NaN_3 for indirect ELISA (AgF- and TAS-ELISA).

h. Penicillinase, 1000 units, ~ 6 mg (Sigma P0389 or similar).

i. PBS: 0.05 M PB, pH 7.2, 8.5 g NaCl/l – see Appendix 2, p. 247.

j. Glutaraldehyde, 25% aqueous, electron microscopy grade, in ampoules (Merck 112179 or similar). **Caution:** toxic, handle with care.

k. Bovine serum albumin (BSA) (Sigma A7638 or similar).

l. Sodium azide (NaN_3), 2% stock solution – see Appendix 2, p. 247. **Caution:** highly toxic.

Substrate/BTB

m. pH-meter, of good precision.

n. Penicillin G, sodium or potassium salt.

o. Bromothymol blue (BTB) dye (e.g. Merck 103026).

p. NaOH, 0.2 M/distilled water.

q. HCl, 5 M. **Caution:** strong acid, handle with care – see Appendix 2, p. 247.

r. HCl, 0.1 M/distilled water.

PROCEDURE.

Conjugate

1. In a hydrophobic test tube dilute the IgG in *PBS without NaN_3* to give 1 mg/ml.

2. Add and dissolve penicillinase powder to give 2 mg/ml.

3. Transfer the solution to a dialysis tubing presoaked in *PBS*.

4. Dialyse against *PBS* (500 ml) in a beaker or Erlenmeyer flask for 1 h at room temperature under slow stirring.

5. Replace the PBS with PBS containing *0.06% glutaraldehyde* (1 ml of 25% glutaraldehyde in 420 ml PBS) and dialyse for 3–4 h at room temperature under slow stirring.

6. Dialyse for 18 h (overnight) at 4°C under stirring against at least three times 500 ml PBS containing *0.02% NaN_3*; change approx. 1, 3 and 6–8 h after start. See **Note** below.

7. Transfer the conjugate to a glass vial. Add *BSA* to give 5 mg/ml and store at 4°C. Do not freeze. Shelf life at least 1 year at 4°C when undiluted.

8. Determine optimal concentration for ELISA by a chequerboard test as described under Section 5.2.3, p. 103, or Section 5.2.4, p. 111.

Substrate/BTB

9. Dissolve 20 mg of *BTB dye* in 50 ml of *0.2 M NaOH*.

10. Neutralize the solution with *5 M HCl*.

11. Make up the volume to 100 ml with distilled water. Stable for at least 9 months at 4°C.

12. Add *penicillin G* at 0.5 mg/ml for the volume needed. The mixture is stable for at least 2 weeks at 4°C, but the pH at room temperature must be checked before each use.

13. Adjust with a freshly calibrated pH-meter the pH of the substrate/BTB to exactly 7.20 at room temperature with 0.1 M HCl or 0.1 M NaOH.

PNC-based ELISA

Conjugate

1. Use the *IgG-PNC conjugate* instead of IgG-AP conjugate. IgG-PNC can generally be used at higher dilutions (1 : 2000 or higher) than IgG-AP.

Substrate/BTB mixture

2. Load the plate with the *substrate/BTB mixture*, after adjusting to *pH 7.20* with a calibrated pH-meter. If stored cold, the mixture must be brought to room temperature before pH measuring.

3. After incubation at room temperature, observe the colour change from blue to greenish-yellow or deep yellow for positive reactions (depending on the amount of penicilloic acid produced). Weakly positive reactions remain greenish-yellow, and negative reactions blue.

4. If desired, read the absorbance at 620 nm (see Section 5.2.3, p. 103, step **18**).
5. With this enzyme/substrate system, no stopping of the reaction can be applied.

N In another procedure (Dijkstra and de Jager, 1998), further $MgCl_2$ and $ZnCl_2$ are added before storing of the conjugate. This can be done by dialysing against a PBS buffer containing 2 mM $MgCl_2$ and 0.1 mM $ZnCl_2$ in the last change.

5.2.7. Other variants of ELISA

Among the many ELISA procedures that have been developed, Van Regenmortel and Dubs (1993) found 11 to be most widely used in plant virology. The 11 procedures vary primarily in reagent type and step sequence, but variants using different enzyme/substrate also exist. Some of these procedures and variants are briefly described below.

Use of F(ab′)$_2$ fragments
Indirect assays using the F(ab′)$_2$ fragments as trapping antibody require only a single virus-specific antiserum and not, for instance, a rabbit IgG as trapping agent and a Mab as the detecting antibody. The F(ab′)$_2$ fragments obtained by enzymatic digestion of virus-specific IgG (see Section 5.1) is used for coating of the plate, and the intact form of the same IgG are used for detection. Only the intact IgG is recognizable by an anti-rabbit IgG/enzyme conjugate or by a protein A/enzyme conjugate (see Use of protein A, below) reactive to the Fc portion but not the F(ab′)$_2$ fragments (Fig. 5.7).

Biotin/streptavidin reaction
To enhance ELISA sensitivity, the antibodies can be coupled to biotin (a water-soluble compound within the vitamin B group) rather than to an enzyme. Biotin has a strong affinity to streptavidin (derived from a bacterium), which can be conjugated with enzyme. Biotinylated antibodies, mostly for use as secondary

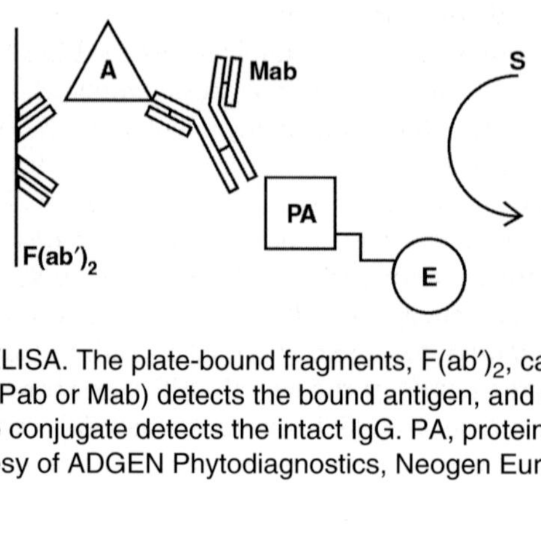

Fig. 5.7. F(ab′)$_2$-ELISA. The plate-bound fragments, F(ab′)$_2$, capture the antigen (A). The intact IgG (Pab or Mab) detects the bound antigen, and – in this case – a protein A–enzyme conjugate detects the intact IgG. PA, protein A; E, enzyme; S, substrate. Courtesy of ADGEN Phytodiagnostics, Neogen Europe Ltd, UK.

antibodies, and streptavidin/enzyme conjugates are commercially available (Sigma™ and others), but can also be prepared in the laboratory. Because of the very high binding affinity of streptavidin to biotin, smaller amounts of antigen are detectable than by an antibody/enzyme conjugate (Converse and Martin, 1990). In differentiation of serologically divergent isolates of two cowpea viruses, Huguenot *et al.* (1993) found the sensitivity of Mab–biotin/streptavidin–AP superior to five other ELISA systems tested. A DAS-ELISA system with the use of a Mab as coating antibody and a biotinylated Mab as the detecting antibody enabled Konate and Neya (1996) to reliably detect one CABMV-infected seed in a group size of 500 seeds. The advantage of the increased sensitivity of the biotin/streptavidin system should be weighed against the extra step(s) involved and the added cost, e.g. could be used primarily to retest particular samples producing questionable (very weak) reactions.

Use of protein A

Protein A (PA), derived from the bacterium *Staphylococcus aureus*, has a strong affinity for the Fc portion of antibodies. This ability is utilized in some indirect ELISA variants by using a PA–enzyme conjugate instead of an anti-first animal IgG–enzyme conjugate. When used in $F(ab')_2$ ELISA or AgF-ELISA, this conjugate will only recognize and bind to the detecting antibody, as only this antibody possesses the Fc portion (Fig. 5.7). PA–enzyme conjugation is described by Clark *et al.* (1986) and the conjugate is commercially available (e.g. Sigma™, DakoCytomation™). PA is also used in ELISA as the first layer in order to bind and orient the subsequent, coating anti-viral IgG. Due to the PA specificity to the Fc portion of the IgG, use of whole unfractionated antiserum for coating of PA-precoated and subsequently blocked plates might be possible. For coating purposes, the PA is diluted in carbonate buffer and the other reactants are diluted in PBST + additives (Van Regenmortel and Dubs, 1993).

The main advantage of using PA is that the assay requires only a single antiserum. However, sometimes difficulties have been experienced with respect to reduction of the non-specific background, while maintaining sufficient assay sensitivity (Van Regenmortel and Dubs, 1993).

Alternative enzyme/substrate systems

Apart from AP and penicillinase (PNC), the enzymes horseradish peroxidase (HRP) and urease are also suitable for use in ELISA. Of these, HRP is the most widely used. Its substrate is the chromogenic 3,3′,5,5′-tetramethyl benzidine dihydrochloride (TMB) or *o*-phenylenediamine (OPD). Note, however, that the latter may be mutagenic. The substrate TMB indicates a positive reaction with development of a blue colour, which can be evaluated at an absorbance of 450 nm in an ELISA reader. The reaction can be stopped by addition of 25 μl 2 M sulphuric acid per well (at 100 μl reactant volume). It should be noted that the enzymatic activity of HRP is blocked by sodium azide, the use of which, therefore, should be avoided in HRP-ELISA. Converse and Martin (1990) describe some details of the use of HRP, and more can be found in the manufacturers' instructions. HRP is cheaper than AP, but may not always work well with plant extracts, due to their possible content of polyphenol oxidases or peroxidases, which can hydrolyse the HRP's substrate and thus lead to false positives in ELISA (Converse and Martin, 1990).

Special applications

A method for specific detection by ELISA of embryo-located PSbMV in seeds has been developed by Masmoudi *et al.* (1994). During seed maturation, a part of the capsid protein of the seed-coat-located virus becomes deleted, while the capsid protein of the embryo-located virus remains intact. An antiserum specific to the deleted capsid protein part only was prepared and used for specific detection of transmissible virus in whole mature pea seeds.

Mabs for ELISA detection of all members of at least one virus genus (*Potyvirus*) have been prepared and are commercially available from, for example, Agdia™ and the American Type Culture Collection, both USA (Appendix 3).

Recombinant single-chain variable fragment antibodies (scFv) (Section 5.1) have been used successfully for DAS-ELISA and TBIA (Section 5.4) detection of *Citrus tristeza virus* (CTV, *Closterovirus*) by Terrada *et al.* (2000). The scFv were genetically fused with alkaline phosphatase and diagnostic reagents were produced by expressing these fusion proteins in bacterial cultures.

Alternative solid phase

ELISA (indirect AgF) for detecting different plant viruses was carried out in polystyrene Petri dishes instead of microtitre plates (Abdalla and Albrechtsen, 2001). The reactants were applied and incubated as several 50 μl drops per Petri dish, effectively separated by hydrophobic wax boundaries. After the washings, remnants of washing liquid were completely removed from the dish before the next reactant drops were added. The sensitivity was similar to or higher than that of the microtitre plate ELISAs run in parallel. Due to the low amount of reagents needed and the inexpensive Petri dishes, the total cost of assay was reduced.

A technique similarly involving drop reactions in Petri dishes, but where a mixture of the substrate, BCIP/NBT (see Section 5.3), and a 40–50°C warm agar, is finally poured into the Petri dish so as to cover the whole dish, produced positive reactions in the form of bluish spots in the agar. The sensitivity in detection of plant viruses equalled that of the above method (Abraham and Albrechtsen, 2000).

5.2.8. Remarks on ELISA systems, including protocol for plate reuse

Trouble shooting

There are a number of variable factors involved in ELISA procedures and therefore several possibilities for errors, each of which may adversely influence the outcome.

Some of the most common problems and their possible causes and remedies are tabulated in Table 5.5.

Increasing the sensitivity

Low sensitivity of an ELISA system can be dealt with in different ways. One possibility could be the use of a more sensitive ELISA system, such as the *biotin/*

Table 5.5. Trouble shooting in ELISA.[a]

Problem	Possible cause	Possible remedy
Too high background	Inefficient washing (quite common)	• More thorough wash (1, wash more and/or soak longer; 2, increase the distance from nozzle to well to make more turbulence in the wells) • Complete emptying of wells before next reagent
	Lacking or inefficient blocking	• Use a higher conc. of blocking agent or include a separate blocking step
	Contamination from infected material during sample preparation	• Better hygiene when sampling and homogenizing; prepare positive controls last; load samples in the correct sequence
	Antiserum reacts to plant proteins (common in indirect AgF-ELISA)	• Choose a more specific antiserum • Use cross-absorption
	Test samples too concentrated	• Dilute test samples (to $\times$ 100 or more)
	Detecting antibody concentration too high	• Dilute antiserum (indirect ELISA) or conjugate (DAS), e.g. 1 : 2000 may work better than 1 : 500 • Check dilutions of all reagents in chequerboard test
	Substrate too old	• Check that substrate is stored as prescribed • Use another batch
	Contamination with enzyme	• Avoid fingerprints on tools, especially when preparing substrate solution (gloves) • Use of clean glassware free from enzyme remnants • Check by including one to two wells containing substrate solution only
Unexpected high absorbance values in certain wells	Contamination from neighbouring wells, pipettes, pipette tips, or fingerprints	• Be careful when loading plates • Avoid contamination of pipettes and use new tips for every sample • Be careful when emptying plates • Do not touch the tops of plates
	Dirt underneath the plate	• Wipe plate clean underneath

Continued

Table 5.5. (*Continued*)

Problem	Possible cause	Possible remedy
High absorbances in outer wells of the plate	Temperature differences in plate during incubation	• Place plates on an insulating material in the incubator • Do not place the plates on top of each other
	Insufficient washing of outer wells	• Improve the washing
	Quality of plates	• Do not use outer wells, or try another brand
Imprecise plate reading	Dirt in reading chamber	• Clean spectrophotometer
	Scratches on plate underside	• Be careful when handling plates
	Spectrophotometer error	• Check: turn plate 180° and read
Generally too high absorbances, also in wells with healthy samples	Too high conc. of one or more reagents	• Do chequerboard test
	Forgotten or insufficient wash after conjugate incubation	• Check procedure
	Substrate or solution may have been exposed to strong light	• Keep substrate in dark and avoid bright light on the substrate solution
	One of the buffers contaminated with conjugate	• Check procedure and contamination risks
	Incubated with cross-absorbed serum for too long (AgF-ELISA)	• Max. 1–3 h incubation
	The cause may be some of those of 'Too high background' above	• See above
Generally too low absorbances	Too low conc. of one or more reagents	• Do chequerboard test
	Defective reagents, e.g. low titre of antiserum	• Chequerboard test, which may give a hint of the defective reagent; then try another batch
	Wrong buffer, wrong pH or molarity	• Check procedure, check and adjust pH, or prepare new buffer
	Incubation time(s) too short	• Longer incubation time(s)
	Remnants of PBS in wells prior to substrate step	• Wash with H_2O before substrate step
Absorbance completely absent, also in positive controls	Missing or wrong reactant(s)	• Check if conjugate or substrate was forgotten • In indirect ELISA: used a wrong conjugate, e.g. anti-rabbit instead of anti-mouse or anti-rat? • Check if positive controls have lost antigenicity

Table 5.5.

Absorbances of infected samples low but positive controls normal	The antibodies may react weakly to the actual virus or virus strain	• Use more specific antibodies • Check dilutions (chequerboard test)
Readings are below zero	Photometer error or blank well failure	• Check spectrophotometer • Make sure that blank wells contain only buffer when reading
Variation in readings	Photometer error	• Check spectrophotometer/ ensure, if needed, warming up before first use

[a]Partly according to Andersen *et al.* (1991) and *Adgen ELISA Protocols* (2000).

streptavidin ELISA mentioned above under Section 5.2.7. For an all-purpose improvement, use of *antibodies of high sensitivity and specificity* for the virus being tested for and with low plant specificity is an obvious measure. Monoclonal antibodies can be made either highly specific or highly sensitive. It may be helpful to use *cross-absorption* (p. 95) to improve the virus specificity of polyclonal antisera having high plant specificity, especially for use in indirect AgF-ELISA. Use of more efficient *blocking* (p. 95) may reduce the overall background. The detectability of viruses may also be improved by using the *correct dilutions* of reagents. Thus, for detection of *Andean potato latent virus* (APLV, *Tymovirus*) by an indirect ELISA, using $F(ab')_2$ fragments for trapping, raising the dilution of the detecting antibody from 1 : 500 to 1 : 5000 resulted in increased absorbances from virus and reduced absorbances from healthy material (Koenig and Paul, 1982). James and Mukerji (1996) found in indirect AgF-ELISA a doubling of the absorbance value when increasing the dilution of infected crude sap from 1 : 10 to 1 : 100 for detection of Cherry mottle leaf virus (*Trichovirus?*). According to Hobbs *et al.* (1987), PeMoV in groundnut seed and pea tissue could be detected at higher tissue : buffer dilution (up to 1 : 10,000) with AgF-ELISA than with DAS-ELISA, while for PCV in plants the sensitivity of the two formats was almost the same.

Lommel *et al.* (1982) compared extraction methods of plant material for detection of different viruses, such as *Carnation mottle virus* (CarMV, *Carmovirus*) and *Carnation ringspot virus* (CRSV, *Dianthovirus*), by AgF-ELISA. They found that extraction of tissue at 1 : 1 w/v in distilled water (with or without addition of the reducing agent, 0.1% mercaptoethanol or DIECA) prior to dilution in carbonate antigen buffer gave the highest absorbance values. Among other methods they used for initial extraction, 1 : 1 homogenization in PBS with DIECA, for example, resulted in only half the absorbance as compared to water. In an overview by Torrance (1998), different methods of improving ELISA sensitivity are compiled.

Reduction of cost, workload and time

OPTIMAL DILUTIONS. One of the most obvious ways to reduce costs in ELISA is to find the optimal dilutions of the reagents required. A chequerboard test (see Sections 5.2.3 and 5.2.4), may reveal that an antibody dilution at, for example, 1 : 5000 works as well as or better than one at 1 : 500, reducing the cost of antibodies tenfold. Similarly, consumption of conjugate and even substrate may be reduced (in our laboratory, *p*NPP at 0.5 mg/ml or less, instead of the 1 mg/ml usually prescribed, worked well). Koenig and Paul (1982) found for TAS-ELISA that the use of a pre-incubated mixture of detecting antibody and conjugate could save one working step. This approach gave similar or even higher sensitivity as compared with the conventional method, but only when the detecting antibody was present at high dilution in the mixture.

COMPOSITE SAMPLES. As mentioned under Sample preparation, p. 102, testing of groups of samples rather than single samples can reduce workload, time and cost considerably. Before determining the group size, i.e. the maximal number of samples that can be tested together still allowing detection of one infected seed or seedling, an initial trial must be made for the actual virus/crop/reagents. Two extracts are made: known infected extract of the species being tested (I) and known healthy extract of the same species (H). These extracts are diluted equally depending on the ELISA type being used (e.g. 1 : 100). Increasing twofold dilutions of I in H are then made and tested by ELISA using optimal reagent dilutions. An apparently reliable detection of virus in, for example, a 1 : 64 dilution theoretically means that groups of 64 samples can be made. However, to be able to detect the virus in samples with low virus content, the number of samples should be reduced at least fourfold, i.e. in this case no more than 16 seeds/seedlings should be pooled.

The level of infection in a seed sample tested in groups can then be estimated by group analysis based on the number of positive and negative groups (Section 7.2).

MIXED ANTISERA. When samples are to be tested for more than one virus at a time and the purpose is to screen for absence or presence of viruses, detecting antibodies of different specificity can be mixed, and workload, time and cost consequently reduced. Successful use of mixed antibodies has been reported by, for example, Banttari and Franc (1982) and Grimm and Daniel (1984) for simultaneous detection of potato viruses, Etienne *et al.* (1991) for nepoviruses, and Joshi and Albrechtsen (1992) for five cowpea and three soybean viruses. In the latter case, each of the five cowpea viruses and each of the three soybean viruses belonged to different genera, and indirect AgF-ELISA was used.

ELISA FORMAT. Use of penicillinase-ELISA (5.2.6. Protocol) reduces substantially the cost of assay due to the lower cost of enzyme and substrate compared to AP-ELISA.

AUTOMATION. Due to the standard format of microtitre plates and not least due to the medical-clinical laboratory use of ELISA, a variety of equipment for automatic or semi-automatic processing is commercially available. With such equipment, like sample extractors, dispensers, plate washers and data recorders including electronic data processing and storage, thousands of samples can be processed in a short time.

REUSE OF ELISA PLATES: PROTOCOL FOR REGENERATION. A possibility to reduce the cost of ELISA would be reusing, twice or more, the microtitre plates. Different methods to regenerate microtitre plates for reuse have been reported (Bar-Joseph *et al.*, 1979; Banttari and Petersen, 1983; Stobbs and Van Schagen, 1986). These methods, however, resulted in a more or less reduced protein-binding capacity of the plates, thus reducing the ELISA sensitivity, and required in some cases a coating with nitrocellulose (1%) after the regeneration to restore the plate performance.

A simple method for cleaning microtitre plates without disturbance of their protein-binding capacity was developed by Manandhar *et al.* (1996). The method was efficient in cleaning plates (NUNC-ImmunoPlate MaxiSorp F96®), allowing at least three times safe reusing of the plates. This was demonstrated after plate cleaning experiments done after each use of them for indirect AgF-ELISA in detection of eight different seed-transmitted viruses in their respective hosts. Although less well investigated for DAS-ELISA, the same efficacy could be expected for this system and probably also for TAS-ELISA, but additional tests are needed to prove the efficacy of the method for these two systems. Among different types of soap tested as cleaning agent, only clothes washing powder was found effective, possibly due to its content of proteolytic enzymes. The regeneration method was efficient for the tested make and batch of microtitre plates, but its cleaning effect and influence on the binding ability of other plate batches or makes remains to be tested.

Equipment and materials

a. Container with space for ten microtitre plates or more, and adapted to allow vertical positioning and fixing of the plates so that they do not touch each other during shaking.

b. Shaker, reciprocating type suitable for slow horizontal movement of the container.

c. Clothes washing powder, Biotex® (A/S Bluemoeller™, Denmark) or similar. Main components: zelites, 15–30%; anionic tensides, 5–15%; non-ionic tensides, soap, polycarboxylates, phosphonates and enzymes, < 5%; and pH, 9.5.

Procedure

1. Unload the substrate solution from the microtitre plate after finished ELISA, wash under running tap water, and keep it submerged in water until cleaning.

2. For cleaning, place the plates in the container as described above and cover them with a 5% (w/v) solution of clothes washing powder.

3. Place the container on a suitable reciprocating shaker and let it shake slowly at room temperature for 48 h. Thereafter, agitate vigorously by (gloved) hand each plate in the solution.

4. Rinse the plates thoroughly under running tap water followed by thorough rinsing in deionised water, then in two changes of distilled water. After air-drying the plates are ready for reuse.

5.2.9. Recording and interpretation (see also p. 99)

Evaluating the ELISA result is not always straightforward, especially in cases of weak reactions of infected material and high absorbances of healthy material. Therefore, the ELISA procedure must first of all be optimized to maximize the differences between absorbances of negative and positive samples. Appropriate blocking, cross-absorption, correct dilutions of antigens and use of antibodies of correct specificity and sufficient titre, mentioned previously, are some of the important measures to optimize ELISA.

The correct time of reading after addition of the substrate is also important. Before setting a standard substrate incubation period for the virus–host–reagent concerned, the optimal absorbances should be determined. It could be done by choosing among a range of readings, taken for instance every 15 min during substrate incubation, the reading with the maximum proportional difference between the mean of positive controls and the mean of negative controls. For an ELISA to be usable, the negative control (NC) must yield an absorbance value (A) not higher than 1/10 of that of the positive control (PC). In the best systems the A_{NC}/A_{PC} may reach 1/100 or more. Still, cases that are difficult to interpret may occur. For example, infected samples with low absorbances due to low antigen titres may give rise to false negatives. High absorbances from healthy material, on the other hand, may give rise to false positives.

Establishing a positive–negative threshold

Sutula *et al.* (1986) studied this problem and found, for example, that the screened plant pathology literature (1984–1985) – unlike medical literature – did not pay much attention to methods for setting threshold values (TVs) or cut-off values for distinguishing healthy and diseased material. In most cases no methods were stated, while some used two or three times the mean of healthy (negative) controls, and only a few used the *negative mean + two to four times the standard deviation (SD)* among the negative controls for setting a TV. However, the use of TVs on the basis of the NC mean plus SD has increased in the literature since then, and should be regarded as the best approach. Use of such thresholds will, for statistical reasons, require a minimum of four to six healthy controls in each plate, and it is important to ensure coverage of the possible range of healthy values by using a number of different individuals. These individuals should match the unknown test samples as regards cultivar, age, growing conditions and tissue type (Sutula *et al.*, 1986). ELISA systems optimized to give low background levels (e.g. < 0.1 absorbance unit) usually yield fewer questionable samples than systems with high backgrounds. Another important measure is replication of test samples (minimum two wells per sample). The choice of a method to determine a threshold must be based on an acceptable reference (another independent method of defining pathogen presence), taking into account the ELISA results and the ELISA system used, which must be critically evaluated. In other words: *there is no inherent property of ELISA, virus or samples which justifies the use of certain thresholds*; or, expressed differently: the question of how to establish the cut-off value or positive–negative threshold in ELISA data analysis has no absolute answer (Sutula *et al.*, 1986; Converse and Martin, 1990).

In their conclusion, Sutula *et al.* (1986) suggest the following guidelines for adequately reporting ELISA results:

1. Clearly state the positive–negative threshold used.
2 Test enough plants to become familiar with the range of negative (healthy) values involved.
3. Include enough known negative controls in each routine assay to ensure representation of the previously established range of negative background values.
4. Always include a positive control.
5. Match control samples and test samples with respect to host type, tissue type, age and position.
6. Strongly consider replication of test samples.

Like any other detection method, ELISA is not foolproof. In cases of doubtful results, these should be confirmed by retesting the samples by a more sensitive ELISA system, bioassay, PCR or electron microscopy. In large-scale routine testing, periodical control of ELISA results by bioassay should be made, as variants of viruses may occur which are undetectable or reacting weakly to the antibodies used (Maury *et al.*, 1998).

5.3. Dot Immunobinding Assay (DIBA)

DIBA (syn. dot-ELISA or dot blot immunoassay (DBIA)) is another serological detection system that is enzyme-amplified. The main differences between this assay and ELISA are that: (i) the antigen samples are applied as dots on a nitrocellulose membrane (NCM), unprecoated or precoated by antibody; (ii) the subsequent reactions take place by submerging the whole sheet in the reagent dilutions; and (iii) the signal indicating the presence of virus appears as an insoluble coloured product, which binds to the NCM at the sites of dot application, and therefore can be used as a permanent record.

5.3.1. Principles of DIBA

Nitrocellulose membranes have been used for protein blotting since 1979. In protein blotting, proteins separated on a gel are transferred to a sheet of nitrocellulose by electroblotting (Towbin *et al.*, 1979). The transferred proteins irreversibly bound to the membrane are then reacted with antibody and detected (Western blotting). Later, this technique was simplified to a dot immunobinding assay (DIBA) by Hawkes *et al.* (1982) in order to assay monoclonal antibodies. Soon, it came into use for detection of viruses, including plant viruses and other pathogens, applied directly to a membrane, either untreated or precoated with antibody. Since the protein-binding capacity of an NCM is much higher than that of a polystyrene surface, as little as 1/50 or less of the antigen amount required in ELISA can be detected by DIBA. However, DIBA generally gives higher background than ELISA, unless certain precautions are taken. Thus, after application of antigen, the NC sheet must always be blocked in a separate blocking step and a blocking agent is to be added to buffers throughout the serological procedure. To avoid or reduce host-specific reactions, cross-absorption of polyclonal antibodies is normally also a must in DIBA.

5.3.2. General components and steps of DIBA

As most of the steps and ingredients are similar to those of ELISA, only items specific for DIBA are treated in the following.

Solid phase

The most used solid phase is nitrocellulose membranes of 0.45 μm pore size, although use of other types, such as nylon membranes, may be possible. Even certain types of plain paper have been successfully used (Heide and Lange, 1988; Lange *et al.*, 1989). Nylon membranes and paper are normally rather hydrophobic unless treated adequately before use, but they have higher mechanical strength than the relatively fragile NCM. A membrane such as an NCM acts both as a surface filter and as an adsorbing matrix for macromolecules. Some NCMs, e.g. those containing a mixed-ester matrix, are designed to have low binding capacity and should not be used for DIBA (Lazarovits, 1990). Geering and Thomas (1996) compared seven types of membranes, six NCMs and one nylon membrane for use in DIBA and found differences in sensitivity. The nylon membrane gave elevated background reactions. The following is based on the use of NCMs.

Preparation of samples

For DIBA, the samples are generally prepared as described for ELISA. Crude plant or seed samples can, however, pose a problem because of deposits of chlorophyll and particulates in the dots on the membrane. Such deposits may not only remain throughout the process, disturbing the recording, but may also partially block the membrane surface, preventing attachment of antigen. In addition, particulates may be washed off during the subsequent process, leaving potential sites for non-specific interactions (Hammond and Jordan, 1990). They therefore recommend that crude samples be clarified by chloroform treatment and centrifugation before application. Even after clarification, according to the same authors, samples may need dilution at 1 : 100 to 1 : 400, and unclarified samples at 1 : 500 to 1 : 1000, for effective discrimination between healthy and infected samples. A similar positive effect of dilution for some antibody-virus–host combinations has been demonstrated by DIBA experiments in our laboratory (unpublished results). The following method of clarification can be used: extract tissue 1 : 10 in buffer; filter through cheesecloth; pipette 0.8 ml of the filtrate into a 1.5 ml Eppendorf tube, add 0.4 ml chloroform, cap and vortex; centrifuge (approx. 12,000 *g*) for 2 min; pipette 200 µl of the aqueous upper layer into 800 µl of buffer and mix (Hammond and Jordan, 1990). Use of organic solvents for clarification, however, adversely affects the sensitivity in ELISA, according to Lommel *et al.* (1982). For indirect AgF-ELISA, they found the lowest absorbance when using chloroform-butanol treatment of extracts rather than other treatments.

An alternative to the use of the hazardous chloroform was reported by Smith and Banttari (1987), who used heating of extract dilutions for clarification in a double-antibody sandwich (DAS) DIBA for detecting *Potato leafroll virus* (PLRV, *Polerovirus*). Heating to 50 or 70°C for 10 min followed by cooling to room temperature, brief agitation and settling of precipitate before pipetting onto the NCM nearly doubled the sensitivity compared to no treatment. However, heating the extracts to 70°C eliminated the antigenicity of *Banana bunchy top virus* (BBTV, *Nanovirus*) in a triple-antibody sandwich (TAS) DIBA (Geering and Thomas, 1996). In the same report, clarification was as effective with dichloromethane as with chloroform. Remains of organic solvents in the antigen supernatants must be removed or minimized by placing them under ventilation for a while to allow evaporation of the solvent before dotting; otherwise the membrane may be damaged. Use of Triton X-100 in the blocking step to reduce the plant sap background in the dots has also been reported (see Blocking, below). See Buffers for more details about additions to extraction buffers.

Application of samples

Handling of membranes requires strict use of forceps and gloves, as they are very sensitive to fingerprints, which will invariably result in false reactions due to the excretion of alkaline phosphatase from the skin. Before use, the positions of dots are marked on the membrane, either in the form of tiny points or a grid made gently with an extra-soft pencil. Huth (1997) used a stamp to mark a grid on membranes using waterproof ink (Polyethylene-stamp ink No. 337) (see also Dot blotting and RNA fixing, Section 6.1.2). According to Hawkes *et al.* (1982), NCMs with a grid already printed are commercially available (Millipore Corp.™). Marking is not needed if a manifold is used for application.

The first step is application of samples (see Preparation of samples, above) on either:

- an untreated NCM, washed in deionized H_2O or sample buffer and allowed to dry (antigen-first indirect (AgF) method); or
- an NCM presoaked in IgG diluted in carbonate buffer, pH 9.6, washed, blocked and – in some protocols – dried (DAS or TAS method).

The antigens can be applied either: (i) directly by use of a capillary tube or an adjustable pipette (1–10 µl) on premarked positions on the NCM, which is placed on a piece of filter-paper; or (ii) by using a manifold (10–400 µl per dot). Should a manifold be used, the membrane is placed on a wetted filter paper between two identical pieces of perforated Plexiglas, which are bolted together. Manifolds with dot (circular) or slot (rectangular) format are available. The slot-blot format makes the results easier to quantify by densitometry. After pipetting the samples into the upper half's holes, the samples are drained through the membrane with or without vacuum, the manifold is disassembled and the membrane is dried (AgF method) or washed (DAS or TAS method) before being further processed. Details of the manifold application are described by Banttari and Goodwin, (1985) and Hammond and Jordan (1990). Larger volumes of antigen can also be applied directly by repeated dotting on the same spot if the NCM is completely dried between each dotting. The orientation of the surface of the NCM to antigen application affected the sensitivity for the NCM make used by Smith and Banttari (1987), necessitating a preliminary test of each new NCM batch.

Antigen-spotted NCMs are suitable for mailing from modestly equipped laboratories to a central laboratory for processing. This is a better method than sending fresh or dried plant material or antigen-coated ELISA plates, according to a number of reports, and in addition the risk of disseminating infectious virus is less.

Dot blotting of crude extracts or RNA extracts onto membranes is also used in nucleic acid spot hybridization (NASH, Section 6.1.1) and in some variants of polymerase chain reaction (PCR, Section 6.2.2).

Incubation and containers

After dotting the samples onto a sheet of nitrocellulose, no additional individual treatment is possible, as the whole membrane is submerged in reagent solutions during incubations. Incubation times may be 1–4 h at 37°C or longer at room temperature or at 4°C, as for ELISA. The area of the container must be slightly larger than the membrane, allowing the membrane to be manipulated in the container. Glass containers, such as glass Petri dishes, are considered better than containers of polystyrene or other plastic types for incubating the NCM, because of antibody absorption to plastics, especially polystyrene. However, containers of polypropylene, such as the lids of disposable pipette-tip boxes, do not absorb protein and have an ideal shape (Hammond and Jordan, 1990). Use of sealed plastic bags as containers during incubation is also reported (e.g. Pollini *et al*., 1993). Depending on the type of container and shape of membrane, 100 to 200 μl of reagent solution per dot is required. For example, about 5 ml is sufficient for a 5 cm × 6 cm membrane with 25–30 dots in a Petri dish 9 cm in diameter. Also, as little as 25 μl/dot has been used (Hibi and Saito, 1985). During incubation, the container is covered with a lid or sealed and periodically agitated or placed in a gently oscillating shaker. Pollini *et al*. (1993) used 80 oscillations/min. During the reactive processes, the membrane is kept in the container and the reagent solutions are changed by pouring them in and draining off; the washings between each incubation are carried out similarly. However, Heide and Lange (1988) found that changing the container between conjugate and substrate incubation markedly reduced the background when plain paper was used as solid phase.

Blocking

In DIBA, it is imperative to include a separate step to block unoccupied sites after the first treatment (which may consist of soaking in coating antibody (DAS- or TAS-DIBA) or application of samples (AgF-DIBA)). Also, the buffers for the subsequent reagents, except substrate buffer, must contain a blocking agent to further minimize unspecific reaction. Skimmed milk powder has a good blocking effect and is widely used in DIBA at 1–5% w/v for the initial blocking and at 0.1–1% w/v in the buffers used subsequently. Addition of 2% Triton X-100 to the blocking-step solution reduces the green stain from sap components on the membrane, according to Powell (1987). Srinivasan and Tolin (1992) removed green colour by rinsing the membrane in 5% Triton X-100 prior to the blocking step in detection of clover viruses by TBIA (see next section). Parent *et al.* (1985) reported that, although 1% Triton X-100 or 2% Tween 20 added to the blocking solution greatly reduced the background, they also found that some of the antigen was removed from the NCM by these agents.

Polyvinyl alcohol (PVA) (Sigma P8136 or similar) has been reported to be very effective as a blocking agent (Miranda *et al.*, 1993, quoted from Makkouk and Comeau, 1994). PVA used for the initial blocking step at only 1 μg/ml buffer and as little as 1 min of incubation is reportedly sufficient to provide even more efficient blocking than albumins, with the added benefit of reducing the total duration of the DIBA procedure.

Antibodies

In the DAS version of DIBA, coating is usually made with polyclonal IgG and antigen is detected by a polyclonal IgG–enzyme conjugate. In TAS-DIBA, antigen is also trapped by a polyclonal IgG and reacted by an antibody, usually a Mab, whose presence is detected by an anti-mouse (or rat) IgG–enzyme conjugate. In the simpler and most widely used indirect AgF-DIBA, the passively trapped antigen is detected by a crude antiserum or Mab that is detected by an anti-antibody–enzyme conjugate. Dilution of antibodies and conjugate is as for ELISA.

According to several reports, antibody and conjugate dilutions can be stored and reused up to ten times without significant loss in sensitivity.

Cross-absorption

To avoid reaction with healthy material, the use of cross-absorption of polyclonal antisera or IgG is particularly important in DIBA. Powell (1987) even found a positive, although unexplained, effect of cross-absorbing monoclonal antibodies. In indirect AgF-DIBA, unspecific reaction is markedly reduced for Pabs by incubating them with healthy material prior to use, as described for ELISA. Use of healthy seed extracts in cross-absorption normally poses no problem, whereas plant extracts may cause an overall greenish colour of the membrane, which may only partly disappear in the subsequent washings. The background staining can be reduced by centrifugation at 10,000 *g* for 10 min of the extract or the antibody–extract mixture before use, or by use of lower concentration of extract.

Conjugates and substrates

The most commonly used enzyme in DIBA is AP, and the substrate 5-bromo-4-chloro-3-indolylphosphate (BCIP) in combination with nitro blue tetrazolium (NBT). Unlike the *p*NP produced in ELISA, the degradation product of BCIP/NBT is insoluble and binds to the membrane, i.e. to the dots, where the conjugate has been bound. Dots containing virus will then appear with a bluish-purple colour. BCIP and NBT must not be mixed until just before use, but separate stock solutions of BCIP and NBT can be made and stored for several months. Alternatively, but more expensively, tablets containing both components (e.g. Sigma™ Fast BCIP/NBT tablets, B5655) can be used.

Horseradish peroxidase (HRP) is also used, although to a lesser extent than AP. The substrate for HRP is chloronaphthol and hydrogen peroxide (Sherwood, 1987) and for chemiluminescent DIBA it is luminol and hydrogen peroxide. Luminol emits light when oxidized by hydrogen peroxide in the presence of HRP, and the emission can be recorded on X-ray film. Geering and Thomas (1996) found this system (called DIBA-ECL (enhanced chemiluminescent detection)) eight times more sensitive than DIBA using BCIP/NBT for the detection of BBTV in banana. Lazarovits (1990) describes four HRP substrates and their use, and DIBA-ECL is described in detail by Pollini *et al.* (1993), who found the method superior to ELISA or colorimetric DIBA in detection of members of *Closterovirus* in grapevine.

Buffers

The most frequently used buffer in DIBA is TRIS-buffered saline (TBS). Buffers for antibody and conjugate incubation and for washing are often 0.02 M TRIS-HCl, pH 7.5, with 0.15–0.5 M NaCl. This buffer, as well as PBS, pH 7.0–7.5, and carbonate buffer, pH 9.6, is reportedly used for sample extraction. The carbonate buffer, however, gave higher background in AgF-DIBA than PBS, according to Lange and Heide (1986).

Lazarovits (1990) and others recommend extraction buffers that are supplemented with chelating agents, such as ethylenediamine tetra-acetic acid (EDTA) (0.01 M), and antioxidants, such as ascorbic acid (0.2%, w/v), sodium sulphite (1–10%, w/v), 2-mercaptoethanol (1%, w/v) or diethyldithiocarbamate (DIECA) (0.01 M). These agents both reduce non-specific reactions and enhance the sensitivity by improving the extraction efficiency for the antigen. Extraction of plant tissue in TBS with 0.01 M EDTA, followed by centrifugation for 10 min, greatly improves the sensitivity in DIBA (Lizarraga and Fernandez-Northcote, 1989). DIBA detection of PSbMV in pea, in which a high background was thought to be due to lectin present in the pea extracts, was improved by addition of glucose and mannose, each of 0.5 M, to all buffers (Ligat *et al.*, 1991).

The most common buffer for AP substrate (BCIP/NBT) is 0.1 M TRIS-HCl, pH 9.5, with 0.1 M NaCl, but 1 M diethanolamine-HCl, pH 9.6 (Parent *et al.*, 1985), and 0.1 M ethanolamine-HCl buffer, pH 9.6 (Lange and Heide, 1986), are also used. Buffers with near neutral pH, such as 50 mM TRIS-HCl, pH 7.4, with 200 mM NaCl, are used for the HRP substrate chloronaphthol/hydrogen peroxide.

Recording

Reactions in the form of coloured (BCIP/NBT) or black (DIBA-ECL) dots can be scored visually, or by reflectance densitometry or a digital image analyser. Such instruments may not be available in virus-diagnostic laboratories or be important where the aim is only to determine the presence of virus regarding certification (Pollini *et al.*, 1993). Membranes can be stored for years without loss of the dots if kept dry and in the dark. Thus, membranes can serve as permanent records.

5.3.3. Indirect AgF-DIBA, protocol

According to the literature, the protocols used for DIBA show a wide variation. Apparently, indirect AgF (antigen-first) methods, a version of which is described below, are most widely applied.

After blotting the first reactant, the antigen, on to the NCM, the NCM is reacted with a crude, unfractionated polyclonal antiserum, the presence of which is detected by an anti-rabbit IgG–AP conjugate. BCIP/NBT is used as a substrate, producing a purple-bluish colour in virus-positive dots (Fig. 5.8 shows the principles).

EQUIPMENT AND MATERIALS.

Equipment

a. Glassware, thoroughly cleaned.

b. Mortar and pestle or hand homogenizer (e.g. Bioreba™), *or*

c. automatic homogenizer (optional).

d. A rack made for photographic slides is used for holding samples in polyethylene bags.

e. Test tube rack(s).

f. Centrifuge, minimum 10,000 *g* for 10 ml tubes, and (optionally) a rotor for Eppendorf tubes.

g. Pencil, soft.

h. Adjustable pipettes, single channel, of different sizes, of which a 1–10 µl pipette is well suited for dot application instead of capillary tubes (see below).

i. Glass Petri dish, 9 or 15 cm diameter, *or*

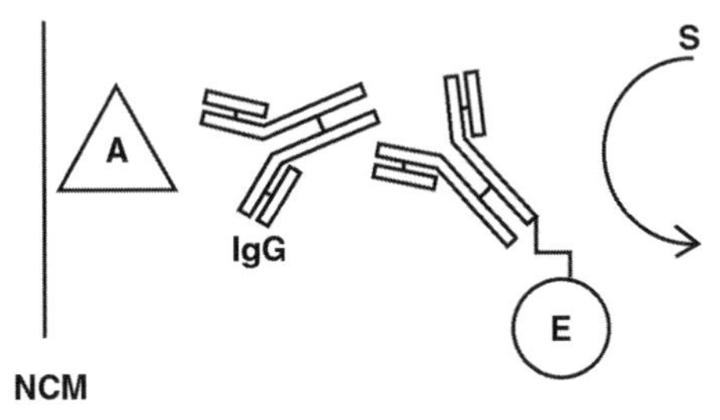

Fig. 5.8. The principles of indirect AgF-DIBA. NCM, nitrocellulose membrane; A, antigen; IgG, virus-specific antibodies in crude antiserum; E, enzyme (AP), which is conjugated to anti-rabbit IgG produced in goat or pig; S, BCIP/NBT substrate. Courtesy of ADGEN Phytodiagnostics, Neogen Europe Ltd, UK.

j. polypropylene container, rectangular (e.g. lid of disposable pipette tips box).

k. Facility for incubation at 37°C (oven with thermostat).

l. Reflectance densitometer or digital image analyser (optional).

Materials

m. Nitrocellulose membrane (NCM), pore size 0.45 μm (Millipore™ HAWP 00010, Sartorius™ 0.45 μm, or similar).

Smith and Banttari (1987) recommend a preliminary test of each new NCM batch so as to determine which of the NCM surfaces have the best DIBA sensitivity.

n. Gloves.

o. Test samples.

p. Controls: known *virus-free* material of the same species, cultivar, age and tissue type as that being tested, and known *virus-containing* material.

q. Known virus-free material for cross-absorption (same as that used for the negative control).

r. Polyethylene bags, sturdy type, approx. 7 cm × 10 cm for preparation of test samples.

s. Test tubes (size 3–4 ml) or Eppendorf tubes, 1.5 ml.

t. Capillary tubes, 10 cm (not needed if an adjustable pipette is used).

u. Pipette tips.

Reagents, solutions and buffers

v. Triton X-100.

w. Blocking agent: skimmed milk powder (SKP) or polyvinyl alcohol (PVA, Sigma P8136 or similar).

x. Virus-specific antiserum, crude (rabbit).

y. Conjugate: anti-rabbit IgG–AP.

z. TBS: 0.02 M TRIS-HCl, pH 7.5, with 0.5 M NaCl – see Appendix 2, p. 249.

aa. Sodium azide (NaN_3), 2% stock solution – see Appendix 2, p. 248. highly toxic.

bb. Ascorbic acid or sodium sulphite (Na_2SO_3).

cc. Antigen buffer (TBS + 0.02% NaN_3 + 0.1 M ascorbic acid D 0.2% Na_2SO_3). Adjust pH to 7.5 with NaOH, if needed.

dd. Tween 20.

ee. Washing buffer: TBST (TBS with addition of 1 ml Tween 20 per litre).

ff. Polyvinylpyrrolidone (PVP), MW 10,000–40,000.

gg. DIBA serum/conjugate buffer (TBST + 2% PVP + 0.2% SKP + 0.02% NaN_3).

hh. Substrate buffer: 0.1 M ethanolamine buffer, pH 9.6 – see Appendix 2, p. 249.

ii. Nitro blue tetrazolium (NBT), Sigma N6876 or similar.

jj. 5-Bromo-4-chloro-3-indolylphosphate (BCIP) Sigma B8503 or similar.

kk. $MgCl_2$.

ll. Substrate: stock solutions of NBT, BCIP and $MgCl_2$ – see Appendix 2, p. 249.

Table 5.6. Flow schedule for indirect AgF-DIBA.

Step (step no. in procedure)		Buffer	Time	Temperature
a. Preparation	(**1–4**)	TBS		
b. Antigen prep.	(**5–6**)	Antigen	≥ 30 min	Ambient
c. Cross-absorption	(**7**)	Serum/conjugate	45 min	37°C
d. Dotting and drying	(**8–9**)			
e. Blocking	(**10–13**)	Washing	30 min[a]	Ambient
f. Specific antiserum[b]	(**14–15**)	Serum/conjugate	1–3 h, not overnight	37°C
g. Washing	(**16**)	Washing	3–5 min × 5	
h. Conjugate	(**17–18**)	Serum/conjugate	1–3 h or overnight	37°C or 4°C, respectively
i. Washing	(**19**)	Wash → substr. buffer	3–5 min × 5	
j. Substrate	(**20–23**)	Substrate	5–20 min	Ambient
k. Recording	(**24**)			

[a]One minute, if PVA is used.
[b]Cross-absorbed.

PROCEDURE (see also Section 5.3.2). Before starting the test, the DIBA flow schedule (Table 5.6) should be checked in order to plan the work, e.g. overnight incubation is not to be used for specific antiserum (step **15**).

Preparation

1. Prepare the loading pattern in a test schedule for each NCM. Indicate both antigens (serial number, etc.) and detecting antibody in their respective, predetermined dilutions. Test samples are to be loaded in at least duplicate dots. Leave dots for negative controls in each NCM (four or more) and positive controls (two).

2 Wear gloves or use clean forceps when handling the membranes. Fingerprints on NCMs should be strictly avoided, as these will lead to unspecific colour development due to AP from the skin.

The NCM is cut with a scalpel or a sharp knife to a size slightly less than the area of the container (for a 9 cm Petri dish, about 5 cm × 6 cm, which can hold up to 30 dots). For convenient handling, membranes should not normally be larger than a size corresponding to a microtitre plate.

3. Mark with a soft pencil the positions of the dots by tiny points 1 cm apart (mark on the NCM's optimal surface; see Materials). Alternatively a grid with 1 cm × 1 cm squares can be made (see Application of samples, Section 5.3.2). Make an orientation mark or cut a corner to indicate the schedule followed.

4. Immerse the membrane in *TBS* so as not to entrap air. After 5 min soaking, remove the membrane and leave it on a filter paper for complete drying (30 min). A pot label of soft plastic (approx. 2 cm × 12 cm) is ideal for transferring wet NCMs.

Antigen

5. Only 10 µl of the final dilution is required per dot. A representative amount of tissue must be sampled and homogenized.

6. Prepare antigens as described under Sample Preparation, pp. 102–103 and 130–131, in the following sequence: negative controls – test antigen samples – positive controls. Homogenize 1 part of tissue in 4 parts of *antigen buffer* (plant tissue); for seed material, use 1 : 10. Transfer some of each homogenate to a test tube to make 1 ml of a final dilution of 1 : 100 or 1 : 200 in the same buffer. Mix, and allow the tubes to stand for 30 min or more for precipitating cell constituents. Alternatively, centrifuge the diluted extracts in Eppendorf tubes at 10,000 *g* or more for 10 min. While waiting, start the cross-absorption.

Cross-absorption

7. Cross-absorption of Pabs for use in DIBA is generally recommended and can be started before the application of dots to the membrane.

(a) Healthy plant or seed material from the same species as that being tested is ground up in *DIBA serum/conjugate buffer* to give a final dilution of 1 : 20–1 : 40 (w/v). Grind first with 4 ml buffer/g tissue (seed material: more buffer) and then dilute further. The volume before filtration is made about 30% larger than the required end volume, which is about 200 µl per dot.

(b) Filter the extract through several layers of gauze, and

(c) use the calculated amount of filtrate to dilute the antiserum into a predetermined dilution.

(d) Stir thoroughly and let the mixture incubate for 45–60 min at 37°C. Meanwhile, go to step **8** below.

(e) Centrifuge at 10,000–15,000 *g* for 10 min to clarify the mixture.

Dotting

8. Place the TBS-treated, dry membrane on a filter paper. Do not shake the test tubes with the antigen solutions. With a capillary tube or an adjustable pipette, take out the antigen supernatants to make dots of approx. 0.5 cm in diameter on the premarked positions. Each dot usually holds 5–10 µl, but a pre-assessment of the volume should be made on a separate piece of membrane. Apply the samples in the following sequence: negative controls – test samples – positive controls.

9. Leave the membrane on the filter paper until it is completely dry. NCMs dotted with antigens can be stored dry at room temperature for several weeks or mailed for later detection. While the NCM dries, prepare for the next step.

Blocking

10. When the NCM is completely dry, transfer it to the container.

11. Triton X-100 treatment (optional step): Prior to blocking, the membranes can be treated with 1% Triton X-100 in *washing buffer* for 10 min to reduce green colour in the dots if plant extracts are being tested. This treatment is usually not needed in the case of seed extracts.

Pour the solution, 200–400 µl/dot, over the membrane and leave it for 10 min.

12. Drain off the Triton solution and blot remaining liquid with a piece of filter paper. Cover the membrane with the blocking solution: 2% SKP in washing buffer. Use 200–400 µl solution per dot and allow to incubate for 30 min at room temperature.
Alternatively, use
13. PVA at 1 µg/ml in *washing buffer* as blocking solution, but then the incubation time is only *1 min*.

Specific antiserum
14. Drain off the blocking solution. Do not wash, but directly add the cross-absorbed, centrifuged antiserum solution (step **7**), approx. 200 µl per dot.
15. Cover with lid and incubate for 1–3 h at 37°C (not overnight). Make sure that the container is placed horizontally. Agitate the container periodically or place the container on a shaker with gentle oscillation.

Conjugate
16. Drain off the antiserum solution and collect it, if desired, for storing at 4°C and later reuse. Wash the membrane by pouring a generous amount of *washing buffer* into the container, soak for 3–5 min with periodical agitation, and drain off. Repeat the washing four times.
17. Add conjugate adequately diluted in *DIBA serum/conjugate buffer* (usually 1 : 1000 or 1 : 2000), approx. 200 µl per dot.
18. Incubate under lid for 1–3 h at 37°C or overnight at 4°C, and otherwise as for antiserum incubation.

Substrate
19. Drain off the conjugate solution (collect for later reuse, if desired) and wash four times with *washing buffer* as in step **16** above. Finally, cover the membrane with *substrate buffer*.
20. Prepare the substrate solution just before use. When preparing and handling the substrate, use gloves (skin contact must be avoided, due both to the risk of AP contamination from skin and to the chemicals' toxicity). For treating about 50 dots, mix the following in a measuring cylinder: NBT stock solution, 1 ml; BCIP stock solution, 0.15 ml; $MgCl_2$ stock solution, 0.02 ml; add *substrate buffer* to 10 ml.
21. Drain off the buffer and add the substrate solution.
22. Observe the development of colour in the dots from virus-containing extracts, which usually takes place within 5–20 min at room temperature.
23. When the whole membrane starts turning to a purple colour, drain off the substrate solution (use gloves) into a chemical waste container, and add deionized H_2O in its place. Change the water two to three times. Used substrate solution should not be dropped into the sink, but treated as chemical waste.

Recording
24. Transfer the membrane to a filter-paper and let it dry. Score the reactions visually or by means of a reflectance densitometer or a digital image analyser. Dry membranes can be kept as a permanent record for years if stored in the dark under dry conditions or, alternatively, be photocopied.

5.4. Tissue Blotting Immunoassay (TBIA)

Tissue blotting immunoassay (TBIA) (syn. tissue print immunoassay (TPIA) and tissue print immunoblotting (TPIB)) is a detection method similar to DIBA, except that it does not involve tissue extraction. Instead of dotted extracts, freshly cut tissue surfaces are printed directly onto nitrocellulose membranes. The antigens trapped in the tissue blots are then reacted with antibodies, conjugate and substrate in the same way as in DIBA. The method was first used for fresh plant material by Lin *et al.* (1990) and has gained popularity for many purposes, not only due to its simplicity, eliminating the need for an extraction step, but also due to its high sensitivity, comparable with or in some cases even higher than ELISA and DIBA.

Leaves from plants or seedlings are tested by rolling a leaf tightly, making a cut, leaving a single-plane cut surface, and pressing the cut surface onto a dry NCM with a firm but gentle pressure for a few seconds. For stems and petioles, a similar printing is made from transverse cuts. Excess exudate occurring in some plants is removed by draining the cut surface on tissue paper just prior to printing. After the blotted membrane is completely dry, it is blocked and further processed as in DIBA. As for DIBA, reactants can be reused. Thus reuse of conjugate five to ten times without significant loss in sensitivity has been reported (Makkouk and Comeau, 1994; Huth, 1997). Figure 5.9 shows a typical result of a TBIA in which cut leaf rolls have been printed on an NCM.

Since its first use in 1990, a number of reports on the successful application of TBIA have been published. Higher sensitivity than ELISA and DIBA has been reported for detection of *Lily symptomless virus* (LSLV, *Carlavirus*) in lily bulbs (Hsu *et al.*, 1995) and a phloem-limited *Closterovirus* in pineapple (Hu *et al.*, 1997). *Barley yellow dwarf virus* (BYDV, *Luteoviridae*) (Makkouk and Comeau, 1994) and *Soybean dwarf virus* (SbDV, *Luteoviridae*) (Makkouk *et al.*, 1997), two other phloem-limited viruses with low concentrations in plants, were also readily detected by this method. A *Begomovirus*, *Tomato yellow leaf curl*

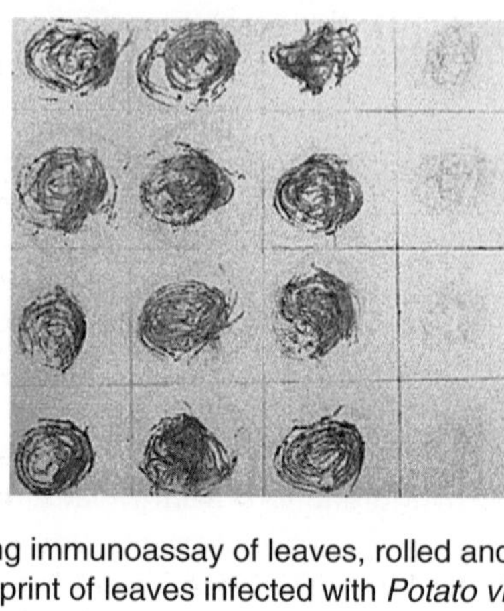

Fig. 5.9. Tissue blotting immunoassay of leaves, rolled and cut before printing. Columns 1–3 from left: print of leaves infected with *Potato virus X* (PVX, *Potexvirus*); column 4: print of healthy leaves.

virus (TYLCV), was detected in tomato, even after storing the dried blots for 1–6 months (Abou-Jawdah *et al.*, 1996). Makkouk *et al.* (1997) used mixed antisera for simultaneous detection of three seed-transmitted viruses, BBSV, PSbMV and BYMV, in lentil. Samples from lentil plants were tested in groups of 25, held together, cut and printed collectively on the membrane. Unsuccessful detection by TBIA has been reported for *Tomato spotted wilt virus* (TSWV, *Tospovirus*) in certain plant species, because of their high anthocyanin content, which masked the purple colour development in the case of infected tissue (Hsu and Lawson, 1991).

Huth (1997) successfully detected 13 viruses in *Poaceae* plants by TBIA, and tried different ways to improve its sensitivity and versatility. The author found no effect of buffer treatment of the NCM, but printing on a slightly moistened membrane gave the best results. However, soaking of the membrane in a 0.02 M Na_2SO_3 solution prior to printing could prevent oxidative brown staining of sap from certain plant species, which otherwise could mask the substrate colour reaction. For other methods to remove plant sap colour, see Section 5.3.2. Duration of printing beyond 0.5–1 s did not increase the signal. Although turgid plant parts are preferable for printing, the same author indicated that plant tissue could also be desiccated and used later for printing after soaking in water or a humidity chamber. Use of polyvinyl alcohol (PVA, MW 30,000–70,000) for blocking instead of albumin or skimmed milk reduced test duration by 1 h (see also DIBA).

Potentially, the method should be well suited for printing membranes at one location and mailing them dry to another, such as a central laboratory, for processing and detection. In contrast to ELISA and DIBA, TBIA seems to be better suited for detection of phloem-limited viruses and other viruses present in low concentration in plants, as reported by the above authors. Makkouk and Comeau (1994) even detected BYDV in wheat spike peduncles of infected plants after 5 years of storage at room temperature. Furthermore, the method is fast; thus the same authors completed the detection in 3 h. TBIA is also superior to ELISA economically. Huth (1997) found that for the same price at least ten times more samples could be tested by TBIA than by ELISA when reusing the conjugate solution five times. The above-mentioned method reported by Makkouk *et al.* (1997) for testing multiple plant samples simultaneously could further reduce costs. It is routinely used in seed-health testing for virus at the International Center for Agricultural Research in the Dry Areas (ICARDA) (K. Makkouk, ICARDA, 2002, personal communication).

Hull (2002) mentions three advantages of TBIA: (i) detailed information on the tissue distribution of a virus can be obtained; (ii) extraction of sap from leaves in which a virus has limited tissue distribution leads to the dilution of the virus with sap from uninfected cells: since the technique samples the content of each cell on the cut surface individually, there is increased sensitivity; and (iii) the technique is easily applicable to field sampling as tissue printings can be made in the field without the need to collect leaf samples for sap extraction in the laboratory.

Tissue printing on membranes is also used in a variant of nucleic acid spot hybridization (Section 6.1.1).

A special procedure similar to TBIA, the direct immunostaining assay (DISA), for non-destructive testing of pepper seed for tobamoviruses is described after the following protocol.

5.4.1. Tissue blotting immunoassay (TBIA), protocol

Below is described a simple procedure for TBIA of the AgF type with the use of crude polyclonal antiserum, anti-rabbit IgG–AP conjugate and BCIP/NBT as substrate. The procedure is as described by Lin *et al.* (1990), with minor modifications, and most of the components and steps are identical to those of DIBA (Section 5.3.2, p. 130).

EQUIPMENT AND MATERIALS. Equipment and materials are as for DIBA, except for the following:

Equipment, further
- Razor blade, or sharp scalpel or knife.
- Stereomicroscope.

Equipment not needed
- Homogenizer.
- Reflectance densitometer/digital image analyser.

Materials, further
- Parafilm® or similar.

Materials not needed
- Test tubes (and racks).
- Capillary tubes or pipette for sample application, except for positive controls.
- Polyethylene bags, small, except for holding positive control samples.
- Antigen buffer – except for preparing positive controls.

PROCEDURE (see also flow schedule, Table 5.7).

Preparation

1. Prepare test schedule and membrane as described in *DIBA steps* ***1–3***, p. 137. For TBIA, a grid may be better than points by easing the positioning of the prints (see Section 5.3.2).

2. Pretreatment: pretreatment with TBS, as in DIBA, is not needed. However, in the case of tests of certain plants (e.g. *Vicia faba*) whose sap becomes dark due to oxidation, and therefore disturbs the reading, soaking the NCM in 0.02 M Na_2SO_3 followed by drying is recommended.

3. The seedling or plant material to be tested must be freshly harvested or, if stored for a short time, be protected against drying (humidity chamber). Seeds have to be soaked in deionized water for some hours or overnight before testing.

Tissue from non-turgid plant parts is difficult to handle and may leave too little antigen on the NCM, although the detectability of virus may not be much affected.

Table 5.7. Combined flow schedules for indirect AgF-DIBA and TBIA.

DIBA					TBIA		
Step (step no. in procedure)		Buffer	Time	Temperature	Step (step no. in procedure)[a]		Time
a. Preparation	(**1**–**3**)				a. Preparation ←	(**1**)	
b. NCM treatment	(**4**)	TBS	5 min		b. NCM treatment*	(**2**)	
c. Antigen prep.	(**5**–**6**)	Antigen	≥ 30 min	Ambient	c. Antigen prep.*	(**3**–**4**)	
d. Cross-absorption	(**7**)	Serum/conjugate	45 min	37°C	d. Tissue blotting*	(**5**–**7**)	
e. Dotting and drying	(**8**–**9**)				e. Cross-absorption ↖	(**7**)	
f. Blocking	(**10**–**13**)	Washing	30 min[b]	Ambient	f. Blocking* (←)	(**8**)	60 min[b]
g. Specific antiserum[c]	(**14**–**15**)	Serum/conjugate	1–3 h, not overnight	37°C	g. Immunoassay... ←	(**9**)	
h. Washing	(**16**)	Washing	3–5 min × 5		h.*		10 min × 3
i. Conjugate	(**17**–**18**)	Serum/conjugate	1–3 h or overnight	37°C or 4°C, respectively	i.		
j. Washing	(**19**)	Washing → substr. buffer	3–5 min × 5		j.*		10 min × 3
k. Substrate	(**20**–**23**)	Substrate	5–20 min	Ambient	k.		
l. Recording	(**24**)				l. Recording*	(**10**)	

[a]Steps with * are different from DIBA.
[b]One minute if PVA is used.
[c]Cross-absorbed.

4. Group testing: A number of freshly harvested seedlings may be tested together, provided that their morphology so allows and that virus is detectable in the tissue to be printed (it must be done with the hypocotyls or stems, in order to distinguish the reaction from individual seedlings).

Tissue blotting

5. Place the prepared NCM on a filter paper.

(a) Seedlings or plants: a leaf or a leaf part is taken fresh from the seedling or plant being assayed. The leaf is rolled into a tight core 2–3 cm long and about 0.5 cm wide. A single-plane cut surface is made by cutting the roll transversely with a razor blade.

If petioles or stems are to be tested, a transverse cut is made in a similar way.

(b) Seed: seeds soaked for some hours or overnight are cut transversely or longitudinally as in (a).

(c) Groups of seedlings: a bundle of ten or more seedlings is held tightly together with a piece of Parafilm® or similar. A single-plane, transverse cut is made through the bundle of hypocotyls/stems.

(d) Controls: negative controls are made by printing cuts from known healthy individual plants or seeds of the same cultivar and age.

Positive controls: prints from known infected plants/seeds the same as those being tested, *or* dots made from other known infected tissue prepared as described under *DIBA, step* ***6***.

Note. Print in the following sequence: negative controls, test samples, positive controls, and rinse the knife between samples in a virus-degrading compound, e.g. 10% trisodium phosphate, followed by thorough wiping with a clean tissue.

6. Immediately after cutting, press the cut surface with a firm but gentle force on to a marked point on the NCM for 1–3 s. If the tissue contains excess exudates, drain off on a piece of filter paper before blotting. Make two to three prints, either on the same spot or beside each other on the NCM by repeated cuttings and blotting from the same sample.

7. Allow the membrane to dry completely. Meanwhile, start *cross-absorption* of the detecting antibody, as described in *DIBA, step* ***7***. This step may be omitted if a Mab is used as the detecting antibody.

Blocking

8. When the NCM is completely dry, proceed with the blocking step, as described for *DIBA, steps* ***10–13***, but block with 2% SKP for *60 min* instead of the 30 min. If using PVA, block for only *1 min* (DIBA, step **13**).

Specific antiserum, conjugate and substrate

9. Proceed with (cross-absorbed) specific antiserum, conjugate and substrate as described for *DIBA, steps* ***14*** *to* ***23***. However, use three times *10 min* soaking in each wash.

Recording

10. Allow the processed membrane to dry on a filter paper and score the result. In some cases, it may be necessary to use a stereomicroscope at low magnification to examine the fine structure of the tissue blots in order to observe the purple-coloured areas, indicating presence of virus in the sap deposited. Blots from healthy plant tissue should appear colourless or weakly greenish, and colour from healthy seed is often absent. A test result (in black and white) is shown in Fig. 5.9.

Dry membranes can be kept as permanent records for years if stored dry in the dark.

Non-destructive testing of seeds for tobamoviruses by the DISA method

A direct immunostaining assay (DISA) was developed by Takeuchi *et al.* (1999) as a technique for detecting tobamoviruses on pepper seeds. In this technique the seeds are used as the solid phase in an immunoassay by which infected seeds become stained.

Pepper seeds were soaked in a polyclonal antiserum solution for 1 h at room temperature, washed in phosphate buffered saline with addition of Tween-20 (PBST), incubated with anti-rabbit IgG–AP conjugate for 1 h at room temperature and washed with PBST and distilled water, before being incubated with NBT/BCIP. After 30–60 min, infected seeds then stained dark purple, while healthy seeds did not. To test whether the detected virus was infective, stained seeds were homogenized and assayed on a hypersensitive tobacco cultivar (Section 4.4.3). The DISA treatment influenced neither the biological detectability of the viruses nor the seed viability.

5.5. Other Serological Test Methods

Serological techniques other than EIAs include precipitin, gel-diffusion and agglutination tests in which the antigen-antibody reaction is mainly directly visible. These methods are simple to carry out, but generally low in sensitivity. Such detection methods in different variations were the only ones available until the invention of ELISA in the 1970s, but some of them are still in use for certain studies. A form of agglutination test with sensitivity close to that of EIAs, virobacterial agglutination, was developed in the 1980s. Immunosorbent electron microscopy (ISEM) is an assay with sensitivity equal to that of EIAs (Matthews, 1993).

Precipitin reaction

When a virus antigen and homologous antibodies are mixed in optimal proportions, the antibodies bind to and, due to their two reactive arms, can link the virions, resulting in aggregates that are visible as a white precipitate. The test is usually performed by mixing a drop each of clarified sap and diluted antiserum on a hydrophobic surface (*microprecipitin test*), followed by incubation at 37°C in a moist atmosphere, or oil-covered, for 0.5–2 h, and recording. The method can be used to test antiserum quality as indicated by a microprecipitin titre, i.e. determination of the highest antiserum dilution still producing a precipitate. Such a titre, however, is 10- to 100-fold lower than that determined by EIAs. The sensitivity of precipitin tests is too low for use in routine detection.

Agglutination tests

Agglutination tests are also based on direct binding between antibodies and virions in liquid phase, involving initial sediments that co-precipitate, forming large complexes (clumping) easily visible with the naked eye. One such assay is *chloroplast agglutination*, in which a drop of crude plant extract is mixed with a drop of diluted antiserum. In the case of virus-containing extract, antibodies bind to virions adsorbed to the chloroplasts, linking these organelles together in green aggregates. The assay is low in sensitivity, requiring relatively large amounts of antiserum, and is best suited for viruses with elongate particles present in infected plants at high concentrations. A much higher sensitivity, though still lower than that of EIAs, is obtainable with agglutination tests involving either latex particles, as used in *latex-agglutination* (LA) (Torrance, 1980), or bacterium cells, *virobacterial agglutination* (VBA) (Chirkow *et al.*, 1984; Walkey *et al.*, 1992). In LA, polystyrene latex particles (80 μm) are first coated with protein A and then with antibodies that are bound due to the specific Fc–protein A affinity (Section 5.1). When the coated latex particles are mixed with clarified virus-containing sap and incubated, virion–antibody linking causes the latex particles to aggregate, resulting in an easily visible granulation. In VBA, a similar reaction is produced, but between virions and killed cells of *S. aureus*, previously coated with antibodies, bound due to naturally occurring protein A in the bacterial cell walls. The VBA was found simple to use, and had higher sensitivity than that of LA and only slightly lower than that of ELISA (Walkey *et al.*, 1992).

Gel diffusion

Virus–antibody binding also occurs in an agar gel, in which the reaction becomes visible as a white band. The *Ouchterlony gel double-diffusion test* (ODD) (Ouchterlony, 1958) is most commonly used. In ODD the two reactants are placed in opposing wells cut in a 3–5 mm thick agar gel in a plastic Petri dish. After 24 h or more, the reactants will have migrated through the gel and, where a virus meets it homologous antibody, their precipitate forms a white line, best observed in indirect light. When two wells containing different virus isolates are placed opposite to a well containing antibodies against one of the isolates, the pattern of bands can reveal whether the isolates are serologically identical, partially related or unrelated. ODD is not suitable for routine detection due to its low sensitivity, equal to that of precipitin tests, but is still used for determining serological relations between viruses and virus isolates. ODD is best suited for detection of isometric virions, whereas most elongated virions require degradation, either by sodium dodecyl sulphate (SDS) or (better) by sonication (Hull, 2002), to be detectable.

Immunosorbent electron microscopy

Among techniques that combine electron microscopy and serology, immunosorbent electron microscopy (ISEM), developed by Derrick (1973), is a highly sensitive technique for detection or identity verification (see Section 7.1.1).

The electron microscopy (EM) specimen carrier, the grid, whose support film is first coated with specific antibodies, is exposed to a virus-containing extract for a few min, followed by washing off unbound host-plant material and contrast staining. Thus, the virions are trapped specifically instead of passively, resulting in an increased concentration of virions per unit area of the grid examined in the EM. The technique enables detection of viruses that occur at low concentrations in their hosts or of viruses in extracts from pooled samples. By a further step, called 'decoration', the virions trapped on the grid film are coated with a layer of specific antibodies, which, after rinsing and staining of the grid, appear in the EM as a dark halo around the virions. In this way, both the identity of the virus can be confirmed and the presence of mixed infection can often be revealed.

Crude, unfractionated antiserum can be used, and the antiserum consumption is low. Presence of host-specific antibodies is less critical here than for EIAs. Also, the quality of the plant or seed extract is not critical.

Each preparation requires only a few min of manual work; however, when adding the time needed for EM examination, fewer samples per time unit can be tested than by EIAs. Moreover, an EM is a costly instrument, not accessible in most plant virus testing laboratories.

Principles and protocols of serological techniques, including chloroplast agglutination, microprecipitin, ODD and several other liquid-phase immunoassays, as well as ISEM, are excellently described in detail in Hampton *et al.* (1990). Detailed protocols for chloroplast agglutination, microprecipitin, ODD and ISEM can also be found in Dijkstra and de Jager (1998).

References

AAB Descriptions of Plant Viruses (n.d.) Nos. 1–409. Available at: http://www.dpvweb.net

Abdalla, M.E. and Albrechtsen S.E. (2001) Petri dish-ELISA, a simple and economic technique for detecting plant viruses. *Acta Phytopathologica et Entomologica Hungarica* 36, 1–8.

Abou-Jawdah, Y., Soubra, K.H. and Shebaro, W.A. (1996) Evaluation of the reaction of tomato genotypes to tomato yellow curl geminivirus infection in Lebanon. *Phytopathologia Mediterranea* 35, 91–99.

Abraham, A. (1995) Studies on alternative enzyme immunoassay techniques and detection of seed-borne viruses from legume crops of Ethiopia. MSc thesis, Danish Government Institute of Seed Pathology for Developing Countries and Royal Veterinary and Agricultural University, Copenhagen, Denmark.

Abraham, A. and Albrechtsen, S.E. (2000) Petri dish-agar dot immunoenzymatic assay (PADIA) – a new and inexpensive method for the detection and identification of plant viruses. *International Journal of Pest Management* 46, 161–164.

Abraham, A. and Albrechtsen, S.E. (2001) Comparison of penicillinase, urease and alkaline phosphatase as labels in enzyme-linked immunosorbent assay (ELISA) for the detection of plant viruses. *Journal of Plant Diseases and Protection* 108, 49–57.

Adgen ELISA Protocols (2000) ADGEN Diagnostics, Neogen Europe Ltd, Auchincruive, UK.

Andersen, H., Sørensen, U.B. and Thomsen, E. (1991) *Immunkemiske Metoder – Teori og Praksis* [Immunochemical methods – theory and practice]. Nucleus, Aarhus, Denmark, 111 pp.

Avrameas, S. (1969) Coupling of enzymes to proteins with glutaraldehyde. Use of the conjugates for the detection of antigens and antibodies. *Immunochemistry* 6, 43–52.

Ball, E., Hampton, R.O., De Boer, S. and Schaad, N. (1990) Antigens, 1. Polyclonal.

In: Hampton, R.O., Ball, E.M. and De Boer, S.H. (eds) *Serological Methods for Detection and Identification of Viral and Bacterial Plant Pathogens, a Laboratory Manual.* APS Press, American Phytopathological Society, St Paul, Minnesota, pp. 33–54.

Banttari, E.E. and Franc, G.D. (1982) Enzyme-linked immunosorbent assay with single or combined antisera for viruses S and X in potato tubers and plants. *American Journal of Potato Research* 52, 375–387.

Banttari, E.E. and Goodwin, P.H. (1985) Detection of potato viruses S, X, and Y by enzyme-linked immunosorbent assay on nitrocellulose membranes (dot ELISA). *Plant Disease* 69, 202–205.

Banttari, E.E. and Petersen, A.C. (1983) A cleaning technique for cuvettes used in enzyme-linked immunosorbent assay. *Plant Disease* 67, 18–20.

Bar-Joseph, M., Moscovitz, M. and Sharafi, Y. (1979) Reuse of coated enzyme-linked immunosorbent assay plates. *Phytopathology* 69, 424–426.

Bharathan, N., Reddy, D.V.R., Rajeshwari, R., Murthy, V.K., Rao, V.R. and Lister, R.M. (1984) Screening of peanut germ plasm lines by enzyme-linked immunosorbent assay for seed transmission of peanut mottle virus. *Plant Disease* 68, 757–758.

Brakke, M.K. (1990) Preparation of antigens, viruses. In: Hampton, R.O., Ball, E.M. and De Boer, S.H. (eds) *Serological Methods for Detection and Identification of Viral and Bacterial Plant Pathogens, a Laboratory Manual.* APS Press, American Phytopathological Society, St Paul, Minnesota, pp. 15–26.

Brattey, C. and Burns, R. (1998) Antibody production. In: Foster, G.D. and Taylor, S.C. (eds) *Plant Virology Protocols: From Virus Isolation to Transgenic Resistance.* Humana Press, Totowa, New Jersey, pp. 279–286.

Brunt, A.A., Crabtree, K., Dallwitz, M.J., Gibbs, A.J. and Watson, L. (eds) (1996) *Viruses of Plants.* CAB International, Wallingford, UK.

Chirkow, S.N., Olovnikov, A.M., Surguchyova, N.A. and Atabekov, J.G. (1984) Immunodiagnosis of plant viruses by a virobacterial agglutination test. *Annals of Applied Biology* 104, 477–483.

Clark, M.F. and Adams, A.N. (1977) Characteristics of the microplate method of enzyme-linked immunosorbent assay for the detection of plant viruses. *Journal of General Virology* 34, 475–483.

Clark, M.F., Lister, R.M. and Bar-Joseph, M. (1986) ELISA techniques. *Methods in Enzymology* 118, 742–766.

Converse, R. and Martin, R. (1990) Enzyme-linked immunosorbent assay (ELISA), 1. Viruses. In: Hampton, R.O., Ball, E.M. and De Boer, S.H. (eds) *Serological Methods for Detection and Identification of Viral and Bacterial Plant Pathogens, a Laboratory Manual.* APS Press, American Phytopathological Society, St Paul, Minnesota, pp. 179–196.

Copeland, R. (1998) Assaying levels of plant virus by ELISA. In: Foster, G.D. and Taylor, S.C. (eds) *Plant Virology Protocols: From Virus Isolation to Transgenic Resistance.* Humana Press, Totowa, New Jersey, pp. 455–460.

Demski, J.W. and Warwick, D. (1986) Testing peanut seeds for peanut stripe virus. *Peanut Science* 13, 38–40.

Derrick, K.S. (1973) Quantitative assay for plant viruses using serologically specific electron microscopy. *Virology* 56, 652–653.

Dijkstra, J. and de Jager, C.P. (1998) *Practical Plant Virology: Protocols and Exercises.* Springer-Verlag, Berlin.

Etienne, L., Clauzel, J.M. and Fuchs, M. (1991) Simultaneous detection of several nepoviruses infecting grapevine in a single DAS-ELISA test using mixed antisera. *Journal of Phytopathology* 131, 89–100.

Falk, B.W. and Purcifull, D.E. (1983) Development and application of an enzyme-linked immunosorbent assay (ELISA) test to index lettuce seeds for lettuce mosaic virus in Florida. *Plant Disease* 67, 413–416.

Foster, G.D. and Taylor, S.C. (eds) (1998) *Plant Virology Protocols: From Virus Isolation to Transgenic Resistance.* Humana Press, Totowa, New Jersey.

Geering, A.D.W. and Thomas, J.E. (1996) A comparison of four serological tests for the detection of banana bunchy top virus in banana. *Australian Journal of Agricultural Research* 47, 403–412.

Grimm, F. and Daniel, G. (1984) Zum Einsatz von Mischseren im ELISA Verfahren: ein Ergebnisvergleich mit Einfachantiseren bei Untersuchungen an Kartoffelblättern [On the use of mixed sera with ELISA: a comparison of results with single antisera in studies with potato leaves] *Potato Research* 27, 13–23.

Hamilton, R.I. (1965) An embryo test for detecting seed-borne barley stripe mosaic virus in barley. *Phytopathology* 55, 798–799.

Hamilton, R.I. and Huguenot, C. (1995) Development and application of serological techniques using monoclonal antibodies for detection of economically important viruses in major African food crops. In: *Final Report (Phase II)*. Pacific Agriculture Research Center, Vancouver Research Branch, Agriculture and Agri-Food Canada, Vancouver, 13 pp.

Hammond, J. and Jordan, R. (1990) Dot immunobinding assay (DIBA), 1. Viruses. In: Hampton, R.O., Ball, E.M. and De Boer, S.H. (eds) *Serological Methods for Detection and Identification of Viral and Bacterial Plant Pathogens, a Laboratory Manual.* APS Press, American Phytopathological Society, St Paul, Minnesota, pp. 237–248.

Hampton, R.O., Ball, E.M. and De Boer, S.H. (eds) (1990) *Serological Methods for Detection and Identification of Viral and Bacterial Plant Pathogens, a Laboratory Manual.* APS Press, American Phytopathological Society, St Paul, Minnesota.

Hau, J. and Van Hoosier, G.L. (2002) *Handbook of Laboratory Animal Science*, Vol I, 2nd edn: *Essential Principles and Practices.* CRC Press, Boca Raton, Florida.

Hawkes, R., Niday, E. and Gordon, J. (1982) A dot-immunobinding assay for monoclonal and other antibodies. *Analytical Biochemistry* 119, 142–147.

Heide, M. and Lange, L. (1988) Detection of potato leaf roll virus and potato viruses M, S, X and Y by dot immunobinding on plain paper. *Potato Research* 31, 367–373.

Hibi, T. and Saito, Y. (1985) A dot immunobinding assay for the detection of tobacco mosaic virus in infected tissues. *Journal of General Virology* 66, 1191–1194.

Higley, P.M., McGee, D.C. and Burris, J.S. (1993) Development of a methodology for non-destructive assay of bacteria, fungi and viruses in seeds of large-seeded field crops. *Seed Science and Technology* 21, 399–409.

Higley, P.M., McGee, D.C. and Burris, J.S. (1994) Effects of non-destructive tissue extraction on the viability of corn, soybean, and bean seeds. *Seed Science and Technology* 22, 245–252.

Hobbs, H.A., Reddy, D.V.R., Rajeshwari, R. and Reddy, A.S. (1987) Use of direct antigen coating and protein A coating ELISA procedures for detection of three peanut viruses. *Plant Disease* 71, 747–749.

Hsu, H.T. and Lawson, R.H. (1991) Direct tissue blotting for detection of tomato spotted wilt virus in *Impatiens. Plant Disease* 75, 292–295.

Hsu, H.T., Kim, J.Y. and Lawson, R.H. (1995) Purification of lily symptomless carlavirus and detection of the virus in lilies. *Plant Disease* 79, 912–916.

Hu, J.S., Sether, D.M., Liu, X.P., Wang, M., Zee, F. and Ullman, D.E. (1997) Use of tissue blotting immunoassay to examine the distribution of pineapple closterovirus in Hawaii. *Plant Disease* 81, 1150–1154.

Huguenot, C., Furneaux, M.T., Thottappilly, G., Rossel, H.W. and Hamilton, R.I. (1993) Evidence that cowpea aphid-borne mosaic and blackeye cowpea mosaic viruses are two different potyviruses. *Journal of General Virology* 74, 335–340.

Hull, R. (2002) *Matthews' Plant Virology*, 4th edn. Academic Press, London.

Huth, W. (1997) Tissue print immunoassay – eine schnelle und zuverlässige Methode für den Routinenachweis von Pflanzenviren am Beispiel der Gramineenviren [Tissue-print immunoassay – a rapid and reliable method for routine evidence of Gramineae viruses]. *Nachrichtenblatt des Deutschen Pflanzenschutzdienstes* 49, 185–191.

James, D. and Mukerji, S. (1996) Comparison of ELISA and immunoblotting techniques for the detection of cherry mottle leaf virus. *Annals of Applied Biology* 129, 13–23.

Johansen, E., Edwards, M.C. and Hampton, R.O. (1994) Seed transmission of viruses: current perspectives. *Annual Review of Phytopathology* 32, 363–386.

Jordan, R. (1990) Antibodies, 2. Monoclonal. In: Hampton, R.O., Ball, E.M. and De Boer, S.H. (eds) *Serological Methods for Detection and Identification of Viral and Bacterial Plant Pathogens, a Laboratory Manual.* APS Press, American Phytopathological Society, St Paul, Minnesota, pp. 55–76.

Joshi, S. and Albrechtsen, S.E. (1992) Use of mixed and individual antisera in ELISA-detection of viruses infecting cowpea and soybean. *Journal of Institute of Agricultural and Animal Sciences* 13, 83–87.

Koenig, R. (1981) Indirect ELISA methods for the broad specificity detection of plant viruses. *Journal of General Virology* 55, 53–62.

Koenig, R. and Paul, H.L. (1982) Variants of ELISA in plant virus diagnosis. *Journal of Virological Methods* 5, 113–125.

Köhler, G. and Milstein, C. (1975) Continuous cultures of fused cells secreting antibody of predefined specificity. *Nature* 256, 495–497.

Konate, G. and Neya, B.J. (1996) Rapid detection of cowpea aphid-borne mosaic virus in cowpea seeds. *Annals of Applied Biology* 129, 261–266.

Kurstak, E. (1986) *Enzyme Immunodiagnosis.* Academic Press, New York.

Lange, L. and Heide, M. (1986) Dot immuno binding (DIB) for detection of virus in seed. *Canadian Journal of Plant Pathology* 8, 373–379.

Lange, L., Jomantor, A. and Heide, M. (1989) Testing seeds for viruses by dot immuno binding (DIB) directly on plain paper. *Danish Journal of Plant and Soil Science* 93, 93–96.

Lazarovits, G. (1990) Dot immunobinding assay (DIBA) 2. Bacteria. In: Hampton, R.O., Ball, E.M. and De Boer, S.H. (eds) *Serological Methods for Detection and Identification of Viral and Bacterial Plant Pathogens, a Laboratory Manual.* APS Press, American Phytopathological Society, St Paul, Minnesota, pp. 249–261.

Ligat, J.S., Cartwright, D. and Randles, J.W. (1991) Comparison of some pea seed-borne mosaic virus isolates and their detection by dot-immunobinding assay. *Australian Journal of Agricultural Research* 42, 441–451.

Lin, N.S., Hsu, Y.H. and Hsu, H.T. (1990) Immunological detection of plant viruses and a mycoplasmalike organism by direct tissue blotting on nitrocellulose membranes. *Phytopathology* 80, 824–828.

Lister, R.M. (1978) Application of the enzyme-linked immunosorbent assay for detecting viruses in soybean seed and plants. *Phytopathology* 68, 1393–1400.

Lister, R.M., Carroll, T.W. and Zaske, S.K. (1981) Sensitive serologic detection of barley stripe mosaic virus in barley seed. *Plant Disease* 65, 809–814.

Lizarraga, C. and Fernandez-Northcote, E.N. (1989) Detection of potato viruses X and Y in sap extracts by a modified indirect enzyme-linked immunosorbent assay on nitrocellulose membranes (NCM-ELISA). *Plant Disease* 73, 11–14.

Lommel, S.A., McCain, A.H. and Morris, T.J. (1982) Evaluation of indirect enzyme-linked immunosorbent assay for the detection of plant viruses. *Phytopathology* 72, 1018–1022.

Makkouk, K.M. and Comeau, A. (1994) Evaluation of various methods for the detection of barley yellow dwarf virus by the tissue-blot immunoassay and its use for virus detection in cereals inoculated at different growth stages. *European Journal of Plant Pathology* 100, 71–80.

Makkouk, K.M., Erskine, W. and Attar, N. (1997) Detection of seed-borne viruses in lentil mother plants as compared to harvested seeds. *LENS Newsletter* 24, 49–51.

Manandhar, H.K., Albrechtsen, S.E. and Mathur, S.B. (1996) Reuse of ELISA plates: development of a simple cleaning method. *Acta Phytopathologica et Entomologica Hungarica* 31, 19–29.

Masmoudi, K., Suhas, M., Khetapal, R.K. and Maury, Y. (1994) Specific serological detection of the transmissible virus in pea seed infected by pea seed-borne mosaic virus. *Phytopathology* 84, 756–760.

Matthews, R.E.F. (1991) *Plant Virology*, 3rd edn. Academic Press, San Diego, California.

Matthews, R.E.F. (1993) Overview. In: Matthews, R.E.F. (ed.) *Diagnosis of Plant Virus Diseases*. CRC Press, Boca Raton, Florida, pp. 1–14.

Maury, Y., Duby, C., Bossenec, J.M. and Boudazin, G. (1985) Group analysis using ELISA: determination of the level of transmission of soybean mosaic virus in soybean seed. *Agronomie* 5, 405–415.

Maury, Y., Duby, C. and Khetarpal, R.K. (1998) Seed certification for viruses. In: Hadidi, A., Khetarpal, R.K. and Koganezawa, H. (eds) *Plant Virus Disease Control*. APS Press, St Paul, Minnesota, pp. 237–248.

McKinney, M.M. and Parkinson, A. (1987) A simple, non-chromatographic procedure to purify immunoglobulins from serum and ascites fluid. *Journal of Immunological Methods* 96, 271–278.

Mernaugh, R., Mernaugh, G. and Kovacs, G. (1990) Antigens and antibodies. In: Hampton, R.O., Ball, E.M. and De Boer, S.H. (eds) *Serological Methods for Detection and Identification of Viral and Bacterial Plant Pathogens, a Laboratory Manual*. APS Press, American Phytopathological Society, St Paul, Minnesota, pp. 3–14.

Michel, J.-P., Gingery, M. and Lavelle, L. (2004) Efficient purification of bromoviruses by ultrafiltration. *Journal of Virological Methods* 122, 195–198.

Miranda, P.V., Brandelli, A. and Tezon, J.G. (1993) Instantaneous blocking for immunoblots. *Analytical Biochemistry* 209, 376–377.

Mowat, W.P. and Dawson, S. (1987) Detection and identification of plant viruses by ELISA using crude sap extracts and unfractionated antisera. *Journal of Virological Methods* 15, 233–247.

Ouchterlony, O. (1958) Diffusion-in-gel methods for immunological analysis. In: Kallos, P. (ed.) *Progress in Allergy*, Vol. 5. S. Karger, Basle, Switzerland, pp. 1–78.

Parent, J.-G., Bélanger, F., Desjardins, S. and Brisson, J.D. (1985) Dot-immunobinding for detection of tomato mosaic and potato virus X infecting greenhouse tomatoes. *Phytoprotection* 66, 53–57.

Pollini, C.P., Giunchedi, L. and Credi, R. (1993) A chemiluminescent immunoassay for the detection of grapevine closteroviruses on nitrocellulose membrane. *Journal of Virological Methods* 42, 107–116.

Porstmann, T. and Kiessig, S.T. (1992) Enzyme immunoassay techniques, an overview. *Journal of Immunological Methods* 150, 5–21.

Powell, C.A. (1987) Detection of three plant viruses by dot-immunobinding assay. *Phytopathology* 77, 306–309.

Premier, R.R., Cox, J.C., Aitken, D.P. and Healy, K. (1985) An evaluation of the use of a pH indicator for the detection of beta-lactamase in enzyme immunoassay. *Journal of Immunological Methods* 83, 371–377.

Reddy, A.S., Hobbs, H.A., Delfosse, P., Murthy, A.K. and Reddy, D.V.R. (1998) Seed transmission of Indian peanut clump virus (IPCV) in peanut and millets. *Plant Disease* 82, 343–346.

Sander, E. and Dietzgen, R.G. (1984) Monoclonal antibodies against plant viruses. *Advances in Virus Research* 29, 131–168.

Sherwood, J.L. (1987) Comparison of a filter paper immunobinding assay, Western blotting and an enzyme linked immunosorbent assay for the detection of Wheat streak mosaic virus. *Journal of Phytopathology* 118, 68–75.

Smith, F.D. and Banttari, E.E. (1987) Dot-ELISA on nitrocellulose membranes for detection of potato leaf roll virus. *Plant Disease* 71, 795–799.

Srinivasan, I. and Tolin, S. (1992) Detection of three viruses of clovers by direct tissue immunoblotting. *Phytopathology* 82, 721 (Abstr.).

Stace-Smith, R. and Martin, R.R. (1993) Virus purification in relation to diagnosis. In: Matthews, R.E.F. (ed.) *Diagnosis of Plant Virus Diseases*. CRC Press, Boca Raton, Florida, pp. 129–158.

Stobbs, L.W. and Van Schagen, J.G. (1986) Reconditioning of ELISA plates. *Phytopathology* 76, 483–486.

Sudarshana, M.R. and Reddy, D.V.R. (1989) Penicillinase-based enzyme-linked immunosorbent assay for the detection of plant viruses. *Journal of Virological Methods* 26, 45–52.

Sutula, C.L., Gillett, J.M., Morrissey, S.M. and Ramsdell, D.C. (1986) Interpreting ELISA data and establishing the positive-negative threshold. *Plant Disease* 70, 722–726.

Suzuki, N., Shirako, Y. and Ehara, Y. (1990) A simple method for elimination of non-specific reactions in non-precoated indirect and electroblot enzyme-linked immunosorbent assay procedures used for detection of Zucchini yellow mosaic virus. *Annals of Phytopathological Society Japan* 56, 337–341.

Takeuchi, S., Hikichi, Y., Kawada, Y. and Okuno, T. (1999) Direct immunostaining assay, a new simplified technique for detection of tobamoviruses from seeds of green pepper (*Capsicum annuum* L.). *Annals of Phytopathological Society Japan* 65, 189–191.

Terrada, E., Kerschbaumer, R.J., Giunta, G., Galeffi, P., Himmler, G. and Cambra, M. (2000) Fully 'recombinant enzyme-linked immunosorbent assays' using genetically engineered single-chain antibody fusion proteins for detection of *Citrus tristeza virus*. *Phytopathology* 90, 1337–1344.

Tijssen, P. (1985) *Practice and Theory of Enzyme Immunoassays.* Laboratory Techniques in Biochemistry and Molecular Biology 15 (Burdon, R.H. and Knippenberg, P.H. (eds)), Elsevier, Amsterdam.

Torok, V.A. and Randles, J.W. (2001) *Tobacco mosaic virus* RNA as an internal control for duplex RT-PCR assay of pea germplasm. *Australasian Plant Pathology* 30, 227–230.

Torrance, L. (1980) Use of protein A to improve sensitisation of latex particles with antibodies to plant viruses. *Annals of Applied Biology* 96, 45–50.

Torrance, L. (1995) Use of monoclonal antibodies in plant pathology. *European Journal of Plant Pathology* 101, 351–363.

Torrance, L. (1998) Developments in serological methods to detect and identify plant viruses. *Plant Cell, Tissue and Organ Culture* 52, 27–32.

Towbin, H., Staehelin, T. and Gordon, J. (1979). Electrophoretic transfer of proteins from polyacrylamide gels to nitrocellulose sheets: procedure and some applications. *Proceedings of the National Academy of Sciences, USA* 76, 4350–4354.

Van Regenmortel, M.H.V. (1982) *Serology and Immunochemistry of Plant Viruses.* Academic Press, New York.

Van Regenmortel, M.H.V. and Dubs, M.-C. (1993) Serological procedures. In: Matthews, R.E.F. (ed.) *Diagnosis of Plant Virus Diseases.* CRC Press, Boca Raton, Florida, pp. 159–214.

Van Regenmortel, M.H.V., Briand, J.P., Muller, S. and Plaue, S. (1988) *Synthetic Polypeptides as Antigens.* Laboratory Techniques in Biochemistry and Molecular Biology 19 (Burdon, R.H. and Knippenberg, P.H. (eds)), Elsevier, Amsterdam.

Van Vuurde, J.W.L. and Maat, D.Z. (1983) Routine application of ELISA for the detection of lettuce mosaic virus in lettuce seeds. *Seed Science and Technology* 11, 505–513.

Walkey, D.G.A., Lyons, N.F. and Taylor, J.D. (1992) An evaluation of a virobacterial agglutination test for the detection of plant viruses. *Plant Pathology* 41, 462–471.

Xu, Z., Chen, K., Zhang, Z. and Chen, J. (1991) Seed transmission of Peanut stripe virus in peanut. *Plant Disease* 75, 723–726.

Yolken, R.H., Wee, S.B. and Van Regenmortel, M.H.V. (1984) The use of beta-lactamase in enzyme immunoassay for the detection of microbial antigens. *Journal of Immunological Methods* 73, 109–123.

Zettler, F.W., Elliot, M.S., Purcifull, D.E., Mink, G.I., Gorbet, D.W. and Knauft, D.A. (1993) Production of peanut seed free of Peanut stripe and Peanut mottle viruses in Florida. *Plant Disease* 77, 747–749.

Zimmermann, D. and Van Regenmortel, M.H.V. (1989) Spurious cross-reactions between plant viruses and monoclonal antibodies can be overcome by saturating ELISA plates with milk proteins. *Archives of Virology* 106, 15–22.

6 Nucleic Acid-based Testing Methods

Unlike serological detection techniques, which build on the specific interaction between antibodies and the protein surface of disease agents, the molecular assay methods rely on the detection of their nucleic acids.

All nucleic acid-based (NAB) detection methods depend upon the specific association of complementary nucleic acid strands, held together by hydrogen bonds between the bases guanine (G) and cytosine (C) and between adenine (A) and thymine (T) or uracil (U). When the hydrogen bonds of a double-stranded DNA (dsDNA) molecule are broken by heat or high pH, the two strands separate, i.e. the dsDNA is denatured. On removal of the heat source or pH extreme, the two single strands of DNA (ssDNA) re-associate, or re-nature, into the dsDNA configuration. When complementary strands from different sources associate, it is called hybridization. Under specified conditions the latter form of association occurs *in vitro* and is utilized in nucleic acid hybridization (NAH) for detection of specific nucleic acid sequences. Complementary association of two ssDNA molecules also occurs *in vitro* in the polymerase chain reaction (PCR) procedure, but only between short DNA strands (oligonucleotides) and a target DNA (called annealing). Two such oligonucleotides, the primers, are designed so as to anneal to their specific binding sites of a DNA strand, defining the beginning and the end of a target-DNA fragment. Subsequently, enzymatic synthesis of complementary copies of the DNA fragment is possible.

Among NAB methods, NAH and PCR are most commonly used for plant virus and viroid diagnosis. NAH was first used for these purposes at the beginning of the 1980s, but, after the invention of the polymerase chain reaction technique in the mid-1980s, this method has become the most-used molecular tool for detection and identification of plant viruses and viroids.

Both methods are remarkable for their detecting specificity, and PCR for its extreme sensitivity compared with serological or biological methods.

Abbreviations and special terms used in this chapter and their meaning are listed in Table 6.1.

Table 6.1. Abbreviations and terms used in Chapter 6.

A – adenine
aa – amino acid
Amplicon – PCR product
AmpliDet RNA – NASBA assay employing molecular beacons
Amplificate – PCR product
AMV RT – RT derived from *Avian myeloblastosis virus*
Annealing – primer binding to target
bp – base pair
C – cytosine
cDNA – complementary DNA/copyDNA
CI – cylindrical inclusion body
Codon – nucleotide triplet coding for either an amino acid or a chain termination
cRNA – complementary RNA/copyRNA
Degenerate primers – mixture of different primers
Denaturation – separation of double strands into single strands
DEPC – diethylpyrocarbonate
DIG – digoxigenin
DMSO – dimethyl sulphoxide
DNA – deoxyribonucleic acid
dNTPs – deoxyribonucleotide triphosphates
d-PCR – duplex PCR
dsDNA – double-stranded DNA
DTT – dithiothreitol
dTTP – deoxy-thymine nucleotide triphosphate
dUTP – deoxy-uridine nucleotide triphosphate
EDTA – ethylenediamine tetra-acetic acid
Exon – translated DNA sequence
Extension – synthesis of DNA strand in PCR
G – guanine
GIT – guanidinium isothiocyanate
IC – immunocapture
Intron – non-translated DNA sequence
kb – kilobases
Klenow fragment – modified polymerase
M-MLV RT – RT derived from *Moloney murine leukemia virus*
Molecular beacon – fluorescence-producing probe
M-PCR – multiplex PCR
mRNA – messenger RNA
NA – nucleic acid
NAB – nucleic acid based
NAH – nucleic acid hybridization
NASBA – nucleic acid sequence-based amplification
NASH – nucleic acid spot hybridization
nt – nucleotide
NTPs – ribonucleotide triphosphates
Oligonucleotide – short nucleotide chain
PAGE – polyacrylamide gel electrophoresis
PCR – polymerase chain reaction
Primer – short nucleotide chains complementary to a specific DNA sequence
Random hexanucleotides – short chains, 6 nts long, with randomly varying sequences
RE – restriction enzyme
RFLP – restriction fragment length polymorphism
RNA – ribonucleic acid
RNase – (ribonuclease) ribonucleic acid-degrading enzyme
RNase H – RNase that degrades the RNA strand of an RNA : DNA hybrid
RT – reverse transcriptase/transcription
RTF-RT-PCR – real-time fluorescent RT-PCR
Scorpion primer – fluorescence-producing probe linked to a primer
SDS – sodium dodecyl sulphate
SSC – standard saline-citrate buffer
SSCP – single-strand conformation polymorphism
ssDNA – single-stranded DNA
ssRNA – single-stranded RNA
T – thymine
T7 RNA polymerase – a polymerase derived from bacteriophage T7
TAE – TRIS-acetate-EDTA buffer
Taq – *Thermus aquaticus* DNA polymerase
TaqMan probe – fluorescence-producing probe
TBE – TRIS-borate-EDTA buffer
TE – TRIS-HCl-EDTA buffer
Tm – melting temperature
tPA, human – human tissue-type plasminogen activator
TPS – true potato seeds
tRNA – transfer RNA
Tth – *Thermus thermophilus* DNA polymerase
U – uracil

6.1. Nucleic Acid Hybridization

Synthesis of DNA that is complementary to (i.e. the reverse transcript of) viral RNA became possible in the 1970s. Complementary DNA (copy DNA or cDNA) recognizes its viral RNA counterpart and hybridizes to it. This specific combining capacity facilitates its use for detection and diagnosis in a way similar to that of serological detection. A DNA or an RNA probe complementary to the RNA of a virus or viroid can be mass-produced by recombinant DNA technology and cloning in bacteria or by PCR. The identifying probe was earlier made radioactive by incorporation of an isotope, e.g. ^{32}P, followed by visualization of the result on an X-ray film. However, the use of the non-radioactive reporter molecules biotin or digoxigenin for labelling of nucleic acid probes, followed by immunological detection and a chemiluminescent or colorimetric visualization, has proved to be as sensitive as radioactive probes.

Originally the term 'hybridization' was proposed to describe DNA–RNA hybrids, but today the term hybridization includes the formation of DNA–DNA, DNA–RNA, or RNA–RNA complexes (Singh and Dhar, 1998).

Three types of molecular hybridization are of interest for virus/viroid detection:

- *Solution hybridization*, in which both probe and target are in solution during the hybridization process, is mostly used in basic studies (Hull, 1993, 2002), although a kind of solution hybridization takes place in some techniques for 'gel-free' analysis of PCR products (Section 6.2.6).
- *Filter hybridization*, where the target is immobilized prior to the hybridization, is the most common technique and will be dealt with in the following.
- *In situ hybridization* allows specific nucleic acid sequences to be detected directly in preserved cells or tissue sections.

In filter hybridization or dot-blot hybridization, also called nucleic acid spot hybridization (NASH), the target nucleic acid is immobilized on a nitrocellulose sheet or, preferably, a nylon membrane and is then hybridized to a labelled, specific nucleic acid probe. As the technique requires certain laboratory facilities and expertise, it may not be applicable for routine use in all seed-health testing laboratories. However, in cooperation with a specialized laboratory, plant or seed extract may be applied onto a membrane either as drops of crude or purified material or as tissue print (as in the first step of DIBA or TBIA). After drying, the membrane is sent to the laboratory for hybridization and detection with a cloned and labelled specific probe of DNA or RNA (riboprobe). Such 'remote' testing is actually practised in some places.

The sensitivity of the test is similar to that of ELISA but the specificity is higher. When using serological detection, only a few per cent of the genomic information of the virus is involved, whereas by nucleic acid hybridization much more of the genetic information of the virus nucleic acid can be utilized. For viroids, due to their lack of a protein coat, NAH would be one of the detection methods.

Details of the principles and techniques of NAH can be found in, for example, Hull (1993, 2002) and Singh and Nie (2002).

6.1.1. General principles and components of nucleic acid spot hybridization (NASH) assays

The principles of NAH are generally the same for the three types mentioned above. In the following, the principles of NASH will be dealt with. These include the steps: preparation of probes, sample preparation (nucleic acid extraction), sample denaturation and its immobilization on a membrane, prehybridization and hybridization, washing of the membrane, and detection of hybridized probes.

The NASH version dealt with here is widely applied. It uses digoxigenin as the reporter molecule and serological detection of hybridized probes, followed by colorimetric or chemiluminescent visualization. Hybridization tests without nucleic acid extraction, such as tissue print (TP) NAH, are also discussed in this section.

Probes

The DNA or RNA probe is a nucleic acid, usually transcribed from the target nucleic acid, which for the majority of plant viruses, and all viroids, is plus-sense single-stranded RNA (ssRNA). Normally a large quantity of probe is produced by cloning or by PCR – for details, see Hull (1993) and Sambrook and Russell (2001). An NAH probe must be labelled with a reporter molecule that can be detected after the hybridization. For many purposes, such as routine detection of plant viruses or viroids, labelling with non-radioactive compounds is preferable. Radioactively labelled probes have to be renewed regularly due to decay of the isotope, special laboratory facilities are required, and storage and disposal of radioactive waste can be problematic.

The most-used non-isotopic labels are biotin (vitamin H) and the hapten digoxigenin (DIG), derived from the *Digitalis* plant. Both compounds are detected by an indirect procedure, using a labelled, specific-binding protein, either streptavidin (for biotin) or anti-digoxigenin serum (Fig. 6.1) and visualized by a chemiluminescent or a colorimetric reaction. The biotin–streptavidin system is very efficient, but false positives or high background may sometimes occur when sap extracts are used, due to endogenous biotin in plant tissue (Hull, 1993). As the *Digitalis* plant is the only natural source of digoxigenin, the anti-DIG antibody does not bind to other biological materials.

The incorporation of a reporter molecule into a probe can be done in different ways (see Hull, 1993; Sambrook and Russell, 2001). The so-called random-primed labelling is efficient and relatively easy, but labelling by means of PCR is now more common (S. Winter, DSMZ-Plant Virus Collection, Germany, 2003,

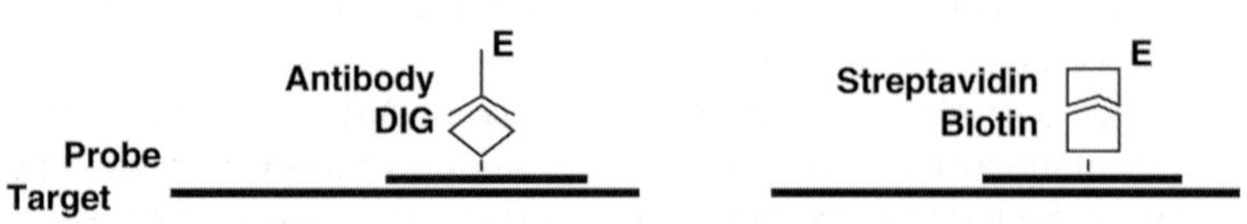

Fig. 6.1. Detection of a hybridized probe, labelled either with digoxigenin (DIG) or biotin, whose presence is detected by antibody–enzyme or biotin–enzyme conjugate. Visualization is by either a colorimetric or a chemiluminescent substrate. E, enzyme.

personal communication). For probe preparation, see Section 6.1.2. Use of biotin or DIG labelling has gained increasing popularity and also made NASH more attractive to less well-equipped laboratories; the sensitivity of systems using these labels equals those using radioactive labels, DIG-labelled probes can be stored for a long time (months or years) at –20°C without losing sensitivity, no facilities for handling radioactive material are required, and the reagents are inexpensive. According to Roche Diagnostics™ (2003, personal communication), the DIG technology is a proprietary technology of this company.

Probe lengths from 200 bases to a few Kb are most commonly used. For viroids, whose genomes are only from 246 to 375 bases long (Hanold, 1993), full-length probes are usually made (Romero-Durbán *et al.*, 1995; Nakahara *et al.*, 1999; Palacio-Bielsa *et al.*, 1999). For viruses, use of probe lengths of size orders 0.5–3 kb has been reported (Singh and Singh, 1995; Saldarelli *et al.*, 1996; Dalmon *et al.*, 2000; Grieco *et al.*, 2000). As already mentioned, probes of cDNA or cRNA can be used. According to several reports cRNA probes are more sensitive than DNA probes and are widely used. On the other hand, DNA probes are cheaper to produce and can be handled without the precautions needed to avoid the degradation of ssRNA (Palacio-Bielsa *et al.*, 1999). The specificity of the probe may be narrow (virus- or strain-specific) if it is made specific to variable parts of the genome, or broad (genus-specific) if the probe is specific to conserved parts. Transcribing and cloning of nucleic acids require some expertise in and facilities for molecular work, and are best carried out in a specialized laboratory. A number of virus- and viroid-specific probes are commercially available from, for example, DSMZ-Plant Virus Collection, Germany or ATCC, USA (Appendix 3). Unlike probe preparation, the labelling of probes is somewhat simpler and can be done in most laboratories when using a commercially available labelling kit, following the manufacturer's instructions. Brief accounts of the DIG-labelling procedures are given below.

DIG PROBE PREPARATION. The digoxigenin label is available as the so-called DIG-dUTP, in which the digoxigenin molecule is linked to the nucleotide deoxy-uridine triphosphate via a spacer arm containing 11 carbon atoms. In the case of cDNA probes, the DIG-labelled nucleotides are normally incorporated into the nucleic acid probes while being synthesized from template DNA by means of one of different types of DNA polymerase. When the DIG-dUTP : dTTP (deoxythymine triphosphate) ratio is optimal in the labelling mixture, the probe synthesized will contain a DIG molecule in every stretch of 20–25 nucleotides. If DIG molecules were closer to each other, steric hindrance would prevent the large anti-DIG antibody from recognizing and binding to the labelled probe (Roche Diagnostics, 2000).

Two methods for preparing labelled DNA probes are commonly used. By *random-primed labelling* the probe is synthesized by a modified polymerase, called the Klenow fragment, in the presence of oligonucleotides of six bases differing in base compositions (hexamers), DIG-dUTP, dTTP and the three other deoxyribonucleotide triphosphates (dNTPs). This labelling method requires a purified DNA template in relatively large quantity. By *PCR labelling*, which has become the preferred method, DIG-dUTP is added to the reaction mixture, from which a large amount of probe is synthesized by a thermostable polymerase. PCR labelling is quick, only a relatively small amount of template-DNA is required and the purity of the template DNA is not as critical as for random-primed labelling. However, according to Singh and Nie (2002), the template should be prepared first by normal PCR and purified, and then be used for the incorporation of the label in a second PCR.

In the case of cRNA probes, DIG-UTP (instead of DIG-dUTP) is incorporated into cRNA during an *in vitro* transcription from a cDNA template by means of a DNA-dependent RNA polymerase. The cDNA template must contain a promoter for the polymerase, and high purity of template is required (Roche Diagnostics, 2000).

Evaluation of labelling efficiency. Before taking a probe into use, the efficiency of the label reaction must be checked. For PCR-labelled probes it is done by electrophoresis; probes labelled by random priming and cRNA probes should be checked by direct detection. By the direct method a series of dilutions prepared from the DIG-labelled probe is spotted directly on a membrane and visualized with the standard DIG detection procedure.

Details and complete protocols for DIG probe preparations and their evaluation are excellently described in the *DIG Application Manual for Filter Hybridization* (225 pp.) (Roche Diagnostics, 2000). Supplementary/basic information can be found in Sambrook and Russell (2001).

Preparation of target RNA

For general aspects of sample preparation, see also Section 5.2.2.

The nucleic acid of the majority, if not all, of the important seed-transmitted viruses and viroids is plus-sense, single-stranded RNA, and therefore only RNA extraction will be presented here.

Extraction of total RNA from seed or plant samples is the most laborious part of the NAH detection procedure. However, procedures without RNA extraction, such as more or less direct spotting of crude plant sap onto membranes or use of tissue-printing techniques (see end of this section) have been developed. But these methods are best suited for tissues with relatively high concentration of virus or viroid.

NAH is best used instead of ELISA for detecting viruses that are poor immunogens or lack a protein coat. As viroids are serologically non-detectable, NASH or PCR is preferred for these pathogens. In order to detect viroids, whose concentration is often low in tissue, such as in seeds, or when testing pooled samples, the best sensitivity is obtained by extracting the RNA from the samples, using a methodology like that described below, before spotting onto membranes.

It should be noted that viruses and viroids located in seeds outside the embryo, and thus not transmissible, are also detected when testing whole seeds by NAH (see Sample preparation, Section 5.2.2).

TOTAL NUCLEIC ACID EXTRACTION. Both during and after extraction from plant material, RNA must be protected against ribonucleases (RNases) by using RNase-free tools and chemicals (see Appendix 2). Handling the material at low temperatures throughout the extraction procedure further improves the result; thus liquid nitrogen is often used to freeze tissue before homogenization with extraction buffer. Many different buffers have been reported for use in plant RNA extraction. Here buffers and solutions used in the protocol following this section will be dealt with.

The extraction buffer (GIT buffer) contains guanidinium isothiocyanate, which is an effective de-proteinizer and RNase inhibitor. By use of this chemical, the use of phenol and chloroform, the two classical and effective but highly toxic chemicals, is avoided. The antioxidant 2-mercaptoethanol, which inactivates RNases irreversibly (Seal and Coates, 1998), and ethylene diamine tetra-acetic acid (EDTA), which is a metal-chelating agent and an enzyme inhibitor, are compounds added to the extraction buffer. The detergent Triton X-100 is added to the homogenate for solubilizing cell membrane components. Before precipitating the RNA, the homogenate is incubated with sodium acetate to reduce co-precipitation of contaminants, e.g. oligosaccharides, and then centrifuged. The resulting supernatant is then mixed with isopropanol and the mixture is allowed to stand for a few min before precipitation of the RNA by centrifugation. Isopropanol precipitation following sodium acetate treatment results in RNA with fewer contaminants than ethanol precipitation (Sambrook and Russell, 2001). Subsequent washings with 80% ethanol further remove contaminants. The ethanol-precipitated pellet is allowed to dry but must not dry completely, as this greatly decreases its solubility (Sambrook and Russell, 2001).

After dissolving the pellets in water, the RNA is ready to be spotted on to a membrane. If needed, RNA dissolved in water can be stored at –80°C. Alternatively, RNA can be stored at – 20°C if dissolved in deionized formamide (Sambrook and Russell, 2001).

Commercially available RNA extraction kits, such as Gene Releaser™ (Bio Ventures Inc.) or RNeasy Miniprep® (Qiagen Inc.), are efficient but add to the costs. A rough estimate of the RNA yield can be made spectrophotometrically. When a solution is measured at 260 nm in a cell of 1 cm path length, one absorbance unit corresponds to an approximate content of 40 μg ssRNA per ml.

Other methods of RNA extraction are described in Sections 6.2.2 and 6.2.3.

DENATURING OF RNA SAMPLE. Denaturing, i.e. separating double-stranded nucleic acids, before the hybridization is not necessary for the majority of plant viruses, the RNA of which is single-stranded already (Rybicki, 1998). However, viroids, which are, in a way, double-stranded due to internal, partial base pairing inside their circular RNA, must be denatured, and can be so by heating in a tube to above their melting temperature, which is 58°C in low salt buffers (Hanold, 1993). A subsequent rapid cooling of the tube in ice water will keep the RNA denatured (the RNA strands remain as open circles) for some time. For RNA viruses, the same treatment may help to increase the hybridization signal (Pallás *et al.*, 1998), probably by breaking down possible strong secondary structures of RNA (see Denaturing in Section 6.1.2).

Membranes

Nylon membranes are superior to nitrocellulose membranes (NCMs), as they bind nucleic acid more effectively. Nucleic acid may leach from NCMs during hybridization and washing and their mechanical strength is less than that of nylon membranes. Unmodified or positively charged forms of nylon membranes are available. The latter type has the highest binding capacity, but also tends to give a higher level of background, which, however, can be reduced by using higher concentrations of blocking agent during prehybridization and hybridization (Sambrook and Russell, 2001). For dot blotting and RNA fixing on to membrane, see Section 6.1.2.

Prehybridization and hybridization

Before hybridization, the membranes are so-called prehybridized to block the sites of the membrane that might bind the probe unspecifically. This is done by incubation in warm hybridization solution *without* probe for a few hours under gentle agitation. Prior to adding the probe to a new portion of hybridization solution, the probe is denatured to be able to react with the target RNA (described in Prehybridization and hybridization, step 3, Section 6.1.2). The membrane is now incubated with the probe in the hybridization solution for several hours (e.g. overnight) under gentle agitation at a fixed temperature to allow the probe to hybridize with the target RNA. An optimal hybrid formation between probe and target depends on certain reaction conditions both during the hybridization and in the subsequent washing steps.

Conditions that influence the rate of association (hybridization) and the stability of the products are the same as those controlling the denaturation (melting) of a double-stranded nucleic acid (Hull, 1993):

- *Temperature*. The melting temperature (Tm) is the temperature at which 50% of the sequences are denatured. The major factors affecting the Tm are the composition of the nucleic acid, the concentration of salt in, and the pH of, the solution, and the presence of materials that can disrupt hydrogen binding, such as formamide.
- *Nucleic acid composition*. G + C base pairs have three hydrogen bands, and are therefore more stable than A + T (or A + U) base pairs (bp), which have two. For perfectly base-paired DNA in 1 × SSC (SSC = 0.15 M NaCl, 0.015 M Na citrate), the Tm is related to the G + C content by:

  $$\mathrm{Tm} = 0.41\ (\%\ \mathrm{G + C}) + 69.3$$

- *Salt and pH*. The salt concentration has a marked effect on the Tm of a duplex. The Tm increases by almost 16°C for each tenfold increase in the concentration of monovalent cation over the range of 0.01 to 0.1 M, but increases less at higher concentrations.

  The Tm is insensitive to pH in the range 5 to 9, but most duplex DNAs are fully denatured at pH 12.
- *Organic solvents*, e.g. formamide and also urea, lower the Tm; thus for each per cent of formamide, the Tm is reduced by 0.7°C.
- *Base-pair mismatch*. Mismatched sequences are less stable than perfectly base-paired duplexes. In nucleic acids of more than 100 bp, a mismatch of 1% reduces the Tm by about 1°C. Thus, mismatching can be assessed by varying the hybridization conditions (see 'stringency' below).

- *RNA : RNA and RNA : DNA duplexes.* The Tm of double-stranded RNA (dsRNA) is significantly higher than that of dsDNA. A general value for the Tm of DNA in 1 × SSC is about 85°C and of RNA is close to 110°C. The Tm of RNA : DNA hybrids is about 4 to 5°C higher than that for DNA : DNA duplexes under the same conditions.

This leads to the concept of *stringency of hybridization*, which relates to the effect of hybridization and/or washing conditions on the interaction between complementary nucleic acids, which may be incompletely matched (Hull, 1993). High stringency allows few mismatches (the Tm goes down), while low stringency allows a higher number of mismatches between the two strands (the Tm goes up).

The stringency increases by lowering the salt concentration and/or increasing the temperature, and decreases by raising the salt concentration and/or lowering the temperature.

During hybridization the optimal association temperature is around 25°C lower than the Tm. Inclusion of formamide or urea in the hybridization solution lowers the Tm, and thereby the association temperature. In the protocol (Section 6.1.2), a so-called high SDS (sodium dodecyl sulphate) buffer is used as hybridization solution. The high SDS content (7%) contributes to suppressing background while preserving a high sensitivity (Sambrook and Russell, 2001).

After hybridization, the membranes are washed, first at low stringency to remove unbound probe and then at high stringency to remove loosely associated (mismatching) probe. Hybridization and washing conditions determine both the sensitivity and the specificity of the test. Low-stringency washing increases the sensitivity, but may also increase the background hybridization. High-stringency washing increases the specificity (Dijkstra and de Jager, 1998).

After hybridization and washing, the membrane may be stored dry until the immunological detection; however, if a reprobing of the target nucleic acid (NA) with another probe is desired after the recording, the first probe cannot be removed from the membrane unless the membrane remains wet at all stages after hybridization and washing.

There are many factors that influence the result of a hybridization test. For details of these factors and theoretical aspects of hybridization, see Hull (1993, 2002) and Sambrook and Russell (2001).

Immunological detection

Presence of a DIG-labelled probe after hybridization and washing, indicating the presence of target RNA, is detected by an anti-DIG–enzyme conjugate and a substrate that results in either a chemiluminescent or a colorimetric visualization. Chemiluminescent visualization is reported to be more sensitive than colorimetric visualization (Pallás *et al.*, 1998).

The detection consists of four steps:

- Washing (with the detection washing buffer) and blocking of the membrane.
- Reaction with anti-DIG-alkaline phosphatase to localize the probe–target hybrids.
- Washing off unbound antibody–enzyme conjugate.
- Detection of DIG by chemiluminescent or colorimetric visualization.

These steps are described in the protocol (Section 6.1.2) and in detail in the *DIG Application Manual for Filter Hybridization* (Roche Diagnostics, 2000).

As mentioned earlier, removing (stripping) of the probe from the membrane after recording and reprobing is possible, provided that the membrane is not allowed to dry. Procedures for stripping after both luminescent and colorimetric recording are described in the *DIG Application Manual.*

Hybridization tests without nucleic acid extraction

Several procedures for hybridization tests without the use of the laborious RNA extraction have been described. Saldarelli *et al.* (1996) used a reduced sample treatment in a large-scale certification programme, where five viruses were detected in tomato with the use of a cRNA probe mixture. The tissue was ground in an NaOH-EDTA solution and the homogenate spotted on to membranes before prehybridization and hybridization.

In tissue-print hybridization (syn. imprint hybridization), the blotting technique developed for tissue blotting immunoassay (TBIA) is employed. Freshly made cuts of plant parts are blotted (printed or squashed) directly on to dry or pretreated membranes instead of RNA extracts, immobilized and then prehybridized and hybridized with the probe. For detection of the hybrids, most workers seem to prefer the DIG system, using either a colorimetric or a chemiluminescent visualization. The tissue-print (TP) method has gained popularity as it saves time and is well suited for large-scale NAH testing. Podleckis *et al.* (1993) printed tissue cuts on SSC-prewetted membranes for the detection of *Potato spindle tuber viroid* (PSTVd, *Pospiviroid*) and *Apple scar skin viroid* (ASSVd, *Apscaviroid*). Más and Pallás (1995) used TP-hybridization for studying plant virus movement in plants. Romero-Durbán *et al.* (1995) detected four viroids by TP-hybridization in seven hosts by printing leaves and stems on nylon membranes, pretreated with $6 \times$ SSPE (standard NaCl-phosphate-EDTA buffer). They concluded that the method could be used for routine indexing of a large number of samples, but also that, like other detection methods, the limit of detection depends on the viroid and viroid-host tissue. Thus, one of the four viroids, *Avocado sunblotch viroid* (ASBVd, *Avsunviroid*), though present in high concentration, was difficult to detect in avocado tissue by TP-hybridization. Similarly, Palacio-Bielsa *et al.* (1999) found TP-hybridization extremely simple for quick analyses of large numbers of *Citrus* samples for viroids, but observed limited sensitivity for tissues of certain hosts.

Stark-Lorenzen *et al.* (1997) reported TP-hybridization as a rapid and sensitive method for PSTVd detection in tomato plants. In comparing ELISA, TBIA and TP-hybridization for detection of *Tomato yellow leaf curl virus* (TYLCV, *Begomovirus*), Dalmon *et al.* (2000) found TBIA to be less sensitive and ELISA and TP-hybridization to be equally sensitive. The authors recommend the latter method because of its ease of sample preparation. Hurtt and Podleckis (1995) tested seedlings raised from pear seeds for seed transmission of ASSVd by TP-hybridization, using prints from seedling petioles. One of the advantages of TP-hybridization is that samples need not be ground and manipulated, thus reducing the risk of cross-contamination.

6.1.2. Non-isotopic nucleic acid spot hybridization, protocol

The protocol described below has been successfully used for detection of PSTVd in true potato seeds (TPSs) (Borkhardt *et al.*, 1994). The procedure involves the use of neither the harmful compounds phenol and chloroform nor radioactive probe labelling. A digoxigenin-labelled cDNA probe is used and the presence of bound probe is detected by an immunological reaction and visualized by a chemiluminescent or a colorimetric assay. The procedure describes the detection of PSTVd in TPS seedlings, following the method of Borkhardt *et al.* (1994) and Roche Diagnostics (2000), with some modifications derived from Pallás *et al.* (1998), Webster and Barker (1998) and Dijkstra and de Jager (1998). By this method, one infected TPSs among 150 TPSs could be detected. The protocol can be adjusted to detect other viroids or viruses in other hosts.

By using chemiluminescent detection, a higher sensitivity is obtained than by use of colorimetric detection. Further, whereas the natural green-brownish colour of leaf sap interferes with colorimetric detection methodology, visible sap-extract residues do not interfere with chemiluminescent detection (Pallás *et al.*, 1998).

Since parts of the following protocol, especially probe labelling and reprobing procedures refer to descriptions in the *DIG Application Manual for Filter Hybridization* (hereafter DIG manual) from Roche Diagnostics (2000), *access to this manual is required.* Among its several and detailed procedures, the manual also contains a comprehensive appendix on trouble shooting.

For other types of viroid assays, see the introduction to Chapter 4 and Sections 6.2 and 7.1.

EQUIPMENT AND MATERIALS.

Equipment

a. Thermal cycler for PCR probe labelling (see Sections 6.2.2 and 6.2.3) (optional).

b. Electrophoresis apparatus for checking PCR-labelled probes (see Section 6.2.4) (optional).

c. Plastic boxes with lids, approx. 25 cm × 25 cm × 10 cm, for germinating seeds (similar to the boxes described in Section 4.4.4 and shown in Fig. 4.8).

d. Mortars and pestles, or another homogenization device.

e. Micro-centrifuge at 15,000 *g*, with cooling (optionally, centrifuge in refrigerator using pre-cooled rotor).

f. Eppendorf tubes, sterile, 1.5 ml.

g. Adjustable pipettes (1 ml, 100 μl, 10 μl) and tips.

h. Water-bath with temperature regulation.

i. Ice-water-bath.

j. Nylon membranes, positively charged (Roche Diagnostics™ or similar). Must be handled with gloves or clean forceps.

k. Hybridization flasks.

l. Hybridization oven in which the flasks rotate (see **Note 1** after procedure).

m. Orbital shaker.

n. Oven with thermostat.

o. UV cross-linker (e.g. Stratalinker™, Stratagene, La Jolla, CA, USA) or an appropriate UV lamp or transilluminator, which would also require an empirically determined exposure time (see **Note 3** after procedure). **Caution:** UV light is dangerous, especially to the eyes; wear protective goggles or mask and shield the UV light source to minimize radiation.

p. Glass dishes for incubating membranes. These and other glassware and containers must be diethylpyrocarbonate (DEPC)-treated and sterilized before use – see Appendix 2, p. 250.

q. Thermal sealer for sealing plastic bags.

r. X-ray photo cassette.

s. Standard laboratory equipment, such as magnetic stir plate, vortexer, glassware, tube racks, ice machine or similar, forceps, gloves (powder-free), filter paper.

Biological material

a. TPS, test sample.

b. TPS, known healthy for negative control.

c. TPS infected with PSTVd or other known PSTVd-containing material, for positive control, e.g. fresh or dehydrated leaf tissue.

Reagents, solutions and buffers

Note that reagents and chemicals must be RNase-free products or be made RNase-free before use – see Appendix 2, p. 250.

General

Double-distilled H_2O (dd H_2O), DEPC-treated – see Appendix 2, p. 250.

Probe template

PSTVd cDNA, cloned in bacteria or PCR-amplified. Commercially available from ATCC, USA, or DSMZ-Plant Virus Collection, Germany (Appendix 3).

DIG labelling by PCR

a. PCR DIG labelling kit containing: *Thermus aquaticus* (*Taq*) polymerase, PCR synthesis mix, PCR buffer, dNTP solution, control template (human tPA), control primer mix for human tPA (Roche Diagnostics™). DNA purification kit: see item a below.

b. Primers specific for amplifying the PSTVd probe template, adjusted to 10–50 μM; obtained custom-built from a commercial laboratory or synthesized on own synthesizer:

Upstream:
5′-GGA TCC CCG GGG AAA CCT GGA GCG-3′ (nucleotides 87–110).
Downstream:
5′-GGA TCC CTG AAG CGC TCC TCC GAG C-3′ (nucleotides 68–92)
(Nie and Singh, 2001).

For the next items (c–g), see Agarose gel electrophoresis protocol, Section 6.2.4, p. 203.

c. DNA molecular weight marker (Roche Diagnostics™ or similar).

d. Agarose, nucleic acid grade.

e. Running buffer: TRIS-acetate-EDTA (TAE) or TRIS-borate-EDTA (TBE).
f. Tracking dye.
g. Ethidium bromide. **Caution:** ethidium bromide is a powerful mutagen; wear double layers of protecting gloves.

Alternative labelling method:

DIG labelling by random priming (Roche Diagnostics™ reagents – see DIG manual)
a. DNA template purification kit (see DIG manual).
b. DIG-High Prime Kit® containing: DIG-High Prime labelling mix (Klenow enzyme, DIG-dUTP, dNTPs and random primer mix).
c. DIG-labelled control DNA.
d. DNA dilution buffer.

Extraction
a. GIT buffer (guanidinium isothiocyanate-TRIS-Mg-EDTA) – see Appendix 2, p. 250. **Caution:** toxic, wear gloves.
b. TRIS-HCl.
c. $MgCl_2$.
d. Na_2-EDTA.
e. 2-Mercaptoethanol. **Caution:** highly toxic, possibly mutagenic, use fume cupboard.
f. Triton X-100.
g. Sodium acetate, 3 M.
h. Isopropanol. **Caution:** toxic, flammable.
i. Ethanol 80%.
j. Denaturing solution: 20 × SSC (standard saline citrate buffer 20 ×), containing 15% formaldehyde. **Caution:** formaldehyde solution and vapours are highly toxic. SSC – see Appendix 2, p. 250.

Hybridization
a. Solution for prehybridization and hybridization ('Hyb solution') – see Appendix 2, p. 251. Items b–h are 'Hyb solution' ingredients. *Alternatively*, use the ready-to-use DIG Easy Hyb® solution from Roche Diagnostics™.
b. Formamide, deionized.
c. 30 × SSC – see Appendix 2, p. 251.
d. Sodium phosphate buffer – see Appendix 2, p. 251.
e. Blocking stock solution (see Immunological and chemiluminescent detection below).
f. Yeast transfer RNA (tRNA), 10 mg/ml (e.g. Ambion 7119).
g. *N*-lauroylsarcosine.
h. SDS (sodium dodecyl sulphate) – **Caution:** harmful, see Appendix 2, p. 251.

Immunological and chemiluminescent detection
a. Maleic acid buffer – see Appendix 2, p. 252.
b. Washing buffer: Maleic acid buffer containing 0.3% Tween 20.
c. Blocking stock solution containing 10% blocking reagent – see Appendix 2, p. 252.

d. Blocking solution: Maleic acid buffer, containing 1% blocking reagent.

e. $MgCl_2$.

f. Anti-DIG–AP (anti-digoxigenin–alkaline phosphatase conjugate, Roche Diagnostics™).

g. Detection buffer – see Appendix 2, p. 252.

h. CSPD®, chemiluminescent substrate (11.6 mg/ml) (Roche Diagnostics™). Store at 4°C in the dark. **Caution:** CSPD may be harmful.

i. X-ray film (Kodak XAR-5® or similar), fitting the photo cassette.

Colorimetric detection

For colorimetric detection, replace items h and i with:

j. NBT/BCIP (nitroblue tetrazolium/5-bromo-4-chloro-3-indolyl phosphate) stock solution (Roche Diagnostics™). Also available as tablets, and from other manufacturers.

k. NBT/BCIP substrate solution: 200 μl of stock solution in 10 ml detection buffer; prepare the day to be used, store in dark.

l. TE buffer (10 mM TRIS-HCl, pH 8.0, 1 mM Na_2-EDTA).

PROCEDURE.

Germination of seeds

1. Place the required number of seeds (see Section 7.2) from the test sample, e.g. on moist filter paper in plastic boxes (see Fig. 4.8 for arrangement) for germination at room temperature for 5–10 days under lid in darkness.

2. Known healthy (100) and known PSTVd-infected (100) TPSs are similarly germinated. Alternatively, other known PSTVd-infected material can be used.

Probe preparation: template DNA. cDNA transcribed from PSTVd RNA and cloned, or PCR-amplified, may be obtained from a specialized laboratory or a commercial source.

For DIG labelling, use one of the two methods below.

DIG labelling by PCR. PCR labelling requires a lesser amount of template than random primed labelling, and purity of the template is less critical. However, Singh and Nie (2002) recommend that the template be prepared first by normal PCR and purified (see random priming below) before being used for incorporation of the label in a second PCR.

1. Follow the PCR labelling procedure in the DIG manual and see also the reverse-transcription polymerase chain reaction (RT-PCR) protocol (Section 6.2.3). Two of the three PCR tubes require the use of specific PSTVd primers. Use 1 μM of each primer per tube (of 50 μl).

2. Check the synthesis of the labelled probe by running a portion (5 μl) of each reaction on an agarose minigel in *TAE* or *TBE buffer*, along with a DNA molecular-weight marker, following the instructions in the DIG manual. See also Section 6.2.4.

3. Evaluate the result as advised in the DIG manual. The PCR-labelled product can be used directly in the hybridization reaction or stored for a short time at 4°C or a long time at –20°C.

DIG labelling by random priming. For random-priming labelling, the purity of the template is critical. It must be free of cloning vector sequences; if not, it can be purified by different methods, e.g. a purification kit, also obtainable from Roche Diagnostics™.

1. Follow the instructions in the DIG manual for template purification, template quantity required and how to carry out random-primed labelling with DIG.
2. Prepare working solutions of: *maleic acid buffer, blocking solution, antibody solution and detection buffer.*
3. To check the synthesis of the labelled probe, prepare serial dilutions of both the labelled probe and the control DIG-DNA in *DNA dilution buffer* (control DIG-DNA and dilution buffer from Roche Diagnostics™).
4. Apply 1 μl spots from each of the dilutions on to a piece of nylon membrane and fix the DNA by UV or baking (DIG manual) (see also Dot blotting and RNA fixing below).
5. Detect the DIG in the spots by following steps **1–9** of Immunological-chemiluminescent detection below. Alternatively, replace steps **5–9** with Colorimetric detection, steps **1–3**.
6. Evaluate the yield of DIG-labelled probe (ng/μl) as guided in the DIG manual. If not immediately using the probe, store it at –20°C.

RNA extraction. During RNA extraction, use gloves and RNase-free glassware and solutions (see Appendix 2, p. 250), and work under general sterile and cold conditions, avoiding dust and draughts. Each composite sample (group) and control sample is treated as follows:

1. Harvest and homogenize seedling material in groups representing 50 seedlings each, in the order: negative control, test samples, positive control material (fewer numbers, especially of control seedlings, may be used and the solution volumes below correspondingly adjusted). For group testing, see Section 7.2.
2. Each batch of 50 seedlings is homogenized in 750 μl of ice-cold *GIT buffer* containing 1 μl *2-mercaptoethanol*, using pre-chilled mortars and pestles or another device for homogenization. Add 2-mercaptoethanol to the GIT buffer just before use.
3. Add 60 μl of 25% *Triton X-100* and homogenize for a further 1 min.
4. Transfer 750 μl of homogenate to an Eppendorf tube and leave it on ice for 15 min.
5. Add an equal amount of *3 M sodium acetate*, mix and leave on ice for another 15 min.
6. Centrifuge the homogenate at 15,000 *g* for 30 min at 4°C and transfer the supernatant to two Eppendorf tubes.
7. Add 0.8 volume of ice-cold *isopropanol* to each tube, mix, leave the tubes on ice for 5 min and centrifuge as before.
8. Carefully decant and discard the supernatant without losing the precipitated pellets.
9. Wash the pellets with cold *80% ethanol*: suspend the pellets, centrifuge at 15,000 *g*, 4°C for 5 min, retain pellets and repeat the procedure once or twice; see **Note 2** below.

10. Air-dry the pellets, dissolve them each in 100 µl *DEPC-treated dd* H_2O and combine the two solutions; see also **Note 2**.

Denaturing

1. Before dotting on membranes, the RNA extract of each group and each control sample is mixed with an equal volume of *denaturing solution*.

2. Heat the mixture for 15 min at 60°C, followed by rapid chilling in ice water.

Dot blotting and RNA fixing

1. Prepare the membrane (use gloves or clean forceps!): cut out a piece of an appropriate size, e.g. 100 cm^2 and, before removing the two layers of protecting paper, mark it into 1 cm × 1 cm squares with a pencil. The membrane below will be sufficiently marked by the pencil's pressure to be visible. For identification of dots, fill in a similar grid on a piece of paper. Cut a corner of the membrane for orientation, and remove the protecting paper.

2. Nylon membranes do not need any prior treatment. Pipette 4 µl per dot of each RNA extract on to the marked squares of the membrane.

3. Fixing of RNA: after the dotted sheet is completely dry, the RNA is fixed to the membrane by (a) baking at 80°C for 2 h or (preferably) (b) UV radiation for 3 min on both surfaces of the membrane (see **Note 3** below).

Prehybridization and hybridization. During prehybridization and hybridization, do not allow the membrane to dry out at any stage.

1. To block the non-specific binding sites, the membrane is prehybridized by placing it in warm '*Hyb solution*', 10 ml/100 cm^2 membrane, in a hybridization flask. Remove air bubbles and close the flask.

2. Incubate under slow rotation of the flasks in the hybridization oven for 4 h at 45°C (see **Note 1** below).

3. Immediately before hybridizing with the DIG-labelled probe, the probe is denatured as follows (for 100 cm^2 membrane):

(a) Withdraw 20 µl (range 5–40 µl) of PCR-labelled probe *or* withdraw random priming-labelled probe corresponding to 200 ng and place it in an Eppendorf tube along with 50 µl of dd H_2O.

(b) Place the tube in boiling water for 5 min, followed by quick chilling in ice water (see also **Note 4** below about reuse of probes).

(c) Immediately add the denatured probe to a tube containing 10 ml of pre-warmed '*Hyb solution*' and mix by inversion. Final concentration of 2 µl/ml (PCR-labelled probe) or 20 ng/ml (random-labelled probe).

4. After the prehybridization step, remove solution completely from the flask and replace immediately with the already prepared mixture of DIG-labelled probe in '*Hyb solution*'. Use 10 ml/100cm^2 membrane, remove air bubbles and allow hybridization by incubation under rotation at 45°C for 16 h (for an alternative incubation set-up, see **Note 1** below).

5. While waiting, prepare the working solutions required for the immunological detection below.

6. Remove the solution from the flask. The hybridization solution can be saved and reused (see **Note 4**). Immediately wash the membrane twice for 5 min with *2 × SSC, 0.1% SDS* at room temperature (low stringency) (10 ml per flask, see **Note 5**).
7. Wash the membrane for 15 min, with 10 ml *0.1 × SSC, 0.1% SDS* at 68°C (high stringency). Add 5 min for warming up time. At the same time, place a sterile flask with another 10 ml of the same solution in the oven.
8. Repeat step **7** with the preheated SSC/SDS solution. Continue immediately with detection of hybridized probe, or store the membrane air-dried before further processing; however, if reprobing after the recording is desired, do not let the membrane dry but continue with detection.

Immunological chemiluminescent or colorimetric detection. The following steps are performed at room temperature with gentle shaking unless otherwise mentioned.

1. Wash the membrane briefly (approx. 2 min) in the *maleic acid washing buffer*.
2. Add *blocking solution* and incubate for 30 min.
3. Dilute anti-DIG–AP conjugate 1 : 10,000 with *blocking solution* and incubate the membrane with 20 ml per 100 cm^2 membrane for 30 min.
4. Unbound conjugate is removed by two washings of 15 min each in *washing buffer*, followed by equilibration for 2–5 min in 20 ml of *detection buffer*.

Chemiluminescent detection
5. Using gloves, dilute the substrate, *CSPD®, 1 : 100 in detection buffer*; 5 ml of freshly made solution per 100 cm^2 membrane are required. Incubate the membrane with the solution in a heat-sealed transparent plastic bag for 5 min in the dark.
6. Cut the plastic bag at the top and pour out the substrate solution (can be reused one to two times when kept sterile (Roche Diagnostics™).
7. Squeeze excess liquid out of the bag, seal again and incubate for 15 min at 37°C. Do not allow the membrane to dry.
8. Make a first exposure to X-ray film at room temperature for approx. 15 min. As luminescence continues for at least 24 h, additional exposures may be taken, if needed.
9. Develop the film and record the results. Positive reactions appear as dark spots (Fig. 6.2).

Store the membranes dry, but keep them wet until reprobing, if this is intended. Procedures for stripping and reprobing are described in the DIG manual.

Colorimetric detection. As an alternative to the chemiluminescent detection (step **5–9** above), a colorimetric detection can be made, as follows (Pallás *et al.*, 1998, slightly modified):

1. After equilibrating (step **4** above), incubate the membrane in a sealed plastic bag in the dark, without shaking, with a freshly made *NBT/BCIP substrate solution*; use approx. 5 ml/100 cm^2 membrane.
2. Observe in dimmed light the development of the bluish-purple colour of positive dots. When the whole membrane starts turning to a bluish-purple colour

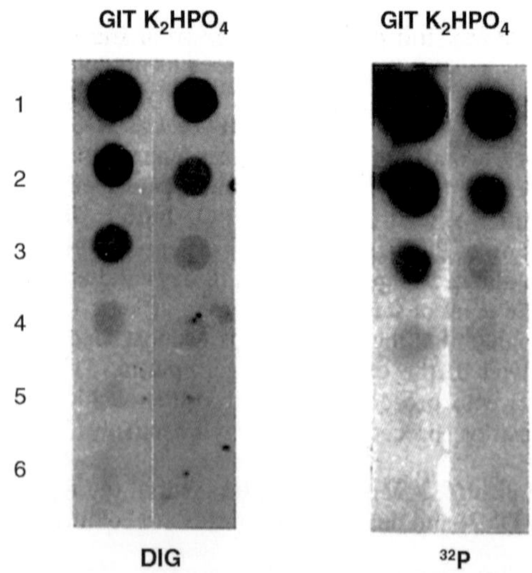

Fig. 6.2. Chemiluminescent NASH detection of seed-borne PSTVd in potato seedlings showing the effects of GIT and K_2HPO_4 extraction procedures and of labelling the probe with DIG (digoxigenin) or ^{32}P (K_2HPO_4 extraction not included in protocol 6.1.2). 1, extracts from PSTVd-infected seedlings (82% infection) diluted 1 : 2 in denaturing agent; 2–5, fivefold serial dilutions of (1) in similarly prepared virus-free extracts; 6, virus-free seedling extracts. From Borkhardt *et al.* (1994) with permission from the copyright holder © Potato Research.

(15 min to 12 h), pour out the substrate solution and wash the membrane with *TE buffer*, which stops the colour reaction.

3. The results can be documented by photocopying or by photographing the wet membrane.

Processed membranes can now be stored dry, provided reprobing is not desired (see step **9** above).

For trouble shooting in the overall procedure, see the DIG manual.

Determination of rate of seed infection. To determine the per cent of infected seeds in the sample, use the *maximum likelihood estimate* (MLE) formula (see Section 7.2).

Notes

1. As an alternative to the use of a commercially available hybridization oven with rotating wheel, membranes can be incubated in sealable plastic bags fixed to a shaking stir device in a water-bath of the desired temperature. This may create problems, however, as plastic bags containing buffers with a high SDS concentration can be very difficult to seal, with the risk of leakages and loss of probes, etc. A hybridization oven with purpose-made flasks, though expensive, has the advantage of being less prone to leak (Sambrook and Russell, 2001).

2. During the ethanol wash, it is advisable to break up precipitated RNA because material with RNase activity is often trapped within the precipitates.

Do not over-dry the pellets, as they may then be more difficult to dissolve in water (Seal and Coates, 1998).
3. Use of UV cross-linking results in a five- to tenfold increase of sensitivity over the standard baking method, as reported for use of nylon membranes and chemiluminescent detection by Pallás *et al.* (1998), and cross-linking on both sides reduces the background (Webster and Barker, 1998). A total of 0.15 J/cm^2 of UV (254 nm wavelength) is appropriate for linking RNA to dry membranes. Over-radiation results in decreased hybridization signal (Sambrook and Russell, 2001).
4. Roche Diagnostics™ recommends as little as 3.5 ml solution per 100 cm^2 membrane for hybridization. The solution containing the probe can be reused several times without loss of sensitivity. If after a few uses, probes are stored at – 20°C in the '*Hyb solution*', they can still give satisfactory results, even after storage up to 2 years (Pallás *et al.*, 1998). Before reusing a probe diluted in hybridization solution, denature at 68°C for 10 min, followed by quick chilling in ice water.
5. During the stringency washings, the membrane must be continuously covered by liquid. The volume of washing buffers depends on the type of container; thus $\geq$ 10 ml may be required.

See also Pallás *et al.* (1998) and Webster and Barker (1998) for several useful tips on NASH detection of plant viral RNA and plant RNA. On the use of NAH for seed health testing, see also Section 7.2.1.

6.2. Enzymatic Nucleic Acid Amplification

Detection and identification by enzymatic nucleic acid amplification, particularly the reverse-transcription polymerase chain reaction (RT-PCR), appears to have great potential for use in seed-health testing for viruses and viroids. The revolutionary technique, PCR, based on an idea of Mullis *et al.* (1986), was made possible thanks to the discovery and utilization of the thermostable enzyme, polymerase, of a hot-water-dwelling bacterium.

Direct detection of target DNA in tissue extracts is difficult unless it occurs in large quantities. And, although DNA can be variously amplified to facilitate detection, PCR is uniquely capable of generating multi-million-fold copies of specific DNA fragments in a matter of hours, even if initially present at extremely low concentrations.

The principle of PCR is, in brief, that, when a thermostable DNA polymerase, specific primers, deoxyribonucleotide triphosphates (dNTPs), the target DNA sample, a reaction buffer, magnesium and optional additives are mixed in a test tube, it is possible, by cycles of specified length and temperature, to double the copies of DNA per cycle, e.g. producing 2, 4, 8, 16, 32 copies, etc. – the 'chain reaction' – finally resulting in an amount of DNA large enough to be detected by agarose gel electrophoresis or by other means. The procedure takes place in a thermal cycling instrument, and the usual number of cycles is 30–40, which can be completed in 2–3 h (Fig. 6.3).

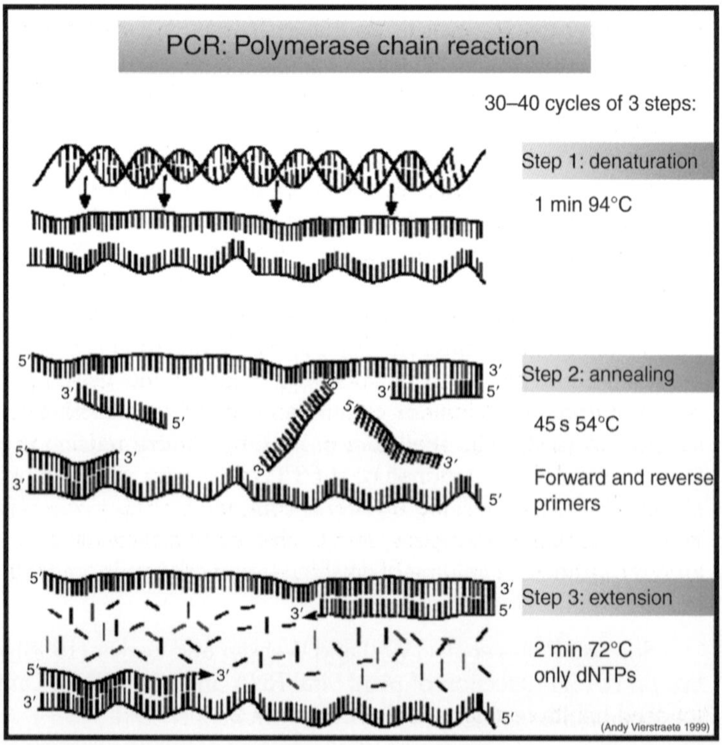

Fig. 6.3. The different steps of PCR with indication of typical cycling conditions. The target strands (step 1) are DNA, most often copy DNA, reverse-transcribed from virus- or viroid-RNA (see p. 180). Forward and reverse primers are also called upstream and downstream primers, respectively (see Fig. 6.5). Reproduced from Vierstraete (2001) with permission.

The thermostable enzyme most often used in PCR is the so-called *Taq* DNA polymerase, derived from the bacterium *Thermus aquaticus*, and capable of withstanding the high temperature (94°C) required in the cycles to denature (melt) DNA. The fact that it is a DNA polymerase requires that RNA – the nucleic acid of most plant viruses and all viroids – be reverse-transcribed (RT) to DNA prior to PCR amplification (Section 6.2.2). In PCR, the two primers, which base-pair to and define each end of the sequence of the target DNA to be amplified, are synthesized specific oligonucleotides (short chains of nucleotides, nts), usually 18–30 nts in length (see Fig. 6.3). The dNTPs are the 'building blocks' required to multiply the DNA. The selected fragment of the target DNA for amplification usually consists of a few hundred base pairs (Henson and French, 1993).

RT-PCR detection has been reported for several plant viruses and viroids (Candresse *et al.*, 1998). Wylie *et al.* (1993) used RT-PCR to detect cucumber mosaic in lupin seed, detecting one infected seed in up to 2000 seeds. Bariana *et al.* (1994) practised simultaneous testing for five seed-transmitted legume viruses by RT-PCR. A way to avoid the often cumbersome and demanding purification of RNA prior to the RT-PCR is to include an immunocapture (IC) step (Sections 6.2.2 and 6.2.3). By IC, the virions in question are initially captured on to a solid surface (usually the wall of the PCR microtube) by a serological procedure similar to the first two steps of DAS-ELISA. Despite the fact that it requires virus-specific antibodies, IC-RT-PCR has gained popularity because of its high sensitivity and simplicity.

The high detection sensitivity of PCR also causes vulnerability to contamination. Thus utmost care is required to exclude contamination, which can lead to false positives (see Section 6.2.3). Also, like serological methods, PCR does not distinguish between non-infective and infective pathogens.

A principal advantage of PCR is that less time is required for preparing a primer set than for preparing pathogen-specific antibodies, particularly monoclonal antibodies. Although comparisons of PCR with ELISA, DIBA, TBIA or NASBA (nucleic acid sequence-based amplification) (see Section 6.2.6) are appropriate, PCR appears to be the best choice for detection and identification of: viroids, viruses lacking a protein coat, virus strains or isolates, viruses with low titres in plants and seeds, and viruses that are difficult to purify or with poor immunogenicity. Accordingly, the following pertains primarily to PCR principles and protocols.

6.2.1. Principle of the polymerase chain reaction (PCR)

The purpose of the polymerase chain reaction is to make a huge number of copies from a specific DNA fragment (for plant viruses most often a fragment of cDNA reverse-transcribed from RNA; see p. 180). The three major steps in a PCR reaction, which are repeated for 30–40 cycles, are shown in Fig. 6.3 and detailed below, essentially as formulated by Vierstraete (2001).

The cycling reactions take place in an automated cycler that can quickly heat and cool the tubes containing the reaction mixtures.

a. Denaturation at 94°C (range 90–96°C): during denaturation, the double strands 'melt' open into single-stranded (ss) DNA; all enzymatic reactions stop (for example, the extension from a previous cycle). Duration, 15 s–2 min.

b. Annealing at temperatures ranging from 45 to 70°C: the primers anneal to (associate with) their complementary nucleotide sequence at the end of each target DNA strand (Fig. 6.3). In this step, the primers are jiggling around, caused by Brownian motion. Hydrogen bonds are constantly formed and broken between the single-stranded primer and the single-stranded template. The more stable bonds last somewhat longer (primers that fit exactly) and the polymerase can attach and start copying the template on that small portion of double-stranded DNA (template and primer). Once there are a few bases built in, the hydrogen bonds are strong enough between the template and the primer for them no longer to break. Duration, 30–60 s.

c. Extension at 72°C: this is an ideal working temperature for the polymerase, which now starts building a strand complementary to the template (the DNA fragment) from the four dNTPs present in the mixture. Primers that are on

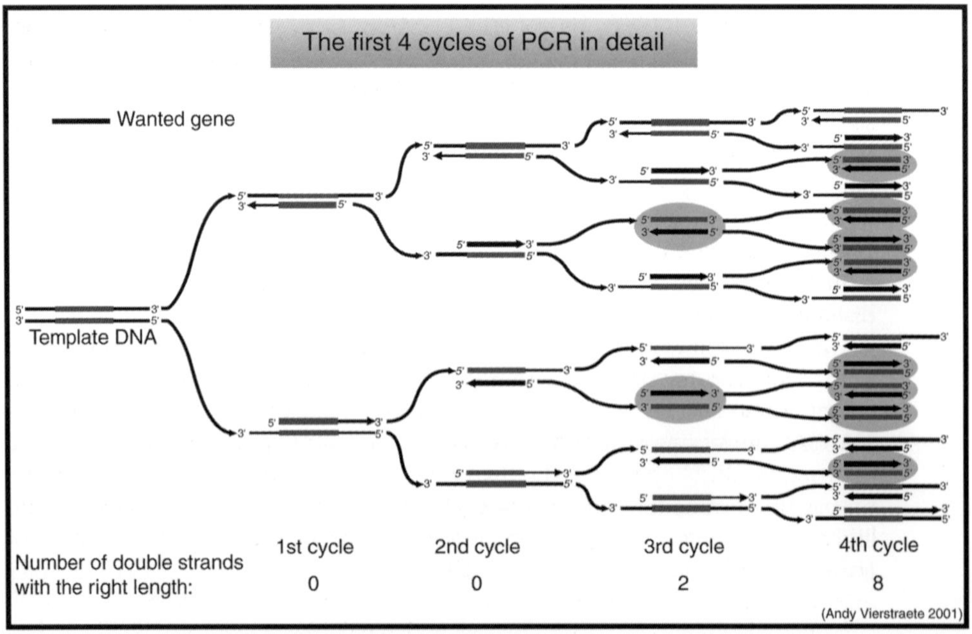

Fig. 6.4. The first four cycles of PCR. Not until the third cycle do DNA copies of the right length start to occur. Reproduced from Vierstraete (2001) with permission.

positions lacking an exact match again detach (because of the higher temperature) and thus do not cause extension of the fragment. Synthesis starts from the 3′ end of each annealed primer and the polymerase adds dNTPs in the 5′-to-3′ direction, reading the template from the 3′ to the 5′ side. Duration, 1–2 min.

Both strands are copied during PCR, thus causing an *exponential* increase in the number of copies of the target DNA (2, 4, 8, 16 . . .). As can be seen in Fig. 6.4, the first synthesized strands are not of the same length as the target DNA. Double strands of the specified length do not occur until the third cycle.

6.2.2. General components of the reverse-transcription polymerase chain reaction (RT-PCR) assay

Segments of the standard RT-PCR assay include test sample preparation, design and preparation of primers, a reverse transcription step, the PCR amplification step, and analysis of PCR products. For detection of DNA viruses, nucleic acid extracts are used directly for PCR without RT.

Sample preparation

For general aspects of sample preparation, see also Section 5.2.2.

A critical factor for successful RT-PCR is the quality of the nucleic acid template. The two major problems are RNA degradation and inhibitory contaminants. Plants contain varying amounts of polysaccharides and phenolic compounds, which are particularly suspected to be inhibitors of enzyme activity

(Candresse *et al.*, 1998; Singh *et al.*, 1998a; Dietzgen, 2002). Tissues from woody plants or tissues from groundnut leaves and seeds, bean leaves and seeds, pepper leaves and potato tubers have been shown to contain PCR inhibitors, whereas leaves of *Nicotiana* species, tomato and cucumber appear to contain less or none of these inhibitors (Dietzgen, 2002). The inhibitory effects of plant compounds can sometimes be avoided simply by diluting the nucleic acid extracts prior to RT-PCR, still leaving sufficient template for amplification but diluting out the inhibitors. However, the ≥ 100-fold dilution often required reduces test sensitivity, which is a drawback with viruses or viroids that produce low titres in plants or when testing composite samples.

In some cases, inhibitors can be removed from plant-tissue extracts. However, many inhibitors appear to be co-purified with nucleic acids, making separation difficult (Seal and Coates, 1998). Many nucleic acid extraction and purification protocols have been reported, of which four different approaches are described in Section 6.2.3. The most effective methods are often laborious and costly, and may involve the use of toxic chemicals, such as phenol, chloroform, lithium salts, etc. Complicated procedures also increase the chance of cross-contamination (Nassuth *et al.*, 2000).

Commercially available RNA extraction kits, such as Flowgene® (Lichfield™, UK), have been shown to be efficient and rapid (Stevens *et al.*, 1997), as has RNeasy® (Qiagen™) according to several reports, but the use of such kits adds to the costs. A so-called silica-capture method, used, for example, by Rott and Jelkmann (2001), Letschert *et al.* (2002), and Menzel *et al.* (2002) was found to be effective in RNA extraction from different plant species. Thompson *et al.* (2003) compared a slightly modified form of RNeasy with silica-capture for nucleic acid extraction from strawberry plants, and found the methods equally reliable. They therefore adopted the silica-capture method because of its lower price per RNA extraction, e.g. ~10% of extraction-kit use (Section 6.2.3).

Candresse *et al.* (1998) stressed that simplification and optimization of sample preparation protocols are needed to develop practical detection assays. Of the following three briefly described simple procedures, **1** may work well for a number of host/pathogen combinations, but is less sensitive than **3**. Procedure **2** has shown promising results for detection of four potato viruses.

1. *Nucleic acid extraction without tissue homogenization.* Such a method has been used successfully for RT-PCR by Thomson and Dietzgen (1995) for different viruses and hosts. A small amount of plant tissue was heated in a TRIS-HCl-KCl-EDTA buffer, pH 7.4 or 8.4, for 10 min at 95°C, and the supernatant, undiluted or slightly diluted, was used directly for RT-PCR. The procedure is described in Section 6.2.3. Nakahara *et al.* (1999) utilized a similar method for viroids in different hosts.

For detection of *Potato leafroll virus* (PLRV, *Polerovirus*) in potato, Singh (1999) released RNA from discs of plant tissue by incubation with a non-ionic detergent (Triton X-405R) at 37°C for 30 min, followed by centrifugation and use of the supernatant for RT-PCR.

One such simple protocol should be attempted before undertaking more complex extraction procedures (Singh, 1998; Dietzgen, 2002).

2. *Membrane-based nucleic acid preparation*. Recently, Singh *et al.* (2004) reported a simple and rapid method for preparing viral RNA from potatoes and aphids. A buffered detergent (Triton XL-80N) solution was used for making tissue extracts, which were clarified and immobilized on a nitrocellulose membrane. Water-eluted extract from cut-out spots from the membrane was used for RT, and cDNA was amplified through PCR. The assay sensitivity was comparable to that of 'long protocols' (phenol extraction, etc.) for detection of four potato viruses, and the method was found especially suitable for remote areas, from where prepared membranes could be mailed to centralized laboratories for detection by RT-PCR.
3. *Immunocapture-RT-PCR* (Wetzel *et al.*, 1992) is a method that usually eliminates both RT and PCR inhibitors and has been shown to be highly sensitive. Prior to the RT, virions are trapped from crude plant extracts on an antibody-coated plastic surface, usually the RT-PCR tubes, by following the initial procedure for DAS-ELISA, including washing after antigen coating to get rid of plant sap contaminants. The method appears to work for any virus, provided that the virus possesses sufficient serological reactivity and that specific antibodies are available (Martin *et al.*, 2000; Dietzgen, 2002). One of the reasons for its high sensitivity may well be the relatively large volume of plant extract (100–200 μl) normally used (Candresse *et al.*, 1998). A protocol for the immunocapture step is described in Section 6.2.3.

SAMPLE PREPARATION FROM SEEDS. Whole, dry seeds often contain compounds inhibitory to RT or PCR and, therefore, the extraction of RNA requires the use of efficient protective procedures, such as phenol/chloroform extraction (Kohnen *et al.*, 1992; Wylie *et al.*, 1993; Bariana *et al.*, 1994; Njeru *et al.*, 1997) or the use of a commercial extraction kit (Gillaspie *et al.*, 2000). IC-RT-PCR was used successfully for testing apricot seeds for PPV (Pasquini *et al.*, 1998) and by Gillaspie *et al.* (2000) for testing groundnut seeds for BCMV-PSt and PeMoV. In RT-PCR detection of CMV in lupin seed flour, Wylie *et al.* (1993) obtained a higher sensitivity (one infected seed in ≤ 2000) by using a phenol–chloroform RNA extraction (Section 6.2.3) rather than a simple, rapid extraction (one infected seed in ≤ 200). (For full names of virus acronyms, see Table 4.2.) The method of Wylie *et al.* (1993) was also used by Njeru *et al.* (1997) for detection of *Subterranean clover mottle virus* (SCMoV, *Sobemovirus*) in clover seeds.

Importantly, the risk of also detecting non-transmissible virus in whole seeds is the same as for NAH and ELISA; see Sample preparation, Section 5.2.2.

To avoid RT-PCR detection of non-transmissible virus in seed-coats when assaying pea seed flour for PSbMV, Torok and Randles (2001) removed the seed-coats prior to RNA extraction. Whole dry seeds were crushed so as to produce a very coarse product consisting of quarter/half seeds, from which the seed-coats could be removed by aspiration. Subsequently, the coat-free seed parts were ground and the RNA extracted.

Lipid content in seeds may cause problems when extracting seeds for IC-RT-PCR, such as those encountered by Gillaspie *et al.* (2000) in groundnut seeds. The problems were overcome by centrifuging the extract and removing only the aqueous portion with a pipette.

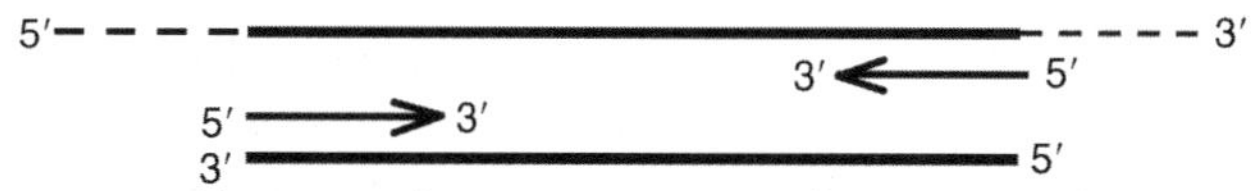

Fig. 6.5. Orientation of and common terms used for primer pairs.

Primers (see also Sections 6.2 and 6.2.1)
DNA polymerases, whether from humans, bacteria or viruses, cannot copy a DNA chain without a short sequence of nucleotides to 'prime' the process (get it started). In the living cell another enzyme, a primase, 'makes' these first few nucleotides of the copy. Thereafter, the polymerase catalyses synthesis of the remaining new chain (Anon., 1992).

The orientation and the common terms used for the two primers are shown in Fig. 6.5. Here below, the terms 'downstream' and 'upstream' are used.

Design of the primers is perhaps the most critical parameter for successful PCR. The result of a poorly designed primer can be little or no product, due to non-specific amplification and/or primer-dimer formation. Among the most critical variables are (Sambrook and Russell, 2001; Prezioso, 2003):

- Primer length.
- Melting temperature (Tm).
- Specificity.
- Complementary primer sequences.
- 3′ end sequence.

Primer length. Primers that are too long or too short impede the amplification. In general, oligonucleotides between 18 and 24 bases are extremely sequence specific. The length of the two primers should not differ by more than three bases.

Melting temperature (Tm). The design of both primers should be such that they have similar melting temperatures (Tm is the temperature at which 50% of the strands in a ds nucleic acid are separated at a given salt concentration; see Section 6.1.1). The Tm of primer partners should not differ by more than 5°C. The Tm of the amplified product should not differ from the Tm values of the primer pairs by > 10°C. A general rule of thumb is to use an annealing temperature that is 5°C lower than the melting temperature. Example: an annealing temperature of 50°C (minimum) corresponds to a primer with a calculated Tm of about 55°C.

The Tm depends on both primer length and its content of the bases C and G. (C and G base-pair with three hydrogen bonds, A and T with only two; hence the differences in Tm.) A rough estimate of the Tm of oligonucleotides can be made using the so-called Wallace formula:

$$\mathrm{Tm} = 2(\mathrm{A} + \mathrm{T}) + 4(\mathrm{G} + \mathrm{C})$$

Using this formula, oligonucleotides with 50% GC content and 18–24 bases long will have Tms of 54–72°C. In general, a Tm within this range gives the best result. Note that inosine used instead of bases (see Degenerate primers below) does not contribute to the Tm (Seal and Coates, 1998). Note also that Tms calculated for very short or longer sequences, using Wallace's rule, are inaccurate; a more precise formula can be found in Prezioso (2003) (available on the Internet).

Specificity. Primers must be chosen so that they have a unique sequence within the template DNA that is to be amplified. According to Seal and Coates (1998) and others, the distance between primers should preferably be a minimum of 200 bp but not longer than 2000 bp. If below 200 bp, the product may be confused with primer-dimers and, if more than 2000 bp, the product may be less efficiently amplified. The sensitivity of the assay is influenced by both the specificity of the primers and the amplified fragment length. Amplification of relatively small rather than large fragments has been shown to improve sensitivity (Rosner *et al.*, 1997; Singh, 1998). Maximum 500 bp fragment length is recommended by M. Nicolaisen (Danish Institute of Agricultural Sciences, 2004, personal communication).

Species-, genus- or even family-specific detection can be performed by use of primers complementary to sequences in a conserved region of a virus genome. As reviewed by Dietzgen (2002), degenerate primer pairs (see below), enabling the detection of all or several members of a virus genus, have been developed for 16 genera, and a degenerate primer pair specific to all members of *Potyviridae* has been designed (Gibbs and Mackenzie, 1997; Gibbs *et al.*, 2003). For species or strain specificity, primers designed from the more or most discriminating regions should be used (see also Section 6.2.5).

Complementary primer sequences. No inverted repeat sequences or self-complementary sequences > 3 bp in length should be present in a primer, because such sequences tend to form hairpin structures, which can prevent the primer from annealing to the target DNA.

The 3′ terminal sequences of one primer should not be able to bind to any site on the other primer. Even weak complementarity between them leads to hybrid formation, resulting in primer-dimers and suppression of target amplification.

3′ end sequence. The 3′ end of the primer is the end that will be elongated, and therefore a G–C-rich region of perfect base pairing will ensure a tight, stable hybridization at this end. If possible, the 3′ end base of each primer should be C or G (but not . . . CG or . . . GC).

Degenerate primers. PCR primers may sometimes be designed on the basis of amino acid sequences rather than directly from the nucleotide (nt) sequence. If a given amino acid (aa) sequence is known to be highly conserved, for example, in all members of a virus genus, a genus-specific primer pair can be designed. However, as most amino acids can be encoded by more than one codon (from one to six), identical aa sequences can correspond to a wide variation of nt sequences. Priming such sequences therefore requires a mixture or a pool of different primers: degenerate primers.

Table 6.2. Symbols used for alternating bases in codons. Most codons alternate only in the 3rd position, but a few do also in 1st and 2nd positions.

Abbreviation letter	Nucleotide represented
R	AG
Y	CT
M	AC
K	GT
W	AT
S	CG
B	CGT
D	AGT
H	ACT
V	ACG
N	ACGT

Example: glutamine (Gln) is encoded by both CAG and CAA, and threonine by ACG, ACA, ACT and ACC. These codons can be abbreviated into CAR and ACN, respectively, where R stands for A or G, and N for A, C, G or T.

Eleven such symbols are used to define degenerate primers (Table 6.2). The degeneracy of the sequence of interest can be very large. If several hundred different primers are required, the number can be reduced by replacing N (A, C, G and T) with inosine (deoxy-inosine, dI, a deoxy-purine ribonucleoside), which base-pairs with all four nucleotides. Also positions with three alternative bases (symbols B, D, H and V, Table 6.2) can be substituted with inosine to further reduce the number of different primers. During primer synthesis, all alternative sequences are synthesized simultaneously and in equal amounts. If the degeneracy is high, the one perfect sequence will be present at a low concentration. Therefore, degeneracy should be minimized as much as possible.

Since a consensus Tm among degenerate primers is impossible, conditions should be used that allow the oligonucleotide with the lowest G + C content to anneal efficiently (Sambrook and Russell, 2001).

Example of a degenerate primer sequence designed after a protein motif (source: http://boneslab.chembio.ntnu.no/DegPCRshortguide.html):

Trp	Asp	Thr	Ala	Gly	Gln
5′ TGG	GAY	ACN	GCN	GGN	CAR 3′

where Trp = tryptophan, Asp = aspartic acid, Thr = threonine, Ala = alanine, Gly = glycine and Gln = glutamine.

Design of degenerate primers for detection and diagnosis can be greatly facilitated by using, for example, the program for CODON USE ANALYSIS, freely accessible on the website of *AAB Descriptions of Plant Viruses* (http://www.dpvweb.net/analysis/codons/index.php).

REMARKS ON PRIMERS. Published primer sequences specific for a great number of viruses and viroids can be found in journal articles, but sometimes one must design usable primers.

It can be rather difficult to design primers manually. Computer programs are available by which one can optimize the design, selection and placement of primers, taking into account the rules and conditions mentioned above. One such software is the GPRIME package of computer programs (Gibbs *et al.*, 1998) and another is the one mentioned above. Many other programs can be found on the Internet, such as under http://www.biocompare.com or via Gateway websites, e.g. http://www.bioinformatics.vg/index.shtml, with many links to information on molecular biology, sequence analyses and tools for primer design. Information on nucleic acid sequences of plant viruses and viroids can be obtained from GenBank (USA), http://www.ncbi.nlm.nih.gov/Genbank/ and EMBL Data Library, http://www.ebi.ac.uk/

Custom-built primers are obtainable from commercial laboratories, or can be made on the spot if a nucleic acid synthesizer is available. Primers are relatively stable when stored as high-concentration solutions in sterile distilled water at – 20°C (Stratford, 1998) or in a TE buffer at 4°C (Holst-Jensen *et al.*, 1998) or lyophilized at 4°C.

In standard reactions, 0.1–1.0 μM of each primer is used.

Reverse transcription

The majority of plant viruses and all known viroids have RNA genomes; the targets to be amplified from these by PCR must therefore be reverse-transcribed into cDNA. The reverse transcription (RT) may be performed as a separate reaction, from which a few μl of the resulting cDNA is transferred to the PCR tube, or in a one-tube RT-PCR. One-tube reactions may be carried out either as two-step or one-step reactions (see One-tube RT-PCR, protocols, Section 6.2.3). For detection of DNA viruses, nucleic acid extracts are used directly for PCR without RT.

RT ENZYMES. The two most commonly used reverse transcriptases (RT, i.e. RNA-dependent DNA polymerases) are AMV RT, derived from *Avian myeloblastosis virus* and M-MLV RT, derived from *Moloney murine leukaemia virus.* Both enzymes possess an RNase H activity, but that of M-MLV is lower than that of AMV RT. Therefore, M-MLV may be preferred for longer cDNAs (> 5 kb). On the other hand, AMV RT is more efficient than M-MLV RT (in terms of units required) and while M-MLV RT is optimal at 37°C, AMV RT can remain active up to 58°C, although normally used at 42°C in two-step RT-PCR. The stability of AMV RT at elevated temperatures makes this enzyme preferable in working with an RNA that has strong secondary structure (McLaren, 1999).

Since remnants of reverse transcriptases inhibit the activity of *Taq* polymerase in the subsequent PCR, the reverse transcriptase should be inactivated by heating to 70°C for 10 min or to a higher temperature for a shorter time.

Another enzyme, the thermo-resistant *Tth* DNA polymerase, derived from the bacterium *Thermus thermophilus*, or its recombinant variant r*Tth* can be used in both PCR and RT, as it possesses an efficient reverse transcriptase activity. For RT, its optimum temperature is 55–70 °C and this elevated temperature minimizes (like AMV RT) difficulties associated with the RT of RNAs containing a high degree of secondary structure. However, for RT-PCR of plant viruses, M-MLV RT or AMV RT appears to be the preferred enzyme. Brief descriptions of RT enzymes can be found on the website http://www.biocompare.com, and facts and details, respectively, about their use are described by McLaren (1999) (available on the Internet) and Sambrook and Russell (2001). The manufacturers of individual RT enzymes provide optimal RT buffers and detailed instructions for optimal RT reaction conditions.

PRIMERS. cDNA synthesis is primed with one primer, either the downstream specific primer, an oligo(dT) primer or random hexanucleotides. The two latter primers are commercially available. Their manufacturer's recommended RT temperature regime should be followed. For viral RNA with a terminal polyadenylated region, such as in the *Potyvirus* genome, a degenerate primer that is a mixture of three 21-mer oligonucleotides consisting of 20 T residues with a 3′-terminal A, G or C, respectively:

5′-TTTTTTTTTTTTTTTTTTTT(A/G/C)-3′

serving both for cDNA synthesis and as the specific downstream primer in the subsequent PCR, was found very useful by Rybicki (2001a). The downstream primer of the primer pair that can be used for RT-PCR amplification of the genome segment shared by all potyvirids, according to Gibbs and Mackenzie (1997), also contains an oligo(dT) sequence.

Most commonly, the downstream primer of a specific primer pair is used for the RT phase. In some cases, such as detection with degenerate primers, Langeveld *et al.* (1991), for example, found that priming in RT with random hexamers, and the use of degenerate primers only during PCR, increased the specificity of RT-PCR. The optimization of multiplex RT-PCR plant virus assays was simplified, when random hexamers or an oligo(dT), or a mixture of both, was used in the RT step in synthesizing a cDNA pool. Target-specific primers were thus (in most cases) not used in RT, but used only in the PCR step to amplify the targets from the cDNA pool (Nie and Singh, 2000, 2001; Rott and Jelkmann, 2001; Thompson *et al.*, 2003) (see also Optimization of RT-PCR, Section 6.2.3).

DNTPS, BUFFER AND ADDITIVES. More dNTPs (up to 500 μM each) are required in RT than in PCR (20–200 μM each), as longer DNA strands are usually synthesized in the RT. For buffers and additives, see the RT-PCR standard protocol, Section 6.2.3.

Polymerase chain reaction

POLYMERASES AND DNTPS. For routine PCRs, *Taq* polymerase (0.5–2.5 units per standard 25–50 μl reaction) is preferable. Apart from *Taq*, a number of other thermostable polymerases with different properties are available, also as mixtures. The properties of the same polymerase sold by different manufacturers may not be identical; therefore PCR should be optimized every time for a new batch (Sambrook and Russell, 2001).

Follow the manufacturer's instruction regarding storing of the dNTPs. They are probably the least stable component, and should be distributed in small tubes that are kept frozen, used a few times and then discarded (Holst-Jensen *et al.*, 1998).

A standard PCR should contain equimolar amounts of the four dNTPs: dATP, dTTP, dCTP and dGTP, at 20–200 μM each (Seal and Coates, 1998).

BUFFER AND ADDITIVES. TRIS-HCl, pH 8.3–8.8 at room temperature, is used as the buffer for standard PCRs. At 72°C (the temperature commonly used for the extension phase), the pH of the reaction mixture drops to about 7.2. An optimal buffer is usually provided by commercial enzyme suppliers.

All thermostable DNA polymerases require free *divalent cations* for suitable activity, usually Mg^{2+}. A concentration of 1.5 mM $MgCl_2$ is routinely used, but higher concentrations may be needed, depending on the concentrations of dNTPs and primers. The correct Mg^{2+} concentration is important and should be optimized for different PCRs.

Standard PCR buffers should also contain *monovalent cations* in the form of KCl, usually 50 mM. KCl facilitates primer annealing but concentrations above 50 mM inhibit the *Taq* polymerase (Seal and Coates, 1998; Rybicki, 2001a). Alternatively, according to Sambrook and Russell (2001), raising the concentration to 70–100 mM often improves the yield of DNA segments shorter than 500 bp.

A number of co-solvents and additives, e.g. dimethyl sulphoxide (DMSO) or glycerol, have been used to improve poor PCR results. According to Sambrook and Russell (2001), however, it is generally far better to optimize the regular PCR components, especially the concentrations of Mg^{2+} and K^+ ions, rather than using such enhancers at the first sign of trouble. For optimal plant virus or viroid detection, Seal and Coates (1998) report, though, that Tween-20 (0.5%), DMSO (5%) or polyethylene glycol 400 (5%) in the PCR buffer has proved useful for eliminating inhibitory effects of some polysaccharides on *Taq* polymerase.

THERMOCYCLERS AND CYCLING PROGRAMME. A wide range of thermocycler brands and types are currently available. The instrument selected should have good uniformity of temperature over the sample block and precision heating and cooling, and should ensure that the block and tube contents quickly reach each programmed temperature. Some instruments require mineral oil on top of the reaction mixture to minimize sample evaporation and condensation on the tube lids; those having heated lids do not. For most thermocyclers the sample tubes are placed in a heating/cooling metal block, but some use air or water for heating and cooling of the reaction mixes, held in thin-walled tubes or glass capillaries. Several real-time detection instruments are also on the market (see Section 6.2.6).

An overview, descriptions and comparisons of thermocyclers are accessible on the Internet at http://www.biocompare.com

A standard cycling programme is typically heating samples at 95°C for 2–3 min (initial denaturation), followed by 30–40 cycles each at 94°C for 10–60 s (denaturation), 45–70°C for 30 s (annealing) and 72–74°C for 30–60 s (extension). Some laboratories conclude with a final (last cycle) extension of 5–10 min;

others do not. The annealing temperature depends on the Tm of the primers, and a good starting point for testing new primers is 5–10°C below the lowest melting temperature of the primers selected. The specificity of primers can be improved by gradually increasing the trial temperature (Seal and Coates, 1998). Fast cycling, i.e. a 30 s duration for each cycle, providing 40 cycles per PCR run in 30 min, is possible with capillary thermocyclers using hot air, because of almost instant temperature shifts.

CONTROLS. As with other detection methods, the use of negative (known healthy) and positive (known infected) controls is crucial in PCR. A PCR assay should also include a tube without a template (a mock sample). In order to avoid false negatives caused by a failed RNA extraction, RT or PCR amplification, an internal control RNA can be included – that is, primer sets for both the target RNA and a control RNA sequence are added to each test tube (a duplex PCR). The control RNA could be a plant genomic messenger RNA (mRNA), as used, for example, by Thomson and Dietzgen (1995), Menzel *et al.* (2002) and Thompson *et al.* (2003) (see Multiplex PCR, Section 6.2.3).

For an internal control, revealing malfunctions in the RT and PCR stages, Torok and Randles (2001) added TMV RNA and a primer set to all tubes, the primers being specific for a 406 nt TMV RNA sequence, for RT-PCR detection of PSbMV in pea seed. The presence of a 406 bp band in the gel but no PSbMV band (1.1 kb) proved that absence of the target was not due to a failure in the RT or PCR phases.

Analysis of PCR products

For analysis of PCR products, the amplificates are probably still most commonly separated by agarose gel electrophoresis, followed by visualization with ethidium bromide staining (Section 6.2.4). Gel-free detection of amplificates (Section 6.2.4) offers some advantages over gel-electrophoresis detection and is increasingly being used, for example, for performing routine assays.

PCR application in plant virology

Candresse *et al.* (1998), Martin *et al.* (2000) and Dietzgen (2002) have reviewed the application of PCR assays in plant virology, and detailed procedures have been described by, for example, Dijkstra and de Jager (1998), Seal and Coates (1998), Singh (1998) and Stratford (1998).

6.2.3. Protocols for RT-PCR assays

This section deals with protocol examples for RT-PCR amplification, while procedures for PCR product analysis are described in Section 6.2.4. For RT-PCR, an extraction of total RNA from plants or seed is usually a prerequisite, and the purity and quality of the extract are critical factors for RT-PCR sensitivity. A large number of nucleic acid extraction and purification procedures have been published; however, simpler and quicker methods have also been reported, such as nucleic acid extraction without tissue homogenization (e.g. Thomson and

Dietzgen, 1995; Nakahara *et al.*, 1999; Singh, 1999). The highly sensitive immunocapture (IC)-RT-PCR requires no RNA extraction, but only sap extraction similar to that used for ELISA. A disadvantage of IC-RT-PCR, however, is the need for specific antibodies.

THE PCR LABORATORY AND EQUIPMENT. As PCR is a highly sensitive assay, the risk of false positive results is also high. Singh (1998) recommends the following precautions as a minimum requirement for eliminating false positives.

1. Performing of (a) sample preparation, (b) reagent preparation, (c) cDNA synthesis, (d) PCR amplification and (e) visualization of amplified products each in their separate rooms or areas.
2. A separate set of adjustable pipettes for each of the rooms or areas (a)–(e), and sterile, disposable, filtered tips.
3. Use of disposable gloves with frequent change is essential, as with strict use of the separate rooms or areas for specific steps.
4. UV light decontamination of pre-PCR reagents and equipment should be practised in the PCR room or area.

Nucleic acid extraction protocols

During nucleic acid (NA) extraction, the work should be conducted under general sterile and cold conditions, avoiding dust and draughts.

Note: All glassware, tools and solutions to be used in RNA extraction, must be made RNase-free by treating with diethylpyrocarbonate (DEPC) – see Appendix 2, p. 250.

Caution: DEPC is mutagenic; wear gloves, work in fume cupboard.

In the following, four NA extraction protocols are described; see also NA extraction in Sections 6.1.1 and 6.1.2.

I. PROTOCOL FOR NUCLEIC ACID EXTRACTION FROM SEED. This procedure has been used in detection of CMV in lupin (*Lupinus angustifolius*) seed by Wylie *et al.* (1993) and of *Subterranean clover mottle virus* (SCMoV, *Sobemovirus*) in seed of subterranean clover (*Trifolium subterraneum*) (Njeru *et al.*, 1997). By this extraction method and the RT-PCR protocol used by Wylie *et al.* (1993), one CMV-infected lupin seed in 2000 healthy seeds could be detected.

Regarding the risk of also detecting non-transmissible viruses when testing whole seeds, see Sample preparation in Sections 5.2.2 and 6.2.2.

Equipment and materials

a. Laboratory mill, easy-to-clean type, fitted with a 2 mm sieve, for grinding of seeds.

b. Microcentrifuge with cooling, 15,000 *g* (optionally, centrifuge in refrigerator using pre-cooled rotor).

c. Microcentrifuge tubes, 1.5–1.6 ml.

d. Vortexer.

e. Adjustable pipettes (0.1–20 μl, 20–200 μl, 200–1000 μl). Use a separate set for extraction work only.

f. Filtered tips for pipettes.

g. Ice/water-bath.

h. Standard laboratory equipment, such as refrigerator, deep freeze, magnetic stir plate, glassware, tube racks, ice machine or similar, forceps, disposable gloves (powder-free), filter paper.

Biological materials

a. Test sample seeds (or plant material).

b. Known healthy material for negative control.

c. Known infected material for positive control.

Reagents, solutions and buffers

a. Extraction buffer (TRIS-HCl-EDTA-NaCl, pH 8.5) – see Appendix 2, p. 252.

b. Phenol, acid (for RNA viruses) – see Appendix 2, p. 252 for details. **Caution:** phenol is highly toxic.

c. Chloroform. **Caution:** highly toxic, avoid inhalation and contact, wear protective clothing and work in a fume cupboard.

d. Isoamyl alcohol. **Caution:** harmful, flammable.

e. Phenol : chloroform : isoamyl alcohol (50 : 50 : 1 (v/v)). Make from stocks or buy commercially. Store in a dark bottle at 4°C. Toxic (see above).

f. Chloroform : isoamyl alcohol (50 : 1). Toxic (see above).

g. Isopropanol. **Caution:** harmful, flammable.

h. TE buffer (TRIS-HCl–EDTA, pH 7.4) – see Appendix 2, p. 253.

i. NaCl, 5 M.

j. Ethanol, 70% – see Appendix 2, p. 253.

Procedure

1. The required number of seeds (for CMV/lupin max. 2000 per group) is ground in a laboratory mill. Before grinding the next sample, the mill must be thoroughly cleaned.

2. Place 100 mg of seed flour in a 1.5 ml microcentrifuge tube and add 500 µl of *extraction buffer* and 500 µl of *phenol : chloroform : isoamyl alcohol (50 : 50 : 1)*.

3. Vortex for 1 min and centrifuge at 13,000 *g* for 1 min.

4. Transfer the aqueous layer to a new tube and add one volume of *chloroform : isoamyl alcohol (50 : 1)*.

5. Vortex for 1 min and centrifuge at 13,000 *g* for 1 min.

6. The aqueous layer is transferred to a new tube and one volume of cold *isopropanol* is added, mixed by inverting and cooled to – 20°C for 3–4 h or overnight.

7. After thawing, the solution is centrifuged at 13,000 *g* for 3 min, the supernatant removed and the pellet air-dried.

8. Resuspend the pellet in a mixture of 100 µl of *TE,* 4 µl *5 M NaCl* and 250 µl *cold ethanol*, and incubate on ice for 20 min.

9. Centrifuge the tube for 3 min at 13,000 *g*, remove the supernatant and let the pellet air-dry (avoid complete drying; see Section 6.1.2, p. 170).

10. Resuspend the pellet in 500 µl *TE*. A rough estimate of the RNA yield can be made spectrophotometrically. When a solution is measured at 260 nm in a cell of 1 cm path length, one absorbance unit corresponds to an approximate content of 40 µg ssRNA per ml.

Use the extract for RT-PCR, or store at –20°C.

II. PROTOCOLS FOR NUCLEIC ACID EXTRACTION FROM PLANT TISSUE

Protocol II A. The following protocol is as described by Spiegel and Martin (1993). The procedure has been used for nucleic acid extraction from potato leaves and tubers for detection of *Potatovirus* Y (PVY, *Potyvirus*) (Nicolaisen *et al.*, 2001) and also for PCR detection of viruses in other plant species (M. Nicolaisen, Danish Institute of Agricultural Sciences, 2003, personal communication). This method has the advantage of not using the highly toxic chemicals phenol and chloroform.

Equipment and materials

a. Plastic bags, sturdy type, approx. 10 cm × 8 cm, for grinding of samples.

b. Roller for grinding samples in the plastic bags, hand model or electric (e.g. Bioreba AG™, Switzerland).

c. Microcentrifuge as for Protocol I, but yielding 20,000 *g*.

d. Equipment as listed for Protocol I, except laboratory mill.

Biological materials

a. Test sample plant material.

b. Known healthy material for negative control.

c. Known infected material for positive control.

Reagents, solutions and buffers

a. Extraction buffer (TRIS-HCl-lithium dodecyl sulphate-LiCl-sodium deoxycholate-Igepal CA-630-EDTA, pH 8.5) – see Appendix 2, p. 253. **Caution:** lithium salts are harmful to eyes and skin, and so is sodium deoxycholate, which is also highly flammable.

b. Potassium acetate, 5 M, pH 6.5.

c. Isopropanol. **Caution:** harmful, flammable.

d. Ethanol, 70% – see Appendix 2, p. 253.

e. dd H_2O.

Procedure

1. Place 0.3 g leaf tissue together with 0.9 ml of *extraction buffer* in a plastic bag and grind the material from outside the bag using the roller.

2. Transfer 500 µl of the extract to a tube, add 750 µl *5 M potassium acetate, pH 6.5*, mix, and let the mixture incubate for 5 min on ice.

3. Centrifuge at 20,000 *g* for 10 min at 4°C.

4. Transfer 600 µl of the supernatant to a new tube and add 500 µl of cold *isopropanol*, mix and cool the tube to – 20°C for 1 h to precipitate the RNA.

5. Centrifuge at 20,000 *g* and retain the pellet.

6. Wash the pellet once or twice with *70% cold ethanol*, air-dry the pellet and resuspend in 20 µl of sterile dd H_2O (details: see Section 6.1.2, p. 170).

7. Estimate the RNA yield spectrophotometrically (see Protocol I) if desired. Store the RNA at –20°C if not used immediately.

Protocol II B. The protocol described below is a silica particle-based extraction method (Rott and Jelkmann, 2001), modified by Menzel *et al.* (2002). The method was used for RT-PCR detection of cherry and apple viruses, and was also used by Letschert *et al.* (2002) for detection of *Tobamoviruses* in pepper and tobacco, and by Thompson *et al.* (2003) for detection of strawberry viruses. According to Rott and Jelkmann (2001) the method was effective and results were considered comparable to those from commercial extraction kits.

Equipment and materials

a. Mortar and pestle or similar for homogenization.
b. Hot-water bath or block heater.
c. Equipment and materials as listed for Protocol I, except laboratory mill.

Biological materials

a. Test sample material.
b. Known healthy material for negative control.
c. Known infected material for positive control.

Reagents, solutions and buffers

a. Grinding buffer: guanidine hydrochloride-sodium acetate-EDTA-potassium acetate-PVP (Appendix 2, p. 253). See **Note** after procedure.
b. Sodium dodecyl sulphate (SDS).
c. SE buffer (optional): PBS, pH 7.4, containing Tween-20, PVP, ovalbumin, BSA and NaN_3 – see Appendix 2, p. 254.
d. Sodium iodide–sodium sulphite solution – see Appendix 2, p. 247.
e. Ethanol, 99.6%.
f. Silica slurry – see Appendix 2, p. 254.
g. Wash buffer: TRIS-HCl-EDTA-NaCl-ethanol – see Appendix 2, p. 255.
h. TE buffer (optional): TRIS-HCl-EDTA – see Appendix 2, p. 253.

Procedure

1. Homogenize tissue in *grinding buffer* (1 ml per 100 mg tissue) and transfer 500 µl of homogenate into a microcentrifuge tube containing 100 µl of 10% SDS. See **Note** below.
2. Let the mixture incubate at 70°C for 10 min with intermittent shaking, place the tube on ice for 5 min, and centrifuge at 15,000 *g* for 10 min.
3. Transfer 300 µl to a new tube, to which is added 300 µl *sodium iodide–Na_2SO_3 solution*, 150 µl *ethanol* (99.6%) and 25 µl *silica slurry*. Mix.
4. Allow the mixture to incubate at room temperature for 10 min with intermittent shaking; then centrifuge for 1 min at 7000 *g*.
5. Remove the supernatant and resuspend the pellet in 500 µl of *wash buffer*. Centrifuge at 7000 *g* for 1 min.
6. Repeat the washing step.
7. Let the pellet dry for several minutes at room temperature before suspending in 150 µl of sterile H_2O or TE buffer. Avoid complete drying of pellet (Section 6.1.2, p. 170).
8. The mixture is incubated at 70°C for 4 min to redissolve the nucleic acids and then centrifuged at 15,000 *g* for 3–5 min.
9. Transfer the supernatant to a new tube and measure the RNA yield spectrophotometrically (see Protocol I) if desired. The total nucleic acid extract can be used directly for RT-PCR (usually 1–5 µl) or stored at –20°C.

Note Instead of guanidine thiocyanate (4 M), used by Rott and Jelkmann (2001), Menzel *et al.* (2002) used the less risky guanidine hydrochloride (6 M) in the grinding buffer, and found it as efficient. According to Thompson *et al.* (2003), the following modification improved the extraction: they homogenized the tissue in an SE buffer (0.3 g in 3 ml) and then added 100 μl of the homogenate to 400 μl of the grinding buffer. This mixture was then used for SDS treatment (step **1**) and steps **2–9** were performed as described.

III. PROTOCOL FOR NUCLEIC ACID EXTRACTION WITHOUT TISSUE HOMOGENIZATION. A simple, single-step plant tissue preparation procedure was developed by Thomson and Dietzgen (1995) and adopted both in their laboratory and several others for routine use of PCR in detection of various viruses in different plant species. A small amount of plant tissue is heated in a TRIS-HCl-KCl-EDTA solution and the supernatant used for RT-PCR undiluted or 5–100 × diluted.

The advantages of the method are that it is rapid, there is less risk of cross-contamination between samples as tissues are not homogenized, and it requires no use of phenol or chloroform. However, a relatively low sensitivity may be expected, especially if high template dilutions are needed to dilute out enzyme inhibitors. Still, for purposes where the highest sensitivity is not required, such as confirmative identification and strain determination, the method should be useful.

Singh (1999) used a similar simple extraction method, and concluded that material cost was up to 100-fold less and time input up to sixfold less compared to use of the chloroform–phenol method.

Equipment and materials

a. Scalpel.
b. Hot-water bath or block heater.
c. Equipment as listed for Protocol I, except laboratory mill.

Biological materials

d. Test sample material.
e. Known healthy material for negative control.
f. Known infected material for positive control.

Reagents, solutions and buffers

a. Extraction solution: 100 mM TRIS-HCl, 1 M KCl, 10 mM EDTA, pH 8.4 – see Appendix 2, p. 255.
b. Sterile distilled H_2O.

Procedure

1. Cut out a 2 mm^2 piece of a leaf or 0.5–1 mg of tissue with a clean scalpel.
2. Place the tissue piece in a sterile 1.6 ml microcentrifuge tube containing 30 μl of *extraction solution* and heat to 95°C for 10 min. See **Note 1** below.
3. Cool and use 1 μl of the supernatant in RT-PCR (with 25 μl reaction volume). The supernatant may be used for RT, undiluted or diluted 5- to 100-fold with sterile water. See **Note 2**.

Notes

1. Fresh, frozen or dried tissue can be used. A disadvantage of the procedure is the small samples, which might be a problem when screening plants with uneven virus distribution. Sampling more pieces per plant treated with a higher volume of extraction solution could possibly alleviate this problem. Homogenization of the tissue (followed by centrifugation) has been shown to increase the assay sensitivity about tenfold (Thomson and Dietzgen, 1995).

2. Depending on the content of substances inhibitory to the RT or PCR, the supernatant may be used undiluted or five- to 100-fold diluted with sterile water. When used undiluted, however, an RT (and PCR) buffer without KCl is required and a compensatory increase in the $MgCl_2$ concentration may be needed.

PROBLEMS IN NUCLEIC ACID EXTRACTION. As indicated in Sample preparation, Section 6.2.2, NA extraction may sometimes be tricky. If none of the above or other published methods works for the pathogen or host concerned, a commercial NA extraction kit could be tried in parallel extractions. If succeeding only with the kit, one might be able to adjust one of the protocols so as to obtain satisfactory results. Another option is to use the immunocapture method described below.

Immunocapturing of virus, protocol

Instead of extracting total nucleic acids for RT-PCR from plants or seeds, capturing the target virions immunologically from crude plant or seed extracts can eliminate the problems of co-extracted PCR inhibitors (Wetzel *et al.*, 1992). Immunocapturing (IC) is usually carried out in the PCR tube and corresponds to the first steps of DAS-ELISA, where a plastic surface is coated with virus-specific IgG, followed by incubation with the crude extract, from which virions present bind to the coated surface. After removal of the extract, followed by washing, RT-PCR is carried out. The following procedure for the IC step is essentially according to Jacobi *et al.* (1998) and Gillaspie *et al.* (2000), who used the method for detection of tobamoviruses and potyviruses, respectively.

Note that, when testing whole seeds, this method also results in detection of both transmissible and non-transmissible virus (see Sample preparation, Sections 5.2.2 and 6.2.2).

EQUIPMENT AND MATERIALS.

a. PCR tubes, 0.2–0.5 ml. See **Note 1** after procedure.
b. Facility for incubation at 37°C (oven with thermostat).
c. Adjustable pipettes (0.1–20 μl, 20–200 μl, 200–1000 μl).
d. Microcentrifuge, 15,000 *g*.
e. Microcentrifuge tubes, 1.5 ml.
f. Polyethylene bags, *c.* 7 mm × 10 mm, sturdy type, or mortars and pestles.
g. Test tubes, 3–4 ml, for making dilutions.
h. Test sample plant or seed material.
i. Known healthy material for negative control.
j. Known infected material for positive control.

Reagents, solutions and buffers

a. IgG, virus-specific, preferably from a broad-spectrum antiserum. (For fractionation of the IgG from whole antiserum, see Section 5.2.3, p. 104).

b. Coating buffer (see Section 5.2.3, p. 108).

c. Washing buffer (see Section 5.2.3, p. 108).
d. Extraction buffer (see Section 5.2.3, p. 108).

PROCEDURE.

Coating with IgG
1. Label the required number of PCR tubes and add 50 µl per tube of trapping antibodies, diluted with *coating buffer* to give 2 µg IgG/ml. See **Note 1** below.
2. Close the tubes and let them incubate for 2–3 h at 37°C or overnight at 4°C. While incubating, go to step **4**.

Washing
3. Pipette out the coating solution. Wash by adding by pipette 200 µl of *washing buffer* to the tube, spinning for 15 s at 5000 rpm and pipetting off the washing buffer.

Extracts
4. Grind the test material (plant tissue, soaked seeds or seed flour) in *extraction buffer*, using tissue:buffer 1 : 5 or 1 : 10. Make a final dilution of 1 : 20 to 1 : 100 with extraction buffer. Seed suspensions are centrifuged at 15,000 *g* for 5 min and the supernatants used.
5. Each tube is loaded with 50 µl of tissue extract. Load in the following sequence: negative control, test sample, positive control.
6. The tubes are closed and incubated for 2–3 h at 37°C or overnight at 4°C.

Washing
7. Pipette off the extracts and wash twice with *washing buffer* as described in step **3**, followed by a final wash with dd H_2O. Blot the tubes dry on a tissue and close them. The tubes may be short-term stored at 4°C.

Disruption of captured virus
8. Before performing the RT-PCR, the attached virions may be treated to separate viral RNA from the protein coat (see **Note 2**).

Notes
1. Gillaspie *et al.* (2000) observed a variation in antibody binding capacity between PCR tube lots. In order to secure a firm attachment of antibodies, therefore, they routinely treated the tubes before use as follows: Soaking in 0.1 M HCl for 15 min, rinsing in *washing buffer*, soaking in 4 M NaOH for 15 min, rinsing in washing buffer, soaking in 95% ethanol, and finally air-drying.
2. Disruption of captured virions prior to RT-PCR has resulted in increased test sensitivity according to Jacobi *et al.* (1998), and this was also done by Gillaspie *et al.* (2000) and others. According to other reports, however, such a treatment did not improve sensitivity (Dietzgen, 2002).

In a one-tube IC-RT-PCR, using the same mix for RT and PCR, Jacobi *et al.* (1998) employed the following treatment for disruption of IC-trapped virions: the reaction mix containing all ingredients, except enzymes, was added to the tubes and then heated to 95°C for 5 min, immediately followed by chilling on ice for 5 min. Subsequently, RT and PCR enzymes were added and RT-PCR performed. In addition to the standard components (see protocols below) this reaction mix contained 0.3% Triton X-100.

As efficacies of virus disruption seem to vary substantially between different viruses, the best conditions for any particular virus are difficult to predict and need to be assessed individually (Jacobi *et al.*, 1998).

RT-PCR standard protocol

Below is described a standard protocol for RT-PCR, essentially as has been used by Singh (1998) for detection of potyviruses and PLRV, but also with elements from Jacobi *et al.* (1998) and Dijkstra and de Jager (1998). The reaction mixtures indicated are typical ones, but concentrations of especially dNTPs, primers and $MgCl_2$ may be adjusted to optimize the assay for different viruses, targets, etc.

For details, see Section 6.2.2; see also Optimization of RT-PCR, this section.

Note in the procedure that the RT reaction mixes used for extracted RNA and for immunocaptured virions are not the same.

PREVENTION OF CONTAMINATION. See the recommendations given at the beginning of this section (p. 184); in addition to these, the following precautions should also be taken to reduce the risk of contamination during PCR amplification (Dijkstra and de Jager, 1998):

1. PCR tubes should be manipulated in a Laminair airflow hood equipped with a UV light source, which is switched on a few hours before beginning work.
2. Place permanently in the hood for PCR use only: a microcentrifuge, a set of pipettes, tips, vortexer, various chemicals and disposable gloves.
3. Prepare dilutions of target nucleic acid in a different area of the laboratory.
4. Open PCR tubes with reagents only after having centrifuged the fluid for a few seconds to minimize the chance of contaminating gloves or pipettes.
5. Always close the PCR tubes immediately after each step.
6. Controls: include (a) a positive control tube, containing the target sequence of interest, (b) a negative control containing nucleic acid from virus-free plant material, and (c) a mock sample containing all PCR components except template nucleic acid.

In addition to these controls, an *internal control* to reveal false negatives may also be included (see p. 183 and Multiplex PCR, p. 197).

EQUIPMENT AND MATERIALS.

a. Thermal cycler.
b. PCR tubes, sterile, 0.2–0.5 ml, fitting the PCR cycler.
c. Adjustable pipettes (0.1–20 μl, 20–200 μl, 200–1000 μl), three sets, for reagent mix preparation, RT and PCR, respectively.
d. Pipette tips with filter.
e. Laminair airflow hood mounted with a UV light source. **Caution:** UV light is dangerous, especially to the eyes; shield the UV light source to minimize radiation, and wear protective goggles or mask.
f. Microcentrifuge (10,000–15,000 *g*).
g. Ice/water bath.

h. Vortexer.

i. Standard laboratory equipment, such as refrigerator, deep freeze, magnetic stir plate, glassware, tube racks, ice machine or similar, forceps, disposable gloves (powder-free), filter-paper.

j. Test samples in the form of RNA extracts or samples from immunocapturing.

k. Sample containing the target of interest, for positive control.

l. Sample containing RNA from virus-free plants or seeds, for negative control.

Reagents, solutions and buffers

a. Buffer for reverse transcriptase, 5 ×. The commercial enzyme supplier usually provides an optimal buffer. (For M-MLV RT, for example: 250 mM TRIS-HCl (pH 8.3 at room temperature), 375 mM KCl, 15 mM $MgCl_2$ and 50 mM dithiothreitol (DTT) (Promega™).

b. dNTP mix, containing 5 mM of each of the four dNTPs. Commercially available. Store in freezer at –20°C in small aliquots. Susceptible to damage by repeated freeze–thaw cycles.

c. Primers, adjusted to 10–50 µM; obtained custom-built from commercial laboratories or synthesized on own synthesizer.

d. RNase inhibitor, e.g. RNasin® (Promega™) 40 U/µl. Follow manufacturer's instructions.

e. RT (reverse transcriptase), M-MLV RT, 200 U/µl or AMV RT 20 U/µl. Commercially available.

f. DTT (dithiothreitol), 0.1 M (e.g. Merck™). **Caution:** harmful, irritant; work in fume cupboard and/or use mask. *Note*: commercial RT buffers may already contain DTT at 50 mM, which is sufficient to give the required concentration of 10 mM in the RT reaction volume below.

g. Triton X-100, 0.8% in sterile dd H_2O – for IC-RT-PCR only.

h. Buffer for PCR, 10 ×. The commercial enzyme supplier usually provides an optimal buffer. (For *Taq* polymerase, for example: 100 mM TRIS-HCl (pH 9.0 at room temperature), 500 mM KCl, 15 mM $MgCl_2$ and 1% Triton X-100) (Promega™).

i. $MgCl_2$, 25 mM, to optimize the Mg^{2+} concentration in the reaction mix. Check the recipe of commercial enzyme buffers for already-added $MgCl_2$. Store at –20°C. Note that $MgCl_2$ solutions form a concentration gradient when frozen: vortex before use.

j. Taq DNA polymerase, 5 U/µl. Commercially available.

k. Mineral oil, if required (light type, RNase-free).

PROCEDURE.

I. *RT of extracted total RNA*

1. Prepare an *RT master mix* in an amount sufficient for the number of sample reactions, including a mock sample. Add the ingredients in the sequence shown. Prepare in slight excess to allow for pipetting errors. The quantities are for each 1 × 20 µl reaction volume (for RT-PCR optimization, see p. 199):

Reaction component	Final concentration (20 μl)
4 μl RT buffer (for M-MLV or AMV RT), 5 ×	1 ×
2 μl DTT, 0.1 M (if needed; see above)	10 mM
2 μl dNTP mix (5 mM each)	0.5 mM each
1 μl downstream primer (10–50 μM)	0.5–2.5 μM
0.5 μl RNasin (40 U/μl) or similar	20 U (RNasin)
1 μl RT: M-MLV RT (200 U/μl) or AMV RT (20 U/μl)	200 U or 20 U
Sterile dd H_2O up to a final volume of 15 μl	

2. Depending on the initial dilution of nucleic acid extracts, extraction method and expected presence of inhibitors, the extracts are used for RT either without further dilution, or diluted to 10^{-1} to 10^{-3} with sterile distilled H_2O.

3. Label the required number of tubes and add 5 μl of nucleic acid extract to each tube. Start with the negative control and the mock sample (5 μl sterile dd H_2O), then add samples, and finally the positive control. Close tubes. See **Note 1** below.

4. Using a preheated block heater or a water bath, denature the RNA by heating for 5 min at 65–70°C, immediately followed by cooling on ice for at least 5 min.

5. Spin samples for a few seconds to remove condensation before opening the tubes.

6. Dispense 15 μl of the *RT master mix* to each tube while placed on ice. Load in the same order as in step **3**, using a fresh pipette tip for each solution, close, mix well by vortexing and centrifuge briefly.

7. Let the samples incubate for 10 min at room temperature, then for 30–60 min at 42°C in the thermocycler. Stop the reaction by heating to 95°C for 2–5 min (to inactivate reverse transcriptase).

8. Proceed with PCR from step **15**, or store the cDNA at –20°C.

II. *RT of immunocaptured virions*

9. Prepare an *RT master mix* as described in step **1** above, but with the difference that the mix below further contains *Triton X-100*, but *no RT enzyme*, which is added later on.

Reaction component	Final concentration (20 μl)
4 μl RT buffer (for M-MLV or AMV RT), 5 ×	1 ×
2 μl DTT, 0.1 M (if needed; see above)	10 mM
2 μl dNTP mix (5 mM each)	0.5 mM each
1 μl downstream primer (10–50 μM)	0.5–2.5 μM
0.5 μl RNasin (40 U/μl) or similar	20 U (RNasin)
7.5 μl Triton X-100 (0.8%)	0.3%

10. To the tubes with captured virions is added 17 μl per tube of the *RT mix* made in step **9**. While placed on ice, add to the tubes in the following order:

negative control tube(s), sample tubes, and positive control tube(s), using a fresh pipette tip for each solution. Include also a mock sample. Close tubes.

11. Using a preheated block heater or a water bath, the mixture is heated for 5 min at 65–70°C (to denature RNA), **or** for 5 min at 95°C (to help separating RNA from virions; see **Note 2** under the immunocapturing protocol); in both cases followed by immediate chilling on ice for at least 5 min.

12. Spin samples for a few seconds to remove condensation before opening the tubes.

13. Dispense to each tube while placed on ice: 3 μl consisting of 1 μl of *M-MLV RT or AMV RT* and 2 μl of *sterile dd H_2O*. Load in the same order as in step **10**, using a fresh pipette tip for each solution, close, mix well by vortexing and centrifuge briefly.

14. Proceed with the RT as described in steps **7–8** above.

PCR

15. Prepare a *PCR master mix* in an amount sufficient for the number of sample reactions. Add the ingredients in the sequence shown. Prepare in slight excess to allow for pipetting errors. The quantities are for each 1 × 50 μl reaction volume (for RT-PCR optimization, see p. 199):

Reaction component	Final concentration (50 μl)
5 μl PCR buffer, 10 ×	1 ×
3 μl $MgCl_2$, 25 mM	1.5 mM
2 μl dNTP mix (5 mM each)	0.2 mM each
1 μl downstream primer (10–50 μM)	0.2–1 μM
1 μl upstream primer (10–50 μM)	0.2–1 μM
0.5 μl *Taq* DNA polymerase (5 U/μl)	2.5 U
Sterile dd H_2O up to a final volume of 45 μl	

16. Dispense 45 μl of the *PCR mix* into each of the required number of labelled tubes, including a mock sample. Place tubes on ice.

17. Add to the PCR mix 5 μl of *cDNA* from the RT tubes in the order: negative control, samples, and positive control. Add 5 μl of sterile distilled H_2O to the mock sample. Close tubes, mix by vortexing and centrifuge briefly to drain the walls.

18. For thermocyclers requiring mineral oil, one drop is layered on top of each sample. Close tubes.

19. PCR-amplify by placing the tubes in the thermocycler, programmed as follows (for RT-PCR optimization, see p. 199):

First cycle: 95°C for 2–3 min (initial denaturation).
Amplification: 30–40 cycles of each at 94°C for 10–60 s (denaturation), 45–70°C for 30 s (annealing) – see **Note 2** – and 72–74°C for 30–60 s (extension).
Last cycle: 72–74°C for 5 min (final extension).

20. Store the samples at 4°C or – 20°C until analysis of the amplified product (Section 6.2.4). Samples can be stored for some hours in the thermocycler if it has a refrigeration function. For trouble shooting, see p. 216.

Notes

1. The manufacturer of RNasin® recommends addition of the RNase inhibitor (0.5 µl) and the downstream primer (1 µl) to each RNA tube prior to step **4**, RNA denaturation. For convenience, these two components can be added before adding the RNAs. The remaining part of the RT mix (13.5 µl) is then added in step **6**. (To ease dispensing, the two components plus 1.5 µl H_2O could be added together, and the remaining RT mix be adjusted to 12 µl.)

2. The annealing temperature depends on the Tm of the primer pair (see Primers, Section 6.2.2).

One-tube RT-PCR, protocols

RT-PCR may be performed as a one-tube reaction, either in two steps or in one step. In a *one-tube, two-step* procedure, the RT reaction is carried out in an RT buffer, i.e. under optimal RT conditions. Thereafter, the PCR is performed in the same tube after adding the PCR mix containing the ingredients shown in step **15** in the above protocol, except the downstream primer.

Another option is to use the same mix for RT and PCR, i.e. a *one-tube, one-step* reaction; however, in the case of difficult targets or low template numbers, it is better to perform RT and PCR separately (Seal and Coates, 1998). For optimal results it is important to follow the enzyme manufacturer's instructions or those of commercial kits for one-tube, one-step RT-PCR, e.g. Titan™ One Tube RT-PCR System (Roche Diagnostics™) or SuperScript™ III One-Step RT-PCR System (Invitrogen™), if used.

Below are two examples of procedures in condensed form. The first is for a one-tube, two-step RT-PCR, essentially as has been used at DGISP, following a protocol from M. Schönfelder and S. Winter (DSMZ, Germany 2001, personal communication). The next is for a one-tube, one step RT-PCR with ranges of reaction mix components and conditions used by Mumford and Seal (1997), Mackenzie *et al.* (1998), Nassuth *et al.* (2000), Nolasco *et al.* (2002) and Olmos *et al.* (2002) for detection of different RNA plant viruses.

ONE-TUBE, TWO-STEP RT-PCR. In this procedure two degenerate primers are used, which enable the detection of all potyviruses with resulting 'amplicon' sizes of 500 to 800 bp.

Note in the procedure that the RT reaction mixes used for extracted RNA and for immunocaptured virions are not the same.

Equipment and materials are as described for the standard protocol above, whereas some items of reagents, solutions and buffers differ as follows (the M-MLV RT and *Taq* polymerase, RT buffer and PCR buffer used are from Gibco™; for other makes, the recommended concentrations and compositions may vary):

b. dNTP mix, containing 25 mM of each dNTP.

c. Primers: A downstream primer with degeneracies at two positions, 'Oligo dT Bam HI anchor': 5′-GCG GGA TCC TTT TTT TTT TTT TTT TVN-3′, and an upstream primer with degeneracies at four positions, 'WCIENGTS', within the conserved WCIENGTS motif of all potyviruses: 5′-TGA GGA TCC TGG TGY ATH GAR AAY GG-3′; 100 μM of each primer. Optimum annealing temperature: 53°C.

d. RNase inhibitor, 10 U/μl (Gibco™).
g. Triton X-100, *0.7%* in sterile dd H_2O – for IC-RT-PCR only.
i. $MgCl_2$, 50 mM (Gibco™).

Procedure

I. *RT of extracted total RNA*

1. Prepare an RT master mix in same way as in the standard protocol above, but with the following components and quantities (per reaction):

Reaction component	Final concentration (20 μl)
4 μl RT buffer, 5 ×	1 ×
2 μl DTT (0.1 M)	10 mM
1.25 μl dNTP mix (25 mM each)	1.56 mM each
1 μl downstream primer (100 μM)	5 μM
0.25 μl RNase inhibitor (10 U/μl)	2.5 U
0.5 μl M-MLV RT (200 U/μl)	100 U
Sterile distilled H_2O up to 15 μl	

2. Continue with steps **2–6** in the standard protocol above.
3. Place the tubes in the thermocycler and allow incubating at 37°C for 10 min, followed by 42°C for 30–40 min.
4. Proceed with PCR from step **11**, or store the cDNA at –20°C.

II. *RT of immunocaptured virions*

5. Prepare a master mix similar to that in step **1**, this protocol, but *without* reverse transcriptase and H_2O. Instead, add 8.5 μl of a *0.7% solution of Triton X-100* (to give a 0.3% concentration in the final volume, i.e. in 20 μl).
6. To the tubes with captured virions is added 17 μl per tube of the RT mix, made in step **5**. While placed on ice, add to the tubes in the following order: negative control tube(s), sample tubes, and positive control tube(s), using a fresh pipette tip for each solution. Include also a mock sample. Close tubes.
7. Using a preheated block heater or a water-bath, the mixture is heated for 5 min at 65–70°C (to denature RNA), **or** for 5 min at 95°C (to help separate RNA from virions; see **Note 2** under the immunocapturing protocol); in both cases follow by immediate chilling on ice for at least 5 min.
8. Spin samples for a few seconds to remove condensation before opening the tubes.
9. Dispense to each tube while placed on ice: 3 μl, consisting of *0.5 μl of M-MLV RT* and *2.5 μl of sterile dd H_2O*. Load in the same order as in step **6**, using a fresh pipette tip for each solution, close, mix well by vortexing and centrifuge briefly.

10. Proceed with the RT as described in steps **3–4** above.

PCR

11. Prepare a PCR mix (per reaction):

Reaction component	Final conc. (in 100 µl)
10 µl PCR buffer, 10 ×	1 ×
4 µl $MgCl_2$ (50 mM)	2 mM
1 µl upstream primer (100 µM)	1 µM
0.5 µl *Taq* DNA polymerase (5 U/µl)	2.5 U
Sterile distilled H_2O up to 80 µl (see **Note** below)	

12. Remove the cDNA synthesis tubes from the cycler and add 80 µl of the *PCR master mix* per tube, giving a 100 µl PCR volume.
13. PCR programme:

> 95°C for 3 min (reverse transcriptase deactivation and initial denaturation).
> Amplification: 35 cycles of each 94°C for 60 s (denaturation), 53°C for 90 s (annealing), and 72°C for 60–90 s (extension).
> Last cycle: 72°C for 10 min (final extension).

14. Store the samples at 4°C or –20°C until analysis of the amplified product (Section 6.2.4). Samples can be stored for some hours in the thermocycler if it has a refrigeration function. For trouble shooting, see p. 216.

Note
A smaller total reaction volume may be used, e.g. 50 µl, by reducing the PCR reaction mix volume to 30 µl and reducing the ingredient quantities, except of polymerase, correspondingly.

ONE-TUBE, ONE-STEP RT-PCR. Table 6.3 shows examples of combined RT-PCR mixes and the cycling conditions applied in the five reports mentioned above.

Multiplex PCR

The simultaneous detection of more than one virus in each PCR assay – multiplex PCR or M-PCR – would be preferable for many purposes and, once developed/optimized, obviously reduce costs and workload. Combined primer pairs have been used to detect simultaneously up to five (Bariana *et al.*, 1994) or six (Bertolini *et al.*, 2001) viruses affecting subterranean clover and olive trees, respectively. Co-amplification of fewer targets, such as duplex PCR (d-PCR), is described in several reports, for example, when an internal control (see below) is incorporated in PCR virus assays.

A successful co-amplification of multiple targets in a single reaction primarily depends on the compatibility of the primers used in the PCR. First of all, these primers must have similar melting temperatures, i.e. similar size and CG content, to ensure that they anneal to and dissociate from their DNA targets at approximately the same temperatures, so that each amplification can proceed at the

Table 6.3. Examples of one-tube, one-step RT-PCR procedures.

	Nolasco *et al.* (2002)	Mumford and Seal (1997)	Nassuth *et al.* (2000)	Olmos *et al.* (2002)	Mackenzie *et al.* (1998)
Reaction mix					
TRIS-HCl	10 mM, pH 8.8	10 mM, pH 8.8	10 mM, pH 8.3	10 mM, pH 8.9	Manuf.[a]
KCl	50 mM	50 mM	100 mM	50 mM	Manuf.
Additive	0.08% Nonidet-P40	0.01% Triton X-100	5 mM DTT + 2% sucrose	0.3% Triton X-100	5 mM DTT
$MgCl_2$	4 mM	1.5 mM	1.5 mM	?	2.5 mM
dNTP, each	200 μM	250 μM	200 μM	250 μM	200 μM
Primer, each	200 nM	20 pM	0.5 μM	1 μM	15–35 pM
RNase inhibitor	3 U	–	–	–	10 U
RTase	7.5 U M-MLV RT	2.5 U AMV RT	0.1 U AMV RT	0.25 U AMV RT	Manuf.
Taq polymerase	1 U	1.25 U	2 U	0.5 U	Manuf.
Template	(IC)	(IC)	2.5 μl	5 μl	1 μl
Final volume	50 μl	50 μl	25 μl	25 μl	25 μl
Reaction conditions					
RT	38°C/45 min	50°C/10 min	54°C/45 min	42°C/45 min	45°C/30 min
Deactiv. of RT/ initial denaturation	94°C/2 min	95°C/4.5 min	n.i.	92°C/2 min	94°C/2 min
PCR cycling: Denaturation	92°C/30 s	95°C/30 s	94°C/30 s	92°C/30 s	94°C/30 s
Annealing	52°C/30 s	55°C/60 s	54°C/45 s	50–60°C/30 s	58°C/45 s
Extension[b]	72°C/30 s	72°C/60 s	72°C/60 s	72°C/60 s	68°C/120 s
No. of cycles	30	35	35	40	35

[a]According to manufacturer's instructions (Titan™, Boehringer Mannheim).
[b]Final extension for 5–10 min in last cycle.
–, None.
n.i., no information.

selected temperature. It is therefore important to select primers with only slightly differing Tms (few °C). The GC content should be within 40–60% and the Tm 50–60°C (Mahony and Chernesky, 1995). The choice of primers should also be such that the resulting amplicons are within the same size range (150–500 bp), allowing each of them to be synthesized effectively and at equal rates. Within this range, the amplicons must differ by at least 40–50 bp in size if they are to be visualized by gel electrophoresis. When many primers are present in a reaction instead of the usual two, the chance of primer-dimer formation is greater, and therefore it is crucial that there is no base pair homology between any of the primers (see also Primers, Section 6.2.2).

Optimal concentrations of both RT and PCR components, especially of Mg^{2+}, are critical, and development of an M-PCR therefore requires considerable effort and time for optimization by the trial-and-error method (see Optimization of RT-PCR below).

For M-PCRs, resulting in many non-specific products obscuring the specific amplicon bands in a gel, it is important, during development of the assay, to verify the identity of amplified products by hybridization to labelled probes (see Section 6.2.4).

INTERNAL CONTROL. Including an internal control in RT-PCR to help exclude false negatives involves the use of a d- (or an m-) RT-PCR. Various reports describe co-amplification of different plant genomic mRNA sequences occurring in the assayed tissue. Presence of an mRNA-specific band and no virus/viroid-specific band indicate that absence of the latter band is not due to RNA degradation during NA extraction or to failures in other phases of the assay, i.e. the result is not a false negative. Use of different plant mRNA sequences is reported, but not all function equally well. Thompson *et al.* (2003) described the application of mRNA-specific internal control primers, which produced a PCR fragment of the expected size in 25 of 27 plant species tested, including both monocot and dicot herbaceous plants and trees. Sequence data from the nicotinamide adenine dinucleotide (NADH) dehydrogenase ND2 subunit (*ndhB* gene) of the complete genome sequence of *Atropa belladonna* L. (Accession No. AJ 316582) was used to design the Atropa Nad2 primers (Schmitz-Linneweber *et al.*, 2002). The *ndhB* gene consists of two exons separated by one intron (Fig. 6.6). The upstream primer was located across the splice junction of the mRNA expressed, with only the last two 3′ end nucleotides being in the second 'exon', and two downstream primers were designed so as to amplify optimally either a short (188 bp) or a longer (571 bp) fragment. Thus one of the primer pairs could be chosen to produce a suitable size of the internal control fragment (Fig. 6.6).

The target amplified is part of the mRNA expressed from the chloroplast DNA of *A. belladonna.* Since the target was shown to be amplifiable from a high number of plant species tested (25 of 27), the internal control primers are assumed to have a potential applicability for a wide range of other species (Thompson *et al.*, 2003). It has been reported that the *ndhB* gene may also have a non-photosynthetic role; it is possible, therefore, that this internal control system is applicable also for RT-PCR detection of virus in non-germinated seeds, but it remains to be tested (J.R. Thompson, USA, 2003, personal communication).

Optimization of RT-PCR

For the maximal specificity and sensitivity of an RT-PCR reaction, the reaction conditions must be optimized. This can be tricky, in particular if RT- or PCR-inhibitory compounds are present in the NA extracts. Inhibitor problems are usually eliminated when using IC-RT-PCR, although its use depends on the availability of a specific antiserum.

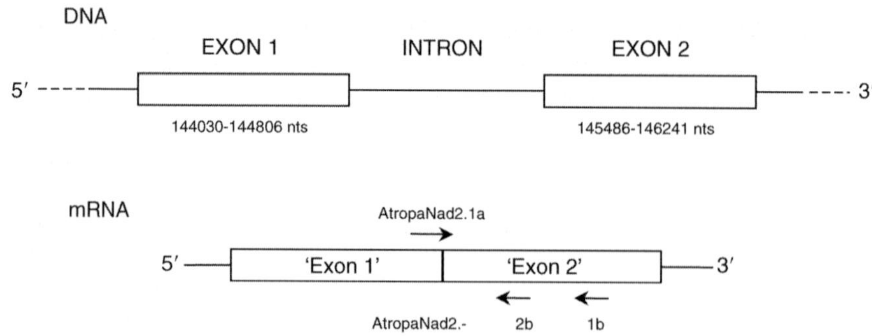

Primer sequences (5′–3′)		Product size	Location in *ndh*B	Acc.No.
AtropaNad2.1a	GGACTCCTGACGTATACGAAGGA**TC**		144784-145487	AJ316582
AtropaNad2.1b	AAACAACGCTTGTAAGGAGTCC	571 nts	146014-146035	
AtropaNad2.2b	AGCAATGAGATTCCCCAATATCAT	188 nts	145627-145650	

Fig. 6.6. Schematic representation of the location of internal PCR control primers in the *ndhB* gene. Uppermost, the *ndhB* gene in chloroplast DNA, consisting of two exons (1 and 2) separated by an intron. Nucleotide positions are indicated. In the middle, the *ndhB* gene in chloroplast mRNA showing the location of the Atropa Nad2 primers across the splice junction (only expressed mRNA is thus detected, not genomic DNA). Below, the primer sequences, size of products and primer locations are shown; only two nucleotides (in bold) of AtropaNad2.1a are located in 'Exon 2'. From Thompson *et al.* (2003) with permission from the first author and the copyright holder © Elsevier.

Optimizing the conditions for an RT-PCR is basically similar to optimizing ELISA conditions by a chequerboard titration, i.e. optimal conditions for each individual reagent are determined in combination with other reagents (Mahony and Chernesky, 1995).

UNIPLEX RT-PCR. With reference to the standard RT-PCR protocol above, reaction conditions can be varied as described below to optimize the assay result for different virus/viroids, targets and hosts.

RT. The standard concentration of dNTPs (0.5 mM each) may be varied, and so may the primer concentration. The amount and quality of RNA template are critical. If the presence of RT- or PCR-inhibitors is suspected, a range of different RNA template dilutions ($10^{-1} - 10^{-3}$) may be tested or a better extraction method used.

PCR. Cycling temperatures and concentrations of Mg^{2+}, primers and DNA polymerase are primary factors that influence the specificity of primer annealing.

An ideal starting point for testing new primers is an annealing temperature 5–10°C lower than the lowest Tm of the primers used. An improved specificity of the primers can then be obtained by gradually increasing the temperature (see cycling conditions below). Too high an annealing temperature, however, may reduce amplification efficiency (Seal and Coates, 1998). To reduce mis-priming and increase the yield of the desired product, *touchdown PCR* can be used. This method has been used, for example, by Seoh *et al.* (1998) in a PCR assay employing degenerate primers. In touchdown PCR, the annealing temperature starts several degrees above the calculated optimal annealing temperature and is gradually lowered in subsequent cycles.

- *Mg^{2+}*: The concentration of Mg^{2+} affects primer annealing, and the *Taq* polymerase requires free Mg^{2+} ions. The amount of free Mg^{2+} is influenced by the concentrations of template, primers and dNTPs, which all chelate these ions. Therefore, for all new template–primer combinations, a titration with varying $MgCl_2$ concentrations between 1.5 and 4 mM should always be made. Note that EDTA, if present, also chelates Mg^{2+}.
- *Primers*: The primer concentration should not be higher than 1 μM each, and use of the minimum concentration that does not limit amplification is preferable, as it reduces non-specific priming. However, for a degenerate primer with a high degree of degeneracy, more than 1 μM may be required.
- *dNTPs*: An optimal concentration of each dNTP may be lower than 200 μM – not higher, as this will result in increasingly poor fidelity (Seal and Coates, 1998). Rybicki (2001a) recommends 50–200 μM of each dNTP.
- *Polymerase*: *Taq* polymerase is generally used at 0.5–2.5 U/50 μl reaction. Too high polymerase concentration can result in non-specific amplification. Enzyme manufacturers recommend optimal concentrations for their polymerases.

Hot start, i.e. adding the polymerase just after the initial denaturation, can improve sensitivity and specificity. To avoid opening the tubes, the hot start can also be accomplished by placing a bead of wax or Vaseline into the tube after adding the PCR mix, except the enzyme, and then putting the enzyme on top of the wax. When the temperature reaches *c.* 80°C, the wax melts, and the enzyme mixes with the rest of the reaction mix. The molten wax floats on top and seals the mix (replacing mineral oil). An alternative is a 'hot-start polymerase', i.e. a polymerase that becomes active only after heating to 95°C for some minutes, and therefore can be added directly to the reaction mix before start of cycling. According to S. Winter (DSMZ, Germany, 2003, personal communication), non-specific products may result, with PCR detection of viroids unless hot start is used.

- *cDNA concentration*: the amount of cDNA may also be adjusted. Too high or too low concentration may negatively influence the result. Great excesses of cDNA can bind all primers in the first cycle, preventing amplification.
- *KCl*: KCl is necessary in PCR to facilitate primer annealing, but concentrations higher than 50 mM usually inhibit *Taq* polymerase.
- *Reagents in general*: low primer, target, *Taq* and dNTP concentrations are preferred because they will normally result in a cleaner product and low background – however, sometimes at the cost of detection sensitivity (Rybicki, 2001b).
- *Cycling conditions*: the initial *denaturation* of a few minutes is necessary to ensure complete separation of the often long cDNA template. When the first PCR products have been synthesized, considerably shorter denaturation times are sufficient (about 30 s at 94°C for up to 500 bp targets). It should be noted that prolonged exposure of *Taq* polymerase to 94°C increases its decay.

  The *annealing* temperature can, as previously mentioned, be set 5–10°C lower than the lowest Tm of the primers. Additional PCRs are then run with gradually (2–5°C steps) increased temperatures until maximal specificity is obtained. The optimal annealing temperature may often be higher than the calculated Tm (Scheinert *et al.*, 2003). The most-often reported optimal duration for the annealing step is 30–60 s.

The most commonly used *extension* temperature (*Taq* polymerase) is 72°C. As a rule of thumb, 1 min extension should be allowed for every 1 kb of the amplification product. Most workers use a longer extension period of 5–10 min in the last cycle. However, according to others, e.g. Sambrook and Russell (2001), such prolonged extension time has no effect.

DUPLEX AND MULTIPLEX RT-PCR. The success of duplex/multiplex RT-PCR is dependent on careful optimization of reaction components and conditions. Before attempting co-amplification of multiple targets, the optimal conditions for each target must be determined in uniplex RT-PCRs.

In optimizing a duplex RT-PCR for detecting PVY and PLRV, Singh *et al.* (2000) found that the RT conditions were more critical than the PCR conditions. In uniplex RT-PCR, a concentration of each dNTP of 0.5 mM in RT was sufficient in yielding a PVY or PLRV band, but, in duplex RT-PCR, a minimum of twice the dNTP concentration (1.0 mM) during RT was required to produce distinct bands in PCR. Similarly, various proportions of downstream primers of the two viruses used during RT affected the outcome of duplex RT-PCR. An interaction of dNTPs and RNA template concentration in RT was also observed; a higher concentration of RNA was required at 0.5 mM dNTP concentration than at 1.0 mM dNTP concentration to yield bands in duplex RT-PCR.

In M-RT-PCR detection of five seed-transmitted viruses, Bariana *et al.* (1994) used the specific downstream primers in the RT step, whereas Nie and Singh (2000), as previously mentioned, used an oligo(dT) as a common cDNA primer in M-RT-PCR to synthesize a cDNA pool. Oligo(dT)s can prime all viruses that have a polyadenylated 3′end, which includes all *Potyvirus* members. Use of a common primer, in this case $(dT)_{21}$, in the RT step when possible, and mainly using specific primers in the PCR step only, reduces optimization work (Nie and Singh, 2000). However, an oligo(dT) primer is best suited for RT when targets to be detected are located in proximity to the virus genome 3′end (Nie and Singh, 2000).

Bertolini *et al.* (2001) simultaneously detected six olive tree viruses by M-RT-PCR, using specific primers and a one-tube, single-step procedure. Thompson *et al.* (2003) detected four strawberry viruses by M-RT-PCR using 10 μM each of a random primer and an oligo(dT) (both commercially available) as primers in the RT step.

Nie and Singh (2000, 2001) found that, for simultaneous detection of up to six targets (five potato viruses and PSTVd) with target lengths of 187–683 bp, when going from uniplex to multiplex RT-PCR, the adjustments of the PCR conditions, shown in Table 6.4, gave optimal results. The synthesis of the six-target cDNA pool was primed by a combination of an oligo(dT) and specific downstream primers, or was primed by random primers.

Henegariu *et al.* (1997) described a step-by-step protocol for optimizing M-PCR. When detecting simultaneously up to nine human chromosomal DNA

Table 6.4. Optimization of the PCR step[a] in multiplex RT-PCR (Nie and Singh, 2000, 2001).

Format	PCR buffer	$MgCl_2$	dNTP each	Primers each	*Taq* polymerase	Cycling (°C/min) Denat.	Anneal.	Extens.
Uniplex	1 ×[b]	1.5 mM	0.2 mM	50 ng	0.625 U	92°/1	57°/1	72°/1
Duplex	1 ×	2.5 mM	0.2 mM	50 ng	0.625 U	92°/1	57°/1	72°/1
Hexaplex	1 ×	4.0 mM	0.5 mM	Balanced	2.4 U	92°/1	60°/1	72°/2

[a]Reaction volume 25 μl.
[b]10 mM TRIS-HCl, pH 8.3, 50 mM KCl.

targets of *c.* 100–500 bp lengths, they found the following parameters especially important (in brief):

- The relative concentrations of primer pairs: use of equimolar concentrations first and then adjusting the concentration of each primer pair.
- The concentration of the PCR buffer: raising the buffer concentration, or only the KCl concentration, up to 100 mM improved the efficiency of M-PCR; and this effect was more important than using any of the adjuvants DMSO, glycerol or BSA.
- The cycling profile: when going from individual PCRs to M-PCR, a lower temperature and a longer time for the extension step, as well as a lower annealing temperature, improved the result.
- The balance between $MgCl_2$ and dNTP concentrations: an increased concentration of dNTPs is required in M-PCR, and was in this case 0.2–0.4 mM each. A higher $MgCl_2$ concentration is necessary at higher dNTP concentration to secure the presence of free Mg^{2+}.

Rybicki (2001a) recommends as a rule of thumb a 0.5–2.5 mM higher Mg^{2+} concentration than the total dNTP concentration.

See also Trouble shooting, p. 216.

6.2.4. Analysis of PCR products and protocol

Analysis of PCR products is probably still most commonly done by gel-electrophoresis in agarose gels, which are ethidium bromide-stained, enabling visualization of the bands in UV light. This system, however, is not well suited for large-scale testing, and therefore various attempts to develop alternative methods, e.g. systems that resemble ELISA formats, have been made (see Gel-free detection of PCR products, this section, and also Section 6.2.6).

Agarose gel electrophoresis protocol

A submarine horizontal gel electrophoresis apparatus is normally used for analysing PCR products on agarose gels. Gels with 1–2% agarose are most common, but 0.8–3% gels may be used for products ranging from very large to very small, respectively, i.e. small products are best seen on a 3% gel that has been run quickly (at relatively high voltage) (Rybicki, 2001b). Sambrook and Russell (2001) recommend gel concentrations of 2.0, 1.5, 1.2 and 0.9% w/v for DNA molecules up to 200, 300, 600 and 700 bp, respectively. Running at 4–10 volts/cm distance between the electrodes is normally used, but too slow (low voltage) migration may cause small products to diffuse, and too fast (high voltage) may cause bands to smear. A commercially available DNA size marker is loaded into the first slot in the gel and, in large gels, also the last slot; or loading only one slot in the middle may be adequate. Size markers consist of DNA digested by restriction enzymes into fragments of increasing lengths, for example, 50–750 bp, producing a 50 bp 'ladder'. Bands are normally visualized by ethidium bromide, which forms complexes with DNA that fluoresce in UV light.

The protocol for PCR product analysis below uses an agarose gel electrophoresis with ethidium bromide visualization, essentially as has been used routinely at DGISP.

EQUIPMENT AND MATERIALS.

a. Gel electrophoresis apparatus (Pharmacia Biotech GNA 200™ or similar).
b. Power supply (Pharmacia Biotech EPS 600™ or similar).
c. Microwave oven or water bath.
d. Adjustable pipette (0.1–20 μl).
e. Pipette tips with filter.
f. Parafilm®.
g. UV light source, preferably a UV transilluminator. **Caution:** use adequate shielding to protect the eyes especially against UV radiation.
h. Photographic equipment (different systems, using instant film or electronic imaging, are available).

Reagents, solutions and buffers

a. Agarose, nucleic acid grade.
b. EDTA, 0.5 M, pH 8.0 – see Appendix 2, p. 250.
c. TBE, electrophoresis buffer: TRIS-borate-EDTA buffer, 5 ×, pH 8.0 (see Appendix 2, p. 256 and **Note 2** after procedure). Alternatively, TAE (TRIS-acetate-EDTA buffer, 10 ×, pH 8.0) can be used (p. 255).
d. TE: TRIS-HCl-EDTA buffer, pH 7.4 – see Appendix 2, p. 253.
e. Ethidium bromide, stock solution: 10 mg/ml in TE buffer – see Appendix 2, p. 256. **Caution:** ethidium bromide is a powerful mutagen; wear double layers of protecting gloves when handling.
f. Gel loading buffer – see Appendix 2, p. 256.
g. DNA size marker, e.g. a 50 or 100 bp ladder, commercially available as a ready-to-use solution or diluted to 100–250 μg/ml in TE buffer.
h. Sterile distilled H_2O.

PROCEDURE. *Note*: opening the PCR tubes and transferring PCR products to the gel must be done in an area away from the PCR preparation area and by use of filtered pipette tips.

1. Mix agarose and *TBE* (1.5 g/100 ml 1 × TBE) in an Erlenmeyer flask. Make a volume sufficient for a gel of 3–5 mm thickness.

2. Melt in microwave oven until boiling. Swirl to ensure that the agarose is completely dissolved. Repeat boiling if necessary. Alternatively, dissolve the agarose in a water bath.

3. Cool to *c.* 60°C.

4. Add *ethidium bromide stock solution* to the agarose solution, 5 μl per 100 ml, to give a final concentration of 0.5 μg/ml. Mix well. **Caution:** wear double layers of protecting gloves. Ethidium bromide is a powerful mutagen and is moderately toxic. (It is possible to stain the gel in a 0.5 μg/ml ethidium bromide staining solution after the run is completed, but see **Note 1** below.)

5. Prepare the gel-casting tray by sealing both ends with tape. Place the combs in the correct position in the tray (0.5–1 mm above the bottom of the tray). Place the tray on a perfectly horizontal surface.

6. Pour the molten agarose slowly into the tray; try to avoid bubbles.

7. Cool for at least 30 min.

8. Meanwhile, add *ethidium bromide stock solution* to 1 × *TBE buffer*, 5 μl per 100 ml, to give a final concentration of 0.5 μg/ml. **Caution** – see step **4**. Pour the solution into the electrophoresis tank (sufficient to cover the gel by about 1–3 mm). See **Notes 1** and **2** below.

9. Remove the tape from the gel casting mould and place the gel in the tank, facing the correct way. Remember gloves! Gently remove the combs, taking care not to damage the slots.

10. Mix 5–10 μl of sample with 1–2 μl tracking dye (= *gel-loading buffer*) on a piece of Parafilm®. Pipette each mixture slowly into separate slots of the electrophoresis gel. Leave slot(s) – the first or the last, both or only the middle – empty for the DNA size marker. Use a new tip for each sample. Be careful not to cross-contaminate slots.

11. Mix 5 μl of *DNA size marker stock solution* (containing 100–250 μg DNA/ml) with 1–2 μl of *gel-loading buffer* per slot (one or two) and load.

12. Place the lid on the electrophoresis tank and connect the cables to the power supply. Run at 4–10 volts/cm distance between electrodes (see introduction to protocol) until the dye reaches the end of the gel (large products) or two-thirds down for small products.

13. When the run is completed, the power is switched off and the leads are removed. Remove the gel carefully from the tank and place it under UV light exposure (a UV transilluminator is preferable), wearing a double layer of disposable gloves and protecting glasses or shield. (If the gel is not already stained with ethidium bromide, it should be stained in a staining tank for 10 min before being placed under the UV light.) See also **Note 3**.

14. Photograph the gel with visualized bands. Examine the photograph for the presence of expected bands, comparing their size with DNA size markers. An example of a gel with bands is shown in Fig. 6.7.

15. Discard stained gels and buffers in a suitable container for hazardous waste.

Notes

1. It is preferable to include ethidium bromide in the gel and also in the buffer. Post-electrophoresis staining can result in band smearing, due to diffusion, and, if there is no ethidium bromide in the buffer, the dye runs backwards out of the

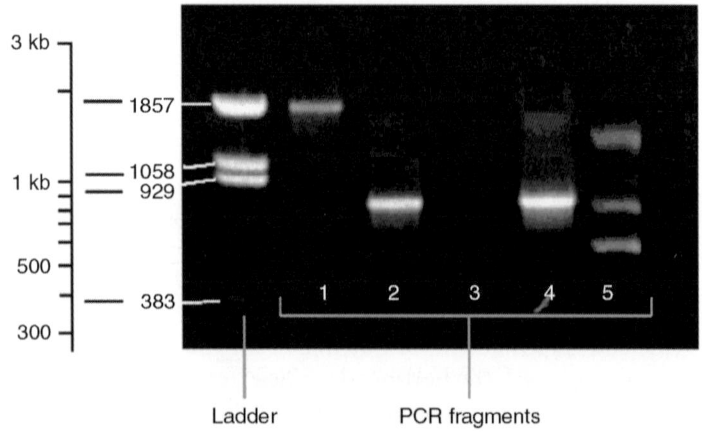

Fig. 6.7. Example of verification of PCR products by gel electrophoresis. Reproduced from Vierstraete (2002) with permission.

gel and smaller bands are stripped of dye and are thus not visible (Rybicki, 2001b).

2. It is important to use the same batch of electrophoresis buffer in both the electrophoresis tank and the gel. Small differences in ionic strength or pH create fronts in the gel that can greatly affect the mobility of DNA fragments (Sambrook and Russell, 2001).

3. Do not expose the gel to UV light longer than necessary – UV light breaks down DNA. Take photographs as soon as possible after running the electrophoresis, as the bands otherwise diffuse and become blurred.

Gel-free detection of PCR products

Among other methods to analyse PCR products are EIA-like systems in which amplified DNA is captured onto a solid surface (wells of a microtitre plate or a nylon membrane) and hybridized to a probe, specific to a part (20–50 nts) of the amplicon. Presence of the amplicon is then detected in different ways involving digoxigenin- or biotin-labelling of either amplicon or probe, and an antibody– or streptavidin–enzyme–substrate reaction, resulting in a colour signal. For details of hybridization techniques, see Sections 6.1.1 and 6.1.2.

Although involving more steps compared to gel electrophoresis, these methods result in a higher sensitivity, are better suited for high-throughput testing work and adaptation to automation, and require no use of the noxious chemical ethidium bromide. Further, in gel electrophoresis, both specific and non-specific amplification, products are detected and, if a non-specific band has the same or nearly the same size as the targeted amplicon, there is an increased risk of false positives. This risk is much less when the amplicon is specifically detected with a probe (Fig. 6.8).

PCR-ELISA. One gel-free method is PCR-ELISA (e.g. Pollini *et al.*, 1997; Rowhani *et al.*, 1998; Shamloul and Hadidi, 1999; Nolasco *et al.*, 2002).

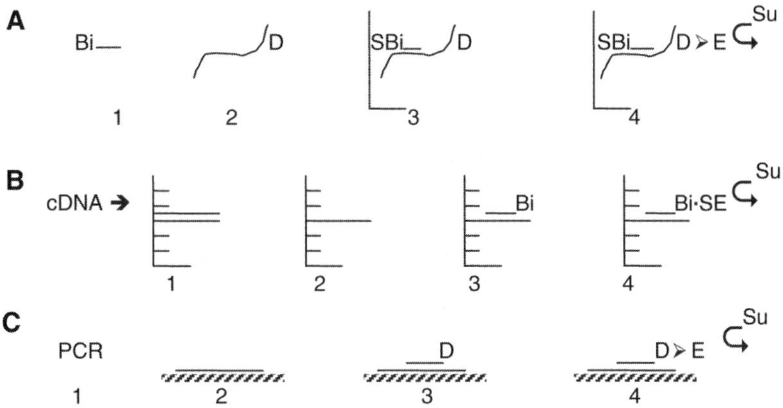

Fig. 6.8. Examples of gel-free amplicon detection systems and their principles: PCR-ELISA (A), DIAPOPS (B) and PCR-NASH (C). A1, biotin-labelled probe; A2, DIG-labelled PCR product; A3, DIG-labelled, denatured PCR product hybridizes with probe and the hybrid is immobilized in a streptavidin-coated microtitre well; A4, immunoenzymatic-colorimetric detection of DIG. B1, the reverse-transcribed target is PCR-amplified, during which the product is immobilized due to a preceding covalent linking of one of the primers to the well surface (short bars represent primer); B2, denaturation of amplicon; B3, hybridization with biotinylated probe; B4, streptavidin-enzyme-colorimetric detection of biotin. C1, amplification; C2, denatured PCR product immobilized on a nylon membrane; C3, hybridization with DIG-labelled probe; C4, immunoenzymatic-colorimetric detection of DIG. Bi, biotin; S, streptavidin; D, digoxigenin (DIG); E, enzyme; Su, substrate.

The PCR-ELISA type used in these four and a number of other reports employs a biotin-labelled oligonucleotide probe, complementary to an internal sequence of the target DNA. The probe is hybridized to the DNA amplicon previously tagged with digoxigenin. The hybridization takes place in a streptavidin-coated microtitre well, resulting in the immobilization of the biotinylated probe and thereby the PCR product.

After hybridization, presence of the amplicon is revealed by an anti-digoxigenin–enzyme–substrate reaction, resulting in a colour signal, which can be analysed on an ELISA reader (Fig. 6.8A).

In a number of reports, including the four mentioned above, a PCR-ELISA kit for semi-quantitative detection of PCR products, using digoxigenin as the reporter molecule, has been used. The digoxigenin technology is a proprietary technology of Roche Diagnostics™, and the steps of the protocol for the DIG-PCR-ELISA kit developed by the same company are, in brief:

1. A digoxigenin (DIG) reporter molecule is incorporated in the amplificate during the PCR assay by employing a kit containing a PCR DIG labelling mix.
2. A capture probe, 17–40 nt long, which specifically recognizes an internal sequence in the amplified target DNA, is 5′ end-labelled with biotin.
3. In each reaction tube, including controls, the PCR product and a denaturation solution are mixed and incubated for 10 min at room temperature.

4. A hybridization solution, containing the biotinylated probe, is added to the blend and mixed.
5. A volume of each mixture from **4** is pipetted into a well of a microtitre plate previously coated with streptavidin, and then allowed to incubate for 1–3 h at 37–55°C (hybridization and immobilization).
6. The solution is removed and the wells washed with a washing solution.
7. An anti-DIG–peroxidase conjugate is added to the wells and incubated for 30 min at 37°C, after which the wells are emptied and washed.
8. After incubation with a substrate solution, the absorbance of the coloured reaction is measured on an ELISA reader.

RT-DIAPOPS. A similar system, involving fewer steps and requiring no transfer of PCR products, is the co-called RT-DIAPOPS (reverse-transcription detection of immobilized, amplified product in a one-phase system) assay (Anon., 1994; Nicolaisen *et al.*, 2001). In this assay the PCR is carried out in the wells of NucleoLink™ strips (NUNC A/S, Denmark). The target is first reverse-transcribed. Prior to the PCR amplification, one of the primers, with an additional 10-thymidine residue linker at its 5′end, is irreversibly bound to the well surface. By a special treatment (NUNC) the primer becomes covalently linked, which causes the amplicon in the PCR to be subsequently immobilized in the well. Presence of the amplicon is then detected by hybridization to an oligonucleotide probe labelled with biotin or digoxigenin, which in turn is detected by streptavidin–enzyme or anti-digoxigenin–enzyme, respectively, followed by a colour-producing substrate and analysis of the signal on an ELISA reader (Fig. 6.8B).

PCR-NASH. Similarly, relatively few steps are required in PCR NASH (e.g. Bertolini *et al.*, 2001; Olmos *et al.*, 2002), in which the PCR product is immobilized on a nylon membrane and subsequently hybridized to an oligonucleotide probe labelled with digoxigenin. Presence of digoxigenin and thereby the specific amplicon is detected by anti-digoxigenin–enzyme and a substrate producing a coloured spot on the membrane (Fig. 6.8C).

OTHER GEL-FREE DETECTION SYSTEMS. PCR products may also be detected in a system where molecular amplification, probe–amplicon hybridization and visualization take place simultaneously: real-time fluorescent RT-PCR and NASBA (Section 6.2.6).

6.2.5. Detection of virus species and strains

For epidemiological investigations, for resistance breeding and for plant virus collections, it is necessary to determine not only the virus species but often also the strain, pathotype or isolate. Identification of viral pathogens at these levels can be done serologically using Mabs or by inoculation to differential plant hosts, but knowledge of the nt sequence of the virus genome, or parts thereof, opens the possibility for more precise and faster identification by PCR.

Diagnosis at the species and subspecies levels by means of PCR is usually done by one of the following methods, and sometimes by combining two of them:

- Use of primers complementary to variable regions of the viral genome.
- Nested PCR.
- Restriction fragment length polymorphism (RFLP).
- Single-strand conformation polymorphism (SSCP).

Primer specificity

For specific identification of viruses, strains or their subtypes, it is best to design primers mainly from the most discriminating regions (Singh, 1998). For example, for the genus *Potyvirus*, variability of the 5′-end-located P1 gene makes it possible, with specific primers, to separate the closely related viruses PVY and *Potatovirus A* (PVA, *Potyvirus*) (Singh, 1998). Two very closely related strains of SMV were distinguished by employing different primer pairs, both located in the CI (cylindrical inclusion) protein region (Omunyin *et al.*, 1996). A specific primer pair located in the coat protein gene of PPV has been used in PCR for direct identification of strain C of this potyvirus (Nemchinov and Hadidi, 1998).

Nested primer PCR

An increased level of specificity can be obtained with nested primer PCR. After amplification with one primer pair, a portion of the product is taken for re-amplification with an internally situated 'nested' primer pair (Fig. 6.9).

Four strains of PPV were differentiated by Szemes *et al.* (2001) using nested PCR. After the RT-PCR, using an external (general) PPV-specific primer pair located within the N-terminal coat protein region, the amplified product (1 μl of a 1000-fold diluted PCR mixture) was used as template for PCRs using different pairs of internal strain-specific primers. In addition, as also experienced by others, RT-PCR followed by nested PCR resulted in a remarkably stronger signal, showing higher sensitivity than RT-PCR alone. One of the purposes of diluting the first PCR product 1000-fold is to avoid the activity of the outer primer pair from the first reaction. In a spot nested RT-PCR, Dovas and Katis (2003a) added 5% of the first-stage product to the nested reaction, where inner primers with a Tm higher than that of the outer primers were used. With the higher Tm, a higher

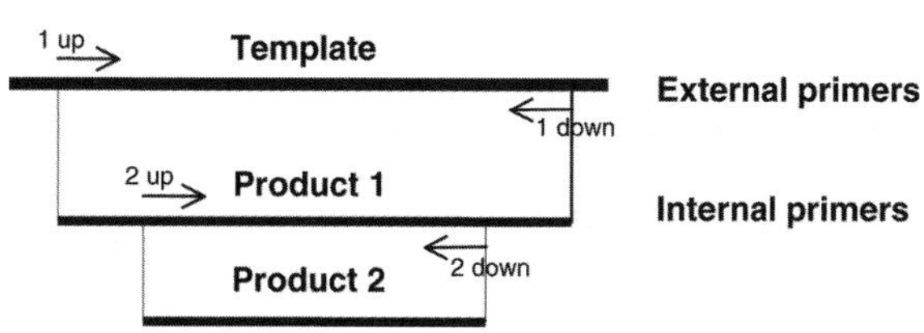

Fig. 6.9. Nested primer PCR, showing the positions of external (outer) upstream and downstream primers on the target cDNA, and the internal (inner) primer pair on the first amplified strand.

annealing temperature could be used in the nested reaction in order to prevent the annealing of the outer primers.

Dovas and Katis (2003b) used a multiplex nested RT-PCR for simultaneous detection and differentiation of clostero-, fovea- and vitiviruses in grapevine.

Semi-nested (hemi-nested) PCR has been used in detection and typing of PPV isolates (Olmos *et al.*, 1997) and as a method to differentiate species and pathotypes of tobamoviruses (Letschert *et al.*, 2002). In this technique, species-specific, external primers are used in the first PCR round, and an external primer and an isolate-specific (internal) primer are used in the second round of PCR.

A simplified form of the nested-PCR, a three-primer PCR, was developed by Weilguny and Singh (1998), which could differentiate the highly aggressive isolate PVY^{NTN} from the PVY^{N} group of strains. The method developed was applicable for large-scale testing of potato samples, where it was performed as a conventional single-primer-pair RT-PCR, but with the use of three primers located in the 214–669 nt sequence of the P1 region. An external downstream primer was used in RT, and an external upstream primer and an internal, nested downstream primer were used in PCR. When the annealing temperature was raised from about 55°C to 63°C, specific amplification of PVY^{NTN} occurred. In addition, the three-primer PCR showed very high sensitivity for NA extracts from leaves and dormant tubers. Singh *et al.* (1998b) further optimized the PVY^{NTN} three-primer PCR.

RFLP

Restriction fragment length polymorphism (RFLP) is a method to differentiate organisms by analysis of patterns derived from cleavage of their DNA. If the DNA of two organisms differs in the distance between sites specific for cleavage by a particular restriction endonuclease enzyme, the length of the fragments produced will differ when the DNA is digested with such an enzyme. Species, strains or isolates may be differentiated from one another, by comparing their fragment-length pattern in gel electrophoresis.

Restriction endonucleases (restriction enzymes) are isolated from a wide variety of bacteria, and they cleave DNA at junctures of specific sequences. The enzymes work only on double-stranded DNA, and restriction sites are often palindromes (same sequence forwards and backwards), for example:

5′ AGCT 3′
3′ TCGA 5′

Restriction enzymes are named from the bacteria they are isolated from. The first letter is for the genus, the two next for the species, the fourth for the strain, and the Roman numerals for the order of discovery. For example, *Eco*RI is the first restriction endonuclease derived from *Escherichia coli* strain RY 13. Restriction enzymes are essential tools in molecular cloning (for basic concepts, see, for example, Watson *et al.*, 1993).

Xu and Hampton (1996) used RT-PCR and restriction enzyme analyses (RFLP) to differentiate BCMV and BCMNV and a number of their pathotypes. Following sequence analyses of the products generated from two virus-specific primer pairs, the four restriction endonucleases *Xba*I, *Eco*RI, *Bam*HI and *Eco*RV were selected for digestion of the products (see brief protocol below).

Three cucumoviruses, CMV, *Peanut stunt virus* (PSV) and *Tomato aspermy virus* (TAV), were differentiated at species and strain level by RT-PCR-RFLP, using one primer pair and 11 restriction enzymes (Choi *et al.*, 1999). To differentiate species and pathotypes of tobamoviruses, Letschert *et al.* (2002) used semi-nested PCR (see above) and RT-PCR-RFLP involving six restriction enzymes.

Each type of restriction enzyme (RE) requires specific conditions for optimal activity. The manufacturer's recommendations for individual enzymes must be followed. The example below of a brief *protocol* for restriction enzyme digestion is as described by Xu and Hampton (1996) for *Xba*I, *Eco*RI, *Bam*HI and *Eco*RV:

- Mix the following:
  - 5 μl PCR product;
  - 0.8 μl 10 × restriction enzyme buffer (specific buffer composition for the enzyme in question);
  - 2.2 μl H_2O;
  - 2–5 U restriction enzyme.
- The mixture is incubated overnight at 37°C (some REs require special incubation temperatures).
- The digestion products are resolved in 1.2% agarose gels by electrophoresis and the bands analysed.

SSCP

Single-strand conformation polymorphism (SSCP) analysis is another method for differentiating strains. When the + and – strands of a dsDNA, often a PCR product, are separated by heat treatment in the presence of formamide, they form sequence-specific secondary, folded structures, which have distinctive electrophoretic mobilities in non-denaturing polyacrylamide gels (Koenig *et al.*, 1995; Dietzgen, 2002). The resulting polyacrylamide gel electrophoresis (PAGE) bands are usually visualized by silver staining. The efficiency of SSCP depends on various conditions, such as fragment length (optimal length < 300 bp (McQuaid, 1997)), temperature during electrophoresis and composition of the gel. Koenig *et al.* (1995) used IC-RT-PCR-SSCP as a rapid method for assigning *Beet necrotic yellow vein virus* (BNYVV, *Benyvirus*) isolates to known strain groups and for detecting mixed infections, minor variants or new strain groups. Sequence variants of *Apple stem grooving virus* (ASGV, *Capillovirus*) were also analysed by IC-RT-PCR-SSCP, as reported by Magome *et al.* (1999).

6.2.6. Variants of molecular amplification assays

A number of techniques and systems exist for molecular amplification and detection of amplificates, and new ones are being developed. The principles of some of the most popular techniques are described below. These and other systems are compared in a brief overview by Schweitzer and Kingsmore (2001).

Real-time fluorescent RT-PCR

Real-time fluorescent RT-PCR (RTF-RT-PCR) is a technique by which detection of the PCR product(s) takes place during amplification. The RT-PCR reaction exploits the 5′ nuclease activity of *Taq* polymerase for cleaving a labelled probe and generating labelled fragments (Holland *et al.*, 1991). The TaqMan® probe consists of an oligonucleotide (25–35 nts) with a reporter dye at the 5′ end of the probe and a quencher dye at the 3′ end. During the reaction, cleavage of the probe separates the reporter dye and the quencher dye, resulting in increased fluorescence of the reporter. Accumulation of PCR products during amplification is detected by directly monitoring the increasing fluorescence of the reporter dye (Fig. 6.10). When the probe is intact, the proximity of the reporter dye to the quencher dye results in suppression of the reporter fluorescence. The 5′ to 3′ nucleolytic activity of the *Taq* polymerase cleaves the probe between the reporter and the quencher only if the probe hybridizes to the target. The probe fragments are then displaced from the target, and polymerization of the strand continues (Fig. 6.10). The 3′ end of the probe is blocked by a phosphate group preventing extension of the probe during PCR. The TaqMan probe, synthesized specifically for a target sequence, is designed to have a Tm 10°C higher than those of the primers, to assure probe annealing before primer annealing. Best results are obtained by amplifying short sequences, i.e. amplicon sizes ranging from 50 to 150 bp (Taqman® Protocol, Applied Biosystems, accessible on http://www.appliedbiosystems.com). The reaction takes place in multi-well plates with 36 to 384 wells, and fluorescence is induced by laser light dispensed to wells via optical fibres, through which the resulting fluorescent emission is also directed to a camera-equipped spectrograph. The sample wells are monitored 8 times/min and the fluorescence data are stored electronically. The instruments monitor fluorescence at several different wavelengths, enabling simultaneous detection of several probes with different reporter dyes. The system is thus well suited to multiplex PCR (http://www.appliedbiosystems.com and http://www.med.unc.edu/anclinic/Tm.htm).

A number of plant viruses, e.g. viruses of potato and sugar cane (Schoen *et al.*, 1996; Mumford *et al.*, 2000a; Korimbocus *et al.*, 2002) have been detected by RTF-RT-PCR. Schoen *et al.* (1996) coupled an immunocapture step to the system by using paramagnetic beads coated with antiserum. Mumford *et al.* (2000a) detected two potato viruses simultaneously by using different fluorescent dyes. A semi-quantitative detection of CMV in lupin seed by this technique has also been reported (Jones, 2001). Salmon *et al.* (2002) found that the use of relatively short probes (< 20 nts) modified with a so-called minor groove binder (MGB) improved detectability of diverse isolates of *Apple stem pitting virus* (ASPV, *Foveavirus*) possessing high genomic variability.

According to the above authors, the main advantage of RTF-RT-PCR over other gel-free techniques is that no post-PCR manipulations are needed, thus reducing the risk of both contamination and false positives, as well as lowering costs of reagents, time and labour. The main disadvantage originally was the sixfold higher price of the combined PCR cycler and fluorescence monitor relative to a conventional cycler. However, a new generation of instruments is now available at about half the price, i.e. about three times the price of a conventional

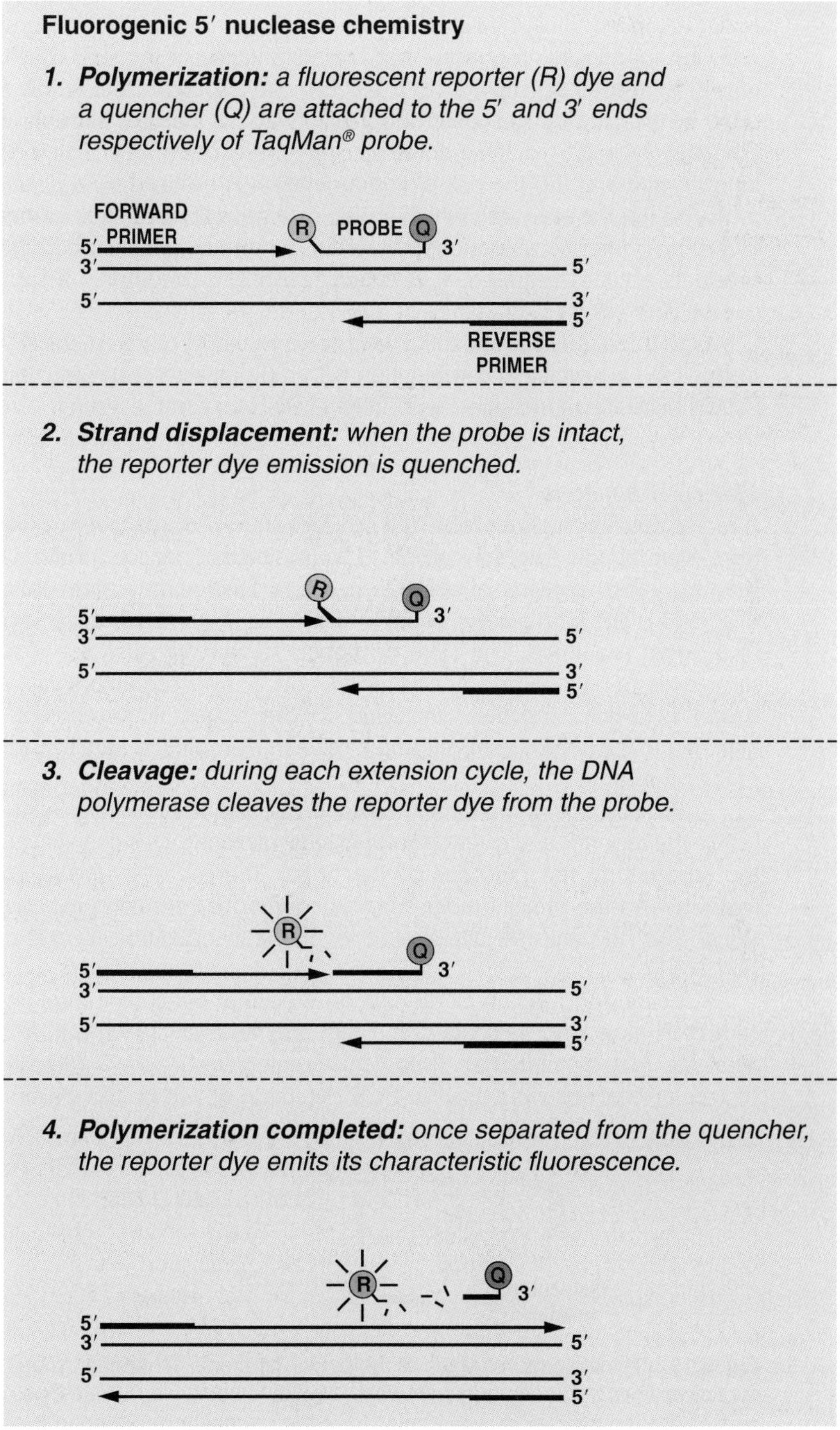

Fig. 6.10. Fluorogenic 5′ nuclease PCR (real-time fluorescent PCR) assay. From Applied Biosystems®, Denmark, with permission.

cycler (Applied Biosystems, 2004, personal communication). Mumford *et al.* (2000b) specifically emphasize that the advantages of the RTF-RT-PCR system are: (i) increased sensitivity; (ii) the accelerated speed of the whole test procedure by precluding gel electrophoresis (with its hazard through the use of ethidium bromide) or chemiluminescent probe detection, both of which are labour-intensive; (iii) the risk of contamination is reduced by a closed-tube format; and (iv) the ease with which multiplex assays fit within this format. Because all probe–primer combinations run under generic conditions and different-sized products are not required, it is easier to optimize multiplex assays to avoid competitive effects.

With these qualities and the lower investment now required, the RTF-RT-PCR technology, especially in combination with rapid, mass-scale extraction methods, should facilitate an increased usefulness of PCR for routine testing.

Molecular beacons

A molecular beacon is a probe that produces fluorescence, but in a way different from that of the TaqMan probe. The molecular beacon probe (Tyagi and Kramer, 1996) consists of a single-stranded DNA with a stem–loop structure (Fig. 6.11). The loop portion consists of a probe sequence that is complementary to a target sequence. The stem portion is formed because the 5′ and 3′ arm sequences anneal. These annealing 5′ and 3′ end sequences are short (5–6 bases) C-G-rich stretches, unrelated to the target. A fluorescent moiety is attached to the 5′ arm terminus and a quenching moiety is attached to the 3′ end of the other arm.

In the absence of the target, the molecule maintains a hairpin structure that brings the reporter and quencher in physical proximity, resulting in quenching of the reporter. In the presence of the target, the probe molecule unfolds and hybridizes to the target (under suitable conditions), resulting in a fluorescence signal from the reporter because of its physical separation from the quencher (Fig. 6.11).

Principles and details of the use of molecular beacons in real-time fluorescent PCR assays are described in Tyagi and Kramer (1996) and in a protocol available on the Internet (http://www.molecular-beacons.org/protocol.html). Molecular beacons were used in PCR detection of two orchid viruses (Eun and

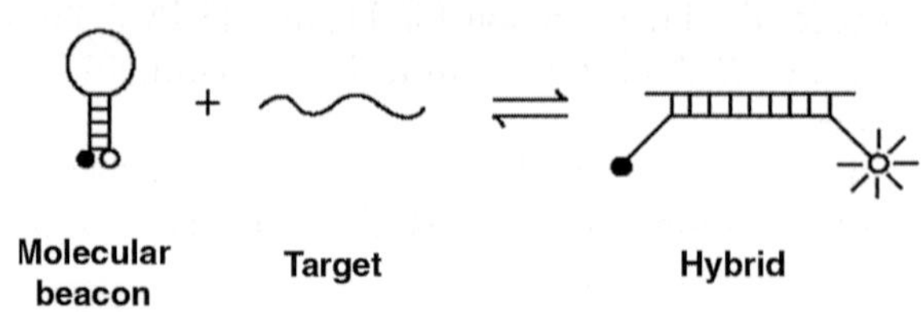

Fig. 6.11. Principle of operation of molecular beacons. When the probe (the loop) hybridizes to its target, the conformational change forces the arm sequences apart and causes the fluorophore to move away from the quencher, resulting in fluorescence. Reproduced with permission from Tyagi and Kramer (1996) and the copyright holder © Nature Biotechnology.

Wong, 2000). However, according to several recent reports on molecular diagnosis, both within the medical–clinical and plant pathological areas, molecular beacons appear to be more often used in combination with amplification by NASBA (see below).

A technique similar to that of molecular beacons is the detection by *scorpion primers* (Scorpions™, DxS, UK, http://www.dxsgenotyping.com). A scorpion also contains a probe in a hairpin structure and also has a fluorophore and quencher attached to the stem, but is linked to the 5′ end of a primer via a PCR stopper. After extension of the primer during the PCR amplification, the specific probe sequence, still linked to the primer and the new DNA strand, is able to bind to its complement within the same strand. When the scorpion is bound to its target, the fluorophore and quencher are separated, resulting in fluorescence. Scorpions were found to perform better in real-time PCR than TaqMan and molecular beacons, especially under fast cycling conditions (Thelwell *et al.*, 2000). Sialer *et al.* (2000) utilized this technique for detection of PPV and *Tomato spotted wilt virus* (TSWV, *Tospovirus*) in plants, and found it at least as sensitive as conventional PCR using gel electrophoresis visualization.

NASBA

Nucleic acid sequence-based amplification (NASBA®) (Kievits *et al.*, 1991) is an amplification technique different from PCR. The resulting product is mainly single-stranded RNA, the reaction requires no cycler as it is isothermal (41°C), the reaction can be completed in 90 min, and no addition of intermediate reagents is required, i.e. it is a closed-tube system. NASBA is a proprietary technology of Organon-Teknika™ (Boxtel, The Netherlands) based on the concurrent activity of three enzymes: AMV reverse transcriptase, RNase H and T7 RNA polymerase. In the presence of nucleotides (dNTPs and NTPs) and two specific primers, one of which contains a promoter sequence for T7 polymerase, a cycling reaction starts, resulting in an exponential amplification. The principle is, in brief (Yates *et al.*, 2001):

1. The primer with the T7 tail binds to its complementary sequence of the single-stranded target RNA and is extended by AMV RT.
2. The RNA strand (the template strand) of the RNA : DNA hybrid is degraded by RNase H (this enzyme specifically degrades only the RNA strand of an RNA : DNA hybrid).
3. The second primer can now bind to the cDNA synthesized in **1**. By means of AMV RT, a dsDNA product with a double-stranded T7 promoter sequence is then formed.
4. From this double-stranded product, the T7 polymerase transcribes multiple copies of single-stranded specific RNA fragments, each of which can then enter a new cycle of amplification.

Illustrations of the principle can be found on http://people.cornell.edu/pages/mbe2/thesis.pdf (page 28 of a thesis) or http://www.devicelink.com/ivdt/archive/98/03/013.html (in the latter called TMA, transcription-mediated amplification).

The procedure is particularly suited for the amplification of ssRNA viruses because the RT step is integrated in the amplification process. Since NASBA is isothermal, any common laboratory incubator can be used. The website http://www.ibi.cc/NASBA%20vs%20PCR.htm lists 23 publications comparing NASBA and RT-PCR for detection of RNA segments of different organisms (viruses in ten cases). In general, NASBA was equal or superior in sensitivity, specificity and simplicity.

As mentioned above, molecular beacons can be used for real-time detection of the amplificate during a NASBA by means of a fluorescence meter. The combination of molecular beacons and NASBA, enabling simultaneous amplification and amplicon detection in a closed tube (a system also called AmpliDet RNA), has been used to detect, for example, PLRV in potato (Leone *et al.*, 1998), ASPV in apple (Klerks *et al.*, 2001) and Sugarcane yellow leaf virus (ScYLV, Luteoviridae?) in sugarcane (Goncalves *et al.*, 2002). These reports provide details on how to set up the NASBA reaction. A method for designing molecular beacon probes for AmpliDet RNA has been proposed by Szemes and Schoen (2003).

6.2.7. Remarks on molecular amplification

PCR and other enzymatic nucleic acid (NA) amplification techniques are powerful tools, unparalleled in sensitivity and specificity, but the techniques are also vulnerable, first of all to contamination. In addition, NA amplification depends on several variables, each of which can compromise the result or spoil the assay, if suboptimal. Unlike PCR assays of fungi, bacteria and DNA viruses, PCR assays of RNA viruses require an RT step, which further adds to the number of variables.

Trouble shooting

Problems often occurring are the following:

A. Amplification products are weak or not visible.
B. Non-specific bands.
C. Smearing.

A, B and C can each have a number of reasons. Some of the most common causes and ways to prevent them, essentially according to the first and second website sources indicated below, are as follows.

- *Inhibitors* (A). Polysaccharides and phenolic compounds are commonly encountered plant substances assumed to inhibit both RT and PCR enzymes (see Sample preparation, Section 6.2.2). Among other inhibitors are chloroform, phenol, EDTA, ionic detergents (e.g. sodium dodecyl sulphate (SDS)) and ethanol.
- *Carry-over contamination* (B). Amplified products transferred from previous amplifications represent potential contaminants in successive amplifications. Due to the high sensitivity of PCR, even a few contaminant molecules can be amplified and spoil the assay. To eliminate or reduce the risk of carry-over contamination, the precautions listed at the beginnings of Section 6.2.3 and RT-PCR, standard protocol should be taken. A non-target negative control should always be included.

- *Primers* (A, B, C). Primers that are non-complementary to the target are obvious causes of failed results. Other errors include self-complementarity or complementarity within a primer pair (see Primers, Section 6.2.2). It is advisable to use computer software to check primer sequences before using them. Primer concentrations that are too low may result in no bands; too high concentrations may cause primer-dimers and/or impeded amplification. Concentrations within 0.1–1.0 μM for each primer are typically best. Concentrations above 1 μM should be used only where there are high degrees of degeneracy. Unlike some of the other components, primers are rather stable and can be stored in TE buffer at 4°C for a long time (Holst-Jensen *et al.*, 1998).
- *dNTPs* (A). The dNTPs are probably the least stable components. They are vulnerable to repeated freeze–thaw cycles, and should therefore be distributed to small tubes before storing at –20°C. After a few uses of each tube, they should be discarded. Too high or too low dNTP concentrations can inhibit the PCR reaction. The optimal range for ordinary PCRs is 20–200 μM of each dNTP, and for RT up to 500 μM each.
- *Mg^{2+}* (A). A suboptimal $MgCl_2$ concentration may result in weak products or their absence. A too low concentration of free magnesium is one of the most common reasons for a usually successful PCR to fail (Seal and Coates, 1998). *Taq* polymerase is dependent on free magnesium ions. However, Mg^{2+} also forms complexes with dNTPs and other components, such as EDTA, which may result in too little free Mg^{2+} to activate polymerase. Therefore, the $MgCl_2$ concentration should be optimized for each new PCR application. A final concentration of 1.5 mM is often used, but the optimal concentration may vary between 1 and 4 mM. It should be remembered that frozen $MgCl_2$ forms a concentration gradient and therefore, after thawing, it must be vortexed before use.
- *Enzyme concentration* (A, C). Too little polymerase results in low yield. Excess *Taq* polymerase may be the cause of a non-specific background smear. Standard concentrations vary from 0.5 to 2.5 units per 25–50 μl reaction.
- *Template concentration* (A). Excess cDNA template may inhibit PCR by binding all the primers in the first cycle, preventing amplification. Too little template, both in RT and PCR, may result in little or no yield.
- *Buffer and master mix* (A). The correct buffer with additives for the enzyme in question should be used. When making the master mix, the reagents should be added in the sequence shown in the standard RT-PCR protocol, p. 191. Adding the enzyme first and the buffer last may cause damage to the enzyme. Adding the reagents in a defined order also reduces the risk of introducing errors.
- *PCR cycling programme* (A, B). An annealing temperature that is too low, results in non-specific amplifications, and excess temperature results in low or no yield (see Optimization of RT-PCR, p. 199). A too-low denaturation temperature (i.e. typically < 90°C) results in low yields or non-specific products. Too high denaturation temperatures and/or long denaturation periods add to the decay of the enzyme.

More on trouble shooting can be seen on the websites http://biology.uio.no/bot/ascomycetes/PCR.troubleshooting.html or http://www.protocol-online.org/prot/Molecular_Biology/PCR/, or via the gateway http://www.bioinformatics.vg/index.shtml

Ways to reduce cost and labour

One of the characteristics of plant disease diagnostics is the low value added for samples being indexed (Candresse *et al.*, 1998). The cost of the detection method used in large-scale indexing of plants or seed is therefore important. Key elements of molecular techniques are often patented, and complex protocols and expensive reagents are usually required (Martin *et al.*, 2000). Further development of systems that are less costly and less labour-intensive is imperative for the practical use of these techniques. Ways to reduce cost and labour currently include the use of multiplex PCR, IC-RT-PCR, one-tube, one-step RT-PCR formats, and testing of as-large-as-practical composite samples. Despite the initial cost involved, the utilization of real-time fluorogenic molecular amplification assays, combined with rapid mass-scale extraction methods, could probably lead to routine use of detection by molecular amplification (Mumford *et al.*, 2000a, b). There is still a need for the development of rapid, simple and cost-effective NA extraction methods that function effectively in tests of inhibitor-rich plants and seeds. See also Section 7.4.2.

Sensitivity, specificity and reliability

The sensitivity of molecular amplification is usually far greater than that of ELISA or dot-blot hybridization assays. Maximum sensitivity depends, however, on optimization of several parameters, including the efficient removal of test-inhibitory plant substances (Candresse *et al.*, 1998). According to the same authors, comparing the sensitivity of PCR assays with those of other techniques should include assays of successively diluted infected-plant extracts with healthy-plant extracts, as this would reflect a 'real life' situation, rather than using plain water as diluent.

Specificity of molecular amplification assays normally depends on the design of primers. Sufficient broad-spectrum specificity is required to ensure detection of all variants of a given virus for quarantine purposes. The reactivity of a primer pair, therefore, must be extensively tested against a variety of viral isolates, and this has not often been done for viruses and viroids for which PCR detection has been reported (Candresse *et al.*, 1998).

For other purposes, such as epidemiological studies that are needed for effective disease management, a narrow specificity is essential to distinguish closely related strains or pathotypes from each other. In such cases highly specific primers, nested PCR or RFLP should be used.

The increased sensitivity of molecular amplification assays over serological assays or even bioassays normally permits greater confidence in test results for both certification and quarantine; however, the high sensitivity also makes these assays less robust by being vulnerable to contamination. Likewise, inhibitors and faulty assay conditions pose potential risks of false negatives and false positives, respectively. To prevent contamination, the precautions and recommendations indicated initially in Section 6.2.3 and initially in the RT-PCR standard protocol (same section) should be strictly maintained in the testing laboratory. Use of internal controls (Section 6.2.3) and uniform testing routines also reduces the risk of false negatives.

Literature on nucleic acid-based methods for plant virus assays

Molecular methods for plant virus detection have been reviewed by, for example, Candresse *et al.* (1998), Martin *et al.* (2000), Dietzgen (2002) and Singh and Nie (2002). Detailed procedures have been described by, for example, Dijkstra and de Jager (1998), Pallás *et al.* (1998), Rybicki (1998, 2001a, b), Seal and Coates (1998), Singh (1998) and Stratford (1998). General information on and protocols for molecular technologies can be found in Sambrook and Russell (2001).

References

Anon. (1992) Polymerase chain reaction – Xeroxing DNA. Available at: http://www.accessexcellence.org/RC/AB/IE/PCR_Xeroxing_DNA.html

Anon. (1994) DIAPOPS using CovaLink™ BreakApart™ modules. *Tech Note* 2 (11), Nunc A/S/Nalge Nunc International. Available at: http://www.nuncbrand.com

Bariana, H.S., Shannon, A.L., Chu, P.W.G. and Waterhouse, P.M. (1994) Detection of five seedborne legume viruses in one sensitive multiplex polymerase chain reaction test. *Phytopathology* 84, 1201–1205.

Bertolini, E., Olmos, A., Martínez, M.C., Gorris, M.T. and Cambra, M. (2001) Single-step multiplex RT-PCR for simultaneous and colorimetric detection of six RNA viruses in olive trees. *Journal of Virological Methods* 96, 33–41.

Borkhardt, B., Vongsasitorn, D. and Albrechtsen, S.E. (1994) Chemiluminescent detection of potato spindle tuber viroid in true potato seed using a digoxigenin labelled DNA probe. *Potato Research* 37, 249–255.

Candresse, T., Hammond, R.W. and Hadidi, A. (1998) Detection and identification of plant viruses and viroids using polymerase chain reaction (PCR) In: Hadidi, A., Khetarpal, R.K. and Koganezawa, H. (eds) *Plant Virus Disease Control.* APS Press, St Paul, Minnesota, pp. 399–416.

Choi, S.K., Choi, J.K., Park, W.M. and Ryu, K.H. (1999) RT-PCR detection and identification of three species of cucumoviruses with a genus-specific single pair of primers. *Journal of Virological Methods* 83, 67–73.

Dalmon, A., Cailly, M. and David, C. (2000) Comparison of serological and molecular techniques for detection of *Tomato yellow leaf curl begomovirus. Bulletin OEPP/EPPO Bulletin* 30, 457–462.

Dietzgen, R.G. (2002). Application of PCR in plant virology. In: Khan, J.A. and Dijkstra, J. (eds) *Plant Viruses as Molecular Pathogens.* Food Products Press/Haworth Press, New York, pp. 471–500.

Dijkstra, J. and de Jager, C.P. (1998) *Practical Plant Virology: Protocols and Exercises.* Springer-Verlag, Berlin.

Dovas, C.I. and Katis, N.I. (2003a) A spot nested RT-PCR method for the simultaneous detection of members of the *Vitivirus* and *Foveavirus* genera in grapevine. *Journal of Virological Methods* 107, 99–106.

Dovas, C.I. and Katis, N.I. (2003b) A spot multiplex nested RT-PCR for the simultaneous and generic detection of viruses involved in the aetiology of grapevine leafroll and rugose wood of grapevine. *Journal of Virological Methods* 109, 217–226.

Eun, A.J.C. and Wong, S.M. (2000) Molecular beacons: a new approach to plant virus detection. *Phytopathology* 90, 269–275.

Gibbs, A. and Mackenzie, A. (1997) A primer pair for amplifying part of the genome of all potyvirids by RT-PCR. *Journal of Virological Methods* 63, 9–16.

Gibbs, A., Armstrong, J., Mackenzie, A.M. and Weiller, G.F. (1998) The GPRIME package: computer programs for identifying the best regions of aligned genes to target in nucleic acid hybridization-based diagnostic tests, and their use with plant viruses. *Journal of Virological Methods* 74, 67–76.

Gibbs, A.J., Mackenzie, A.M. and Gibbs, M.J. (2003) The 'potyvirid primers' will probably provide phylogenetically informative DNA fragments from all species of Potyviridae. *Journal of Virological Methods* 112, 41–44.

Gillaspie, A.G. Jr, Pittman, R.N. and Pinnow, D.L. (2000) Sensitive method for testing peanut seed lots for *Peanut stripe* and *Peanut mottle viruses* by immunocapture-reverse transcription-polymerase chain reaction. *Plant Disease* 84, 559–561.

Goncalves, M.C., Klerks, M.M., Verbeek, M., Vega, J. and Van den Heuvel, J.F.J.M. (2002) The use of molecular beacons combined with NASBA for the sensitive detection of *Sugarcane yellow leaf virus*. *European Journal of Plant Pathology* 108, 401–407.

Grieco, F., Alkowni, R., Saponari, M., Savino, V. and Martelli, G. (2000) Molecular detection of olive viruses. *Bulletin OEPP/EPPO Bulletin* 30, 469–473.

Hanold, D. (1993) Diagnostic methods applicable to viroids. In: Matthews, R.E.F. (ed.) *Diagnosis of Plant Virus Diseases*. CRC Press, Boca Raton, Florida, pp. 295–314.

Henegariu, O., Heerema, N.A., Dlouhy, S.R., Vance, G.H. and Vogt, P.H. (1997) Multiplex PCR: critical parameters and step-by-step protocol. *BioTechniques* 23, 504–511.

Henson, J.M. and French, R. (1993) The polymerase chain reaction and plant disease diagnosis. *Annual Review of Phytopathology* 31, 81–109.

Holland, P.M., Abramson, R.D., Watson, R. and Gelfand, D.H. (1991) Detection of specific polymerase chain reaction product by utilizing the $5' \rightarrow 3'$ exonuclease activity of *Thermus aquaticus* DNA polymerase. *Biochemistry* 88, 7276–7280.

Holst-Jensen, A., Bruns, T. and Gardes, M. (1998) PCR trouble shooting, help, suggestions and advice. *Keys and Services.* Available at: http://biologi.uio.no/bot/ascomycetes/PCR.troubleshooting.html

Hull, R. (1993) Nucleic acid hybridization procedures. In: Matthews, R.E.F. (ed.) *Diagnosis of Plant Virus Diseases*. CRC Press, Boca Raton, Florida, pp. 253–271.

Hull, R. (2002) *Matthews' Plant Virology*, 4th edn. Academic Press, London.

Hurtt, S.S. and Podleckis, E.V. (1995) Apple scar skin viroid is not seed transmitted or transmitted at a low rate in oriental pear. *Acta Horticulturae* 386, 544–547.

Jacobi, V., Bachand, G.D., Hamelin, R.C. and Castello, J.D. (1998) Development of a multiplex immunocapture RT-PCR assay for detection and differentiation of tomato and tobacco mosaic tobamoviruses. *Journal of Virological Methods* 74, 167–178.

Jones, R.A.C. (2001) Developing integrated disease management strategies against non-persistently aphid-borne viruses: a model programme. *Integrated Pest Management Reviews* 6, 15–46.

Kievits, T., van Gemen, B., van Strijp, D., Schukkink, R., Dircks, M., Adriaanse, H., Malek, L., Sooknanan, R. and Lens, P. (1991) NASBA isothermal enzymatic in vitro nucleic acid amplification optimized for the diagnosis of HIV-1 infection. *Journal of Virological Methods* 35, 273–286 (Abstr.).

Klerks, M.M., Leone, G., Lindner, J.L., Schoen, C.D. and Van den Heuvel, J.F.J.M. (2001) Rapid and sensitive detection of *Apple stem pitting virus* in apple trees through RNA amplification and probing with fluorescent molecular beacons. *Phytopathology* 91, 1085–1091.

Koenig, R., Lüddecke, P. and Haeberlé, A.M. (1995) Detection of beet necrotic yellow vein virus strains, variants and mixed infection by examining single-strand conformation polymorphisms of immunocapture RT-PCR products. *Journal of General Virology* 76, 2051–2055.

Kohnen, P.D., Dougherty, W.G. and Hampton, R.O. (1992) Detection of pea seedborne mosaic potyvirus by sequence specific enzymatic amplification. *Journal of Virological Methods* 37, 253–258.

Korimbocus, J., Coates, D., Barker, I. and Boonham, N. (2002) Improved detection of *Sugarcane yellow leaf virus* using a real-time fluorescent (TaqMan) RT-PCR assay. *Journal of Virological Methods* 103, 109–120.

Langeveld, A.S., Dore, J.M., Memelink, J., Derks, A.F.L.M., van der Vlugt, C.I.M., Asjes, C.J. and Bol, J.F. (1991) Identification of potyviruses using the polymerase chain reaction with degenerate primers. *Journal of General Virology* 72, 1531–1541.

Leone, G., van Schijndel, H., van Gemen, B., Kramer, F.R. and Schoen, C.D. (1998) Molecular beacon probes combined with amplification by NASBA enable homogeneous, real-time detection of RNA. *Nucleic Acids Research* 26, 2150–2155.

Letschert, B., Adam, G., Lesemann, D.E., Willingmann, P. and Heinze, C. (2002) Detection and differentiation of serologically cross-reacting tobamoviruses of economical importance by RT-PCR and RT-PCR-RFLP. *Journal of Virological Methods* 106, 1–10.

Mackenzie, A.M., Nolan, M., Wei, K.J., Clements, M.A., Gowanlock, D., Wallace, B.J. and Gibbs, A.J. (1998) Ceratobium mosaic potyvirus: another virus from orchids. *Archives of Virology* 149, 903–914.

Magome, H., Yoshikawa, N. and Takahashi, T. (1999) Single-strand conformation polymorphism analysis of apple stem grooving capillovirus sequence variants. *Phytopathology* 89, 136–140.

Mahony, J.B. and Chernesky, M.A. (1995) Multiplex polymerase chain reaction. In: Wiedbrauk, D.L. and Farkas, D.H. (eds) *Molecular Methods for Virus Detection.* Academic Press, San Diego, California, pp. 219–229.

Martin, R.R., James, D. and Lévesque, C.A. (2000) Impacts of molecular diagnostic technologies on plant disease management. *Annual Review of Phytopathology* 38, 207–239.

Más, P. and Pallás, V. (1995) Non-isotopic tissue-printing hybridization: a new technique to study long-distance plant virus movement. *Journal of Virological Methods* 52, 317–326.

McLaren, B. (1999) Technically speaking, Q & A: reverse transcription and amplification. *Promega Notes* 71, 22–25. Available at: http://www.promega.com

McQuaid, S. (1997) Mutation detection systems – SSCP analysis: MRCPath self help course 21/10/97. Available at: http://www.ich.ucl.ac.uk/cmgs/sscp.htm

Menzel, W., Jelkmann, W. and Maiss, E. (2002) Detection of four apple viruses by multiplex RT-PCR assays with coamplification of plant mRNA as internal control. *Journal of Virological Methods* 99, 81–92.

Mullis, K.F., Faloona, F., Scharf, S., Saiki, R., Horn, G. and Erlich, H. (1986) Specific enzymatic amplification of DNA *in vitro*: the polymerase chain reaction. *Cold Spring Harbor Symposia on Quantitative Biology* 51, 263–273.

Mumford, R.A. and Seal, S.E. (1997) Rapid single-tube immunocapture RT-PCR for the detection of two yam potyviruses. *Journal of Virological Methods* 69, 73–79.

Mumford, R.A., Walsh, K., Barker, I. and Boonham, N. (2000a) Detection of *Potato mop top virus* and *Tobacco rattle virus* using a multiplex real-time fluorescent reverse-transcription polymerase chain reaction assay. *Phytopathology* 90, 448–453.

Mumford, R.A., Walsh, K. and Boonham, N. (2000b) A comparison of molecular methods for the routine detection of viroids. *Bulletin OEPP/EPPO Bulletin* 30, 431–435.

Nakahara, K., Hataya, T. and Uyeda, I. (1999) A simple rapid method of nucleic acid extraction without tissue homogenization for detecting viroids by hybridization and RT-PCR. *Journal of Virological Methods* 77, 47–58.

Nassuth, A., Pollari, E., Helmeczy, K., Stewart, S. and Kofalvi, S.A. (2000) Improved RNA extraction and one-tube RT-PCR assay for simultaneous detection of control plant RNA plus several viruses in plant extracts. *Journal of Virological Methods* 90, 37–49.

Nemchinov, L. and Hadidi, A. (1998) Specific oligonucleotide primers for the direct detection of plum pox potyvirus–cherry subgroup. *Journal of Virological Methods* 70, 231–234.

Nicolaisen, M., Rasmussen, H.N., Husted, K. and Nielsen, S.L. (2001) Reverse transcription-detection of immobilized, amplified product in a one-phase system (RT-DIAPOPS) for the detection of potato virus Y. *Plant Pathology* 50, 124–129.

Nie, X. and Singh, R.P. (2000) Detection of multiple potato viruses using an oligo(dT) as a common cDNA primer in multiplex RT-PCR. *Journal of Virological Methods* 86, 179–185.

Nie, X. and Singh, R.P. (2001) A novel usage of random primers for multiplex RT-PCR detection of virus and viroid in aphids, leaves, and tubers. *Journal of Virological Methods* 91, 37–49.

Njeru, R., Ferris, D.G., Jones, R.A.C. and Jones, M.G.K. (1997) Studies on seed transmission of subterranean clover mottle virus and its detection in clover seed by ELISA and RT-PCR. *Australian Journal of Agricultural Research* 48, 343–350.

Nolasco, G., Sequeira, Z., Soares, C., Mansinho, A., Bailey, A.M. and Niblett, C.L. (2002) Asymmetric PCR ELISA: increased sensitivity and reduced costs for the detection of plant viruses. *European Journal of Plant Pathology* 108, 293–298.

Olmos, A., Cambra, M., Dasí, M.A., Candresse, T., Esteban, O., Gorris, M.T. and Asensio, M. (1997) Simultaneous detection and typing of plum pox potyvirus (PPV) isolates by heminested-PCR and PCR-ELISA. *Journal of Virological Methods* 68, 127–137.

Olmos, A., Bertolini, E. and Cambra, M. (2002) Simultaneous and co-operational amplification (Co-PCR): a new concept for detection of plant viruses. *Journal of Virological Methods* 106, 51–59.

Omunyin, M.E., Hill, J.H. and Miller, W.A. (1996) Use of unique RNA sequence-specific oligonucleotide primers for RT-PCR to detect and differentiate soybean mosaic virus strains. *Plant Disease* 80, 1170–1174.

Palacio-Bielsa, A., Foissac, X. and Duran-Vila, N. (1999) Indexing of citrus viroids by imprint hybridization. *European Journal of Plant Pathology* 105, 897–903.

Pallás, V., Más, P. and Sánchez-Navarro, J.A. (1998) Detection of plant RNA viruses by nonisotopic dot-blot hybridization. In: Foster, G.D. and Taylor, S.C. (eds) *Plant Virology Protocols: From Virus Isolation to Transgenic Resistance*. Humana Press, Totowa, New Jersey, pp. 461–468.

Pasquini, G., Simeone, A.M., Conte, L. and Barba, M. (1998) Detection of plum pox virus in apricot seeds. *Acta Virologica* 42, 260–263.

Podleckis, E.V., Hammond, R.W., Hurtt, S.S. and Hadidi, A. (1993) Chemiluminescent detection of potato and pome fruit viroids by digoxigenin-labeled dot blot and tissue blot hybridization. *Journal of Virological Methods* 43, 147–158.

Pollini, C.P., Giunchedi, L. and Bissani, R. (1997) Specific detection of D- and M-isolates of plum pox virus by immunoenzymatic determination of PCR products. *Journal of Virological Methods* 67, 127–133.

Prezioso, V.R. (2003) General notes on primer design in PCR. Brinkmann Instruments Inc. Available at: http://www.brinkmann.com/PCR_appl_primer.asp

Roche Diagnostics (2000) *DIG Application Manual for Filter Hybridization*. Roche Diagnostics GmbH, Mannheim, Germany.

Romero-Durbán, J., Cambra, M. and Duran-Vila, N. (1995) A simple imprint-hybridization method for detection of viroids. *Journal of Virological Methods* 55, 37–47.

Rosner, A., Maslenin, L. and Spiegel, S. (1997) The use of short and long PCR products for improved detection of prunus necrotic ringspot virus in woody plants. *Journal of Virological Methods* 67, 135–141.

Rott, M.E. and Jelkmann, W. (2001) Characterization and detection of several filamentous viruses of cherry: adaptation of an alternative cloning method (DOP-PCR), and modification of an RNA extraction protocol. *European Journal of Plant Pathology* 107, 411–420.

Rowhani, A., Biardi, L., Routh, G., Daubert, S.D. and Golino, D.A. (1998) Development of a sensitive colorimetric-PCR assay for detection of viruses in woody plants. *Plant Disease* 82, 880–884.

Rybicki, E. (1998) Detection of nucleic acids by hybridization. In: Coyne, V.E., James, M.D., Reid, S.J. and Rybicki, E.P. (eds) *Excerpts from a Molecular Biology Techniques Manual*, 3rd edn. University of Cape Town, Department of Molecular and Cell Biology. Available at: http://www.mcb.uct.ac.za//manual/MolBiolManual.htm

Rybicki, E. (2001a) PCR primer design and reaction optimization. In: Coyne, V.E., James, M.D., Reid, S.J. and Rybicki, E.P. (eds) *Excerpts from a Molecular Biology Techniques Manual,* 3rd edn. University of Cape Town, Department of Molecular and Cell Biology. Available at: http://www.mcb.uct.ac.za//manual/MolBiolManual.htm

Rybicki, E. (2001b) Standard PCR protocol. In: Coyne, V.E., James, M.D., Reid, S.J. and Rybicki, E.P. (eds) *Excerpts from a Molecular Biology Techniques Manual,* 3rd edn. University of Cape Town, Department of Molecular and Cell Biology. Available at: http://www.mcb.uct.ac.za//manual/MolBiolManual.htm

Saldarelli, P., Barbarossa, L., Grieco, F. and Gallitelli, D. (1996) Digoxigenin-labeled riboprobes applied to phytosanitary certification of tomato in Italy. *Plant Disease* 80, 1343–1346.

Salmon, M.A., Vendrame, M., Kummert, J. and Lepoivre, P. (2002) Rapid and homogenous detection of *Apple stem pitting* virus by RT-PCR and a fluorogenic 3′ minor groove binder-DNA probe. *European Journal of Plant Pathology* 108, 755–762.

Sambrook, J. and Russell, D. (2001) *Molecular Cloning: A Laboratory Manual,* 3rd edn. Cold Spring Harbor Laboratory Press, New York.

Scheinert, P., Behrens, B. and Kahle, D. (2003) Optimizing DNA amplification protocols using the Eppendorf Mastercycler®. *Applications On-line,* Brinkmann Instruments Inc. Available at: http://www.brinkmann.com/PCR_appl_protocolsMC.asp

Schmitz-Linneweber, C., Regel, R., Du, T.G., Hupfer, H., Herrmann, R.G. and Maier, R.M. (2002) The plastid chromosome of *Atropa belladonna* and its comparison with that of *Nicotiana tabacum*: the role of RNA editing in generating divergence in the process of plant speciation. *Molecular Biology and Evolution* 19, 1602–1612.

Schoen, C.D., Knorr, D. and Leone, G. (1996) Detection of potato leafroll virus in dormant potato tubers by immunocapture and a fluorogenic 5′ nuclease RT-PCR assay. *Phytopathology* 86, 993–999.

Schweitzer, B. and Kingsmore, S. (2001) Combining nucleic acid amplification and detection. *Current Opinion in Biotechnology* 12, 21–27.

Seal, S. and Coates, D. (1998) Detection and quantification of plant viruses by PCR. In: Foster, G.D. and Taylor, S.C. (eds) *Plant Virology Protocols: From Virus Isolation to Transgenic Resistance.* Humana Press, Totowa, New Jersey, pp. 469–485.

Seoh, M.L., Wong, S.M. and Zhang, L. (1998) Simultaneous TD/RT-PCR detection of cymbidium mosaic potexvirus and odontoglossum ringspot tobamovirus with a single pair of primers. *Journal of Virological Methods* 72, 197–204.

Shamloul, A.M. and Hadidi, A. (1999) Sensitive detection of potato spindle tuber and temperate fruit tree viroids by reverse transcription-polymerase chain reaction-probe capture hybridization. *Journal of Virological Methods* 80, 145–155.

Sialer, M.M.F., Ciancio, A. and Gallitelli, D. (2000) Use of Scorpions for fast and sensitive detection of plant viruses. *Bulletin OEPP/EPPO Bulletin* 30, 437–440.

Singh, M. and Singh, R.P. (1995) Digoxigenin-labelled cDNA probes for the detection of potato virus Y in dormant potato tubers. *Journal of Virological Methods* 52, 133–143.

Singh, R.P. (1998) Reverse-transcription polymerase chain reaction for the detection of viruses from plants and aphids. *Journal of Virological Methods* 74, 125–138.

Singh, R.P. (1999) A solvent-free, rapid and simple virus RNA-release method for potato leafroll virus detection in aphids and plants by reverse transcription polymerase chain reaction. *Journal of Virological Methods* 83, 27–33.

Singh, R.P. and Dhar, A.K. (1998) Detection and management of plant viroids. In: Hadidi, A., Khetarpal, R.K. and Koganezawa, H. (eds) *Plant Virus Disease Control.* APS Press, St Paul, Minnesota, pp. 428–447.

Singh, R.P. and Nie, X. (2002) Nucleic acid hybridization for plant virus and viroid detection. In: Khan, J.A. and Dijkstra, J. (eds) *Plant Viruses as Molecular Pathogens.* Food Products Press/Haworth Press, New York, pp. 443–469.

Singh, R.P., Singh, M. and King, R.R. (1998a) Use of citric acid for neutralizing polymerase chain reaction inhibition by chlorogenic acid in potato extracts. *Journal of Virological Methods* 74, 231–235.

Singh, R.P., Singh, M. and McDonald, J.G. (1998b) Screening by a 3-primer PCR of North American PVY^{N} isolates for European-type members of the tuber necrosis-inducing PVY^{NTN} subgroup. *Canadian Journal of Plant Pathology* 20, 227–233.

Singh, R.P., Nie, X. and Singh, M. (2000) Duplex RT-PCR: reagent concentrations at reverse transcription stage affect the PCR performance. *Journal of Virological Methods* 86, 121–129.

Singh, R.P., Dilworth, A.D., Singh, M. and McLaren, D.L. (2004) Evaluation of a simple membrane-based nucleic acid preparation protocol for RT-PCR detection of potato viruses from aphid and plant tissues. *Journal of Virological Methods* 121, 163–170.

Spiegel, S. and Martin, R.R. (1993) Improved detection of potato leaf roll virus in dormant potato tubers and microtubers by the polymerase chain reaction and ELISA. *Annals of Applied Biology* 122, 493–500.

Stark-Lorenzen, P., Guitton, M.C., Werner, R. and Mühlbach, H.P. (1997) Detection and tissue distribution of potato spindle tuber viroid in infected tomato plants by tissue print hybridization. *Archives of Virology* 142, 1289–1296.

Stevens, M., Hull, R. and Smith, H.G. (1997) Comparison of ELISA and RT-PCR for the detection of beet yellows closterovirus in plants and aphids. *Journal of Virological Methods* 68, 9–16.

Stratford, R. (1998) PCR cloning of coat protein genes. In: Foster, G.D. and Taylor, S.C. (eds) *Plant Virology Protocols: From Virus Isolation to Transgenic Resistance.* Humana Press, Totowa, New Jersey, pp. 269–278.

Szemes, M. and Schoen, C.D. (2003) Design of molecular beacons for AmpliDet RNA assay – characterization of binding stability and probe specificity. *Analytical Biochemistry* 315, 189–201.

Szemes, M., Kálmán, M., Myrta, A., Boscia, D., Németh, M., Kölber, M. and Dorgai, L. (2001) Integrated RT-PCR/nested PCR diagnosis for differentiating between subgroups of plum pox virus. *Journal of Virological Methods* 92, 165–175.

Thelwell, N., Millington, S., Solinas, A., Booth, J. and Brown, T. (2000) Mode of action and application of Scorpion primers to mutation detection. *Nucleic Acids Research* 28, 3752–3761.

Thompson, J.R., Wetzel, S., Klerks, M.M., Vasková, D., Schoen, C.D., Spak, J. and Jelkmann, W. (2003) Multiplex RT-PCR detection of four aphid-borne strawberry viruses in *Fragaria* spp. in combination with a plant mRNA specific internal control. *Journal of Virological Methods* 111, 85–93.

Thomson, D. and Dietzgen, R.G. (1995) Detection of DNA and RNA plant viruses by PCR and RT-PCR using a rapid virus release protocol without tissue homogenization. *Journal of Virological Methods* 54, 85–95.

Torok, V.A. and Randles, J.W. (2001) *Tobacco mosaic virus* RNA as an internal control for duplex RT-PCR assay of pea germplasm. *Australasian Plant Pathology* 30, 227–230.

Tyagi, S. and Kramer, F.R. (1996) Molecular beacons: probes that fluoresce upon hybridization. *Nature Biotechnology* 14, 303–308.

Vierstraete, A. (2001) Principle of the PCR. Available at: http://allserv.rug.ac.be/~avierstr/index.html

Vierstraete, A. (2002) Tif output files. Available at: http://allserv.rug.ac.be/~avierstr/tif.html

Watson, J.D., Gilman, M., Witkowski, J., Zoller, M. and Witkowski, G. (1993) *Recombinant DNA.* W.H. Freeman, New York.

Webster, K.D. and Barker, H. (1998) Detection and quantification of transcript RNA in transgenic plants using digoxigenin-labeled cDNA probes. In: Foster, G.D. and Taylor, S.C. (eds) *Plant Virology Protocols: From Virus Isolation to Transgenic Resistance*. Humana Press, Totowa, New Jersey, pp. 437–445.

Weilguny, H. and Singh, R.P. (1998) Separation of Slovenian isolates of PVY^{NTN} from the North American isolates of PVY^{N} by a 3-primer PCR. *Journal of Virological Methods* 71, 57–68.

Wetzel, T., Candresse, T., Macquaire, G., Ravelonandro, M. and Dunez, J. (1992) A highly sensitive immunocapture polymerase chain reaction method for plum pox potyvirus detection. *Journal of Virological Methods* 39, 27–37.

Wylie, S., Wilson, C.R., Jones, R.A.C. and Jones, M.G.K. (1993) A polymerase chain reaction assay for cucumber mosaic virus in lupin seeds. *Australian Journal of Agricultural Research* 44, 41–51.

Xu, L. and Hampton, R.O. (1996) Molecular detection of bean common mosaic and bean common mosaic necrosis potyviruses and pathogroups. *Archives of Virology* 141, 1961–1977.

Yates, S., Penning, M., Goudsmit, J., Frantzen, I., Van de Weijer, B., Van Strijp, D. and Van Gemen, B. (2001) Quantitative detection of hepatitis B virus DNA by real-time nucleic acid sequence-based amplification with molecular beacon detection. *Journal of Clinical Microbiology* 39, 3656–3665.

7 Epilogue

7.1. Other Detection Techniques

Among other detection techniques commonly used in plant virology are electron microscopy and, for viroid detection, return-gel electrophoresis.

7.1.1. Electron microscopy

Combined with plant-host symptomatology and host range of viruses, electron microscopic examination of plant-sap preparations can quickly categorize viruses according to particle shapes and sizes, e.g. spherical or rod-shaped, modal diameters or particle lengths. Two or more particle shapes/sizes in sap preparations are indicative of mixed infections. As earlier mentioned, virions of members of the genera to which seed-transmitted viruses belong are normally present in high concentration in plant sap and therefore easy to study compared to, for example, luteoviruses and geminiviruses, whose concentration is low.

The principle of an electron microscope (EM) is similar to that of a light microscope, but, because the wavelength of the electron beam is several thousand times shorter than that of light, the resolving power of an EM is great enough (≤ 0.2 nm point to point) to visualize virus particles and other microstructures. There are two types of EM, the transmission electron microscope (TEM) and the scanning electron microscope (SEM). In the TEM, the electron beam passes through and delineates the specimen, whereas the SEM is for examining surfaces. The range of magnification for the TEM is usually $\times$ 20,000–$\times$ 200,000 and $\times$ 4000–$\times$ 40,000 for the SEM. The TEM is best suited for virus studies.

Adjustable electromagnetic lenses are used for EM instead of the glass lenses used for a light microscope. The EM image is visualized either directly on a fluorescent screen or indirectly by a monitor or by photographs taken with a built-in camera. Air within the EM column must first be evacuated before electron illumination of specimens is possible. The specimen carrier for a TEM is a small grid (3 mm in diameter) of fine copper or nickel wire mesh covered with a special ultra-thin plastic film on which the material to be examined is placed. The specimen can either be diluted plant sap mounted as an extremely thin layer on the film (allowed to dry) for examination by TEM, or ultra-thin sections of plant tissue embedded in an epoxy resin or similar material for histological studies. Whereas sap preparations can be completed in less than 2 min, preparation of sections takes several days; the plant tissue must be stabilized by fixing chemicals and then gradually transferred from the aqueous phase into an organic solvent phase before embedding in resin. When the resin is hardened (almost as hard as amber), ultra-thin sections can be cut. The sections have to be ≤ 100 nm in thickness, requiring a precision ultramicrotome. Regardless of specimen types, organic structures can only be observed by TEM when treated with an electron-dense stain, usually solutions of salts of tungsten, molybdenum, uranium or lead.

Because of the high mechanical and electronic precision required for constructing EMs, their costs are ≥ US$200,000, to which must be added the expenses of running, maintaining and servicing the instrument. The use of EM is thus often restricted to research laboratories. However, as currently practised, cooperative work is arranged for identification of viruses between research institutes and seed-health testing laboratories or quarantine stations. When needed, desiccated plant material (see Section 4.5) can be sent over long distances to be examined by EM. This has been routinely practised in seed-health agencies, e.g. at DGISP, where it has served its former students in developing countries.

Useful details on TEM and protocols of preparation techniques can be found in, for example, Griffin (1991). Immunosorbent electron microscopy (ISEM), combining serology and EM, is described in Section 5.5.

Electron micrographs of virions of numerous viruses are accessible on the web – for example, in the *AAB Descriptions of Plant Viruses* (n.d.) (http://www.dpvweb.net), which includes more than 400 viruses.

7.1.2. Return electrophoresis for viroid detection

Due to the absence of coat protein, serological detection of viroids is not possible. Detection requires bioassays, electrophoresis, NA hybridization or RT-PCR (see Section 3.2.2; Chapter 4, Introduction; and Chapter 6). Inoculation to indicator plants has been commonly used for this purpose, but is a time-consuming method (in an extreme case, *Avocado sunblotch viroid* (ASBVd, *Avsunviroidae*), up to 2 years), compared to the other three methods (Singh and Dhar, 1998).

A two-step return-polyacryamide gel electrophoresis (R-PAGE), developed by Schumacher *et al.* (1986), is relatively sensitive. In the first electrophoresis step, nucleic acid extracts are electrophoresed in a non-denaturing polyacrylamide gel, implying that the electrophoretic mobility of the viroid is similar to that of the linear nucleic acid molecules of the host plant. In the second step, 'backwards' electrophoresis is carried out under denaturing conditions, i.e. with the polarity of migration reversed and with a low-salt buffer that is gradually heated to 60–75°C. During this return run under denaturing conditions, the viroid exhibits its circular form, which results in decreased electrophoretic mobility compared to that of linear molecules. Therefore, clear separation is achieved between the viroid band and a smear of plant nucleic acids (Singh and Dhar, 1998; Roenhorst *et al.*, 2000). Despite being less sensitive than molecular methods, R-PAGE is still considered to be a useful tool in phytosanitary testing, particularly because detection of a viroid is not influenced by variations in its nucleotide sequence (Roenhorst *et al.*, 2000). Thus, R-PAGE is useful for detection, but a molecular assay is required for viroid identification.

PAGE has been used in a mobile field laboratory since 1988 to detect *Coconut cadang-cadang viroid* (CCCVd, *Cocadviroid*) (Randles and Rodriguez, 2003).

7.2. Organization and Interpretation of Seed-health Assays

Significant yield losses can be the result of even low seed transmission rates, depending on the virus and the crop. Setting up a seed-health test procedure to provide an accurate estimate of infected-seed incidence in seed lots first requires a seed sample that is representative of the seed lot. Next is an assay with sufficient sensitivity, specificity, reliability and repeatability. Last but essential is defining the optimal seed-sample size for testing, and an appropriate statistical analysis of the outcome.

Which diseases to test for?

Not all seed-transmitted viruses and viroids are of equal importance. Lists of seed-transmitted pathogens have been provided by, for example, Richardson (1990), but none of the lists addresses the economic importance of the pathogen, its distribution or the significance of seed transmission. In the late 1990s a comprehensive source of information became available in the form of the *Crop Protection Compendium* (CPC) on CD-ROM, published by CAB International. The compendium contains a Database on Seed-borne Diseases (DSD) with valuable information, not only on the pathogen, but also on seed-borne aspects, economic impact, geographical distribution, phytosanitary significance and controls for a large number of important diseases. Such readily accessible scientific information on seed-borne diseases should also enable phytosanitary authorities to make more justifiable and rational decisions on regulations (McGee, 1997). The CPC further contains, among other things, a bibliographic database, illustrations and a glossary, as well as efficient facilities for searching, e.g. by crop and by pathogen. The CPC is updated annually; see more regarding the latest edition (Anon., 2004a) on http://www.cabicompendium.org/cpc. Another valuable information source on viruses is the 409 viruses described in *AAB Descriptions of Plant Viruses*, which are also continuously updated with new descriptions and which can be accessed on http://www.dpvweb.net. These descriptions, however, though eminently describing the pathogens, differ from the CPC in not containing information on economic impact, etc. *Viruses of Plants* (Brunt *et al.*,

1996), which is a product of the VIDE (Virus Identification Data Exchange) project, contains data on more than 900 viruses. Though also without data on importance, etc., this source, unlike the *AAB Descriptions*, includes lists of virus non-hosts (Section 4.1). Descriptions and lists from the VIDE Database are accessible also on 'Plant viruses online', http://image.fs.uidaho.edu/vide/refs.htm. Another source is ICTVdb: the Universal Virus Database of the International Committee on Taxonomy of Viruses, http://www.ncbi.nlm.nih.gov/ICTVdb/index.htm. A useful gateway with many links to plant virological topics is 'The Plant Pathology Internet Guide Book', http://www.pk.uni-bonn.de/ppigb/ppigb.htm.

An attempt has been made to list some of the most important seed-transmitted viral diseases (Table 4.1). Fourteen of these 35 virus and viroid species are aphid-transmitted and, among these, ten belong to the family *Potyvirus*. Several members of this very large family cause significant losses. Tobamoviruses are important, especially for tomato and pepper (Section 4.4.3). The other important diseases listed in Table 4.1 belong to the hordei-, como-, peclu-, tobra-, ilar- and sobemoviruses, and one to the pospiviroids. Although many of the nepoviruses are seed-transmitted at high frequencies, they are not generally considered to be of great significance.

7.2.1. Sampling and test methodology

In seed-health testing only a small portion of a seed lot can be tested; it is therefore crucial that this small quantity is representative of the seed lot. As emphasized by the chairman of the Bulking and Sampling Committee of ISTA, Arne Wold (1986, quoted from Mathur and Kongsdal, 2003): 'Incorrect sampling may lead to misleading test results, discarding seed lots of high quality, or to the approval of seed lots of low quality, which may reduce crop yield or even result in complete failure.'

Sampling must be carried out as prescribed in ISTA's *International Rules for Seed Testing* (Anon., 2004b). The chapter (in the 1999 edition) on sampling and its corresponding annex have been summarized by Mathur and Kongsdal (2003) in their book on seed-health testing for fungi. Some important points and definitions from this summary (with slight modifications) are:

- *Object of sampling*. The object of sampling is to obtain from a seed lot a representative sample of a size suitable for tests, in which the probability of a constituent being present is determined only by its level of occurrence in the seed lot.

Definitions:

- *Lot*. A lot is a specified quantity of seeds (max. 1000–40,000 kg depending on species) for which an ISTA International Seed Lot Certificate may be issued.
- *Primary sample*. A primary sample is a small portion taken from one point in the lot.

- *Composite sample.* A composite sample is formed by combining and mixing all the primary samples taken from the lot.
- *Submitted sample.* This is the sample submitted to the testing station. It must be of a size at least two to three times larger than the amount adequate for detecting low rates of seed infection (see Fig. 7.1) if required, and thus contain enough seeds for a possible retesting. The submitted sample may comprise either the whole or a subsample of the composite sample.
- *Working sample.* The submitted sample received by the seed testing station generally needs to be reduced to one or more working sample(s) equal to or greater than the size prescribed for each test.
- *Rules for sampling.* Explicit rules and methods for sampling from each seed lot are described in the ISTA *Rules* and must be strictly followed. In general, a specified number of *primary samples* are drawn from different parts of a seed lot. They are mixed to give a *composite sample*. The composite sample is thoroughly mixed before being properly reduced into the *submitted sample*.
- *Reduction of samples.* Whenever a seed sample is reduced or subdivided, any subsample must be drawn so as to be representative of the original seed lot. Rules, apparatuses and methods for reducing seed samples are described in the ISTA *Rules* and must be strictly followed.
- *Labelling and storing of samples.* How to label, seal and store seed samples is described in the ISTA *Rules*. Labels must contain necessary information, and bags must be sealed by an authorized sampling and sealing agency. Samples, if stored, must be kept in a cool, well-ventilated room.

Testing methodology

Most seed-transmitted viruses can be detected by serological, molecular or biological methods (viroids not by serological assays). Whether whole seeds or seedlings raised from seeds should be tested depends to a large extent on the virus–host combination. For five virus–host systems, LMV/lettuce, PeMoV and BCMV-PSt/groundnut, SMV/soybean and CMV/lupin (see Sample preparation, Section 5.2.2), the determination of infection levels by testing whole seeds has been shown to correspond well with that of testing the seedlings (i.e. gives a good indication of likely risk of transmission). For other virus–hosts, such as PSbMV-infected peas, whole-seed assays may not give a reliable indication (may result in an overestimate) of transmission due to the presence of detectable virus in the testae of a high proportion of seeds with virus-free embryos (Sections 2.1, 3.2.2 and 5.2.2). In such cases, testing of either decorticated seeds or seedlings raised from the seeds must be done. A method for freeing dry pea seeds from the seed-coat is described by Torok and Randles (2001) (see Section 6.2.2).

For tobamoviruses that are able to survive for months or years in or on the seed-coat of, for example, tomato and pepper, a serological test (e.g. by DISA, see Section 5.4) or a molecular test of whole seeds will reveal the absence or presence of virus, but only a bioassay (Section 4.4.3) can reveal whether the detected virus is viable or not.

Not much information is available on the location of seed-transmitted viroids in seeds, but, as far as PSTVd is concerned, similar rates of seed transmission through true potato seeds were found whether dormant seeds, germinated seeds or seedlings were tested (Singh *et al.*, 1988). PSTVd is also transmitted through tomato seed (Diener and Raymer, 1971; Kryczynski *et al.*, 1988).

According to various reports on molecular detection methods for plant viruses and viroids, PCR is generally considered more sensitive than nucleic acid hybridization (NAH). However, despite its lower sensitivity, there is a renewed interest in NAH (Section 6.1), which is becoming the method of choice for small commercial laboratories (Singh and Nie, 2002). Mumford *et al.* (2000) discuss the advantages and disadvantages of RT-PCR, including real-time fluorescent RT-PCR, and NAH (see also Section 6.2.6), and conclude that PCR methods are generally best suited for low sample numbers and NAH for large numbers of samples.

OPTIMAL PROCEDURES AND CONFIRMATIVE TESTS. As mentioned earlier (Sections 5.2.8 and 6.2.3), parameters for optimal assay sensitivity need to be worked out for each virus–host system before initiating routine testing. Critical parameters are optimal extraction methods, optimal dilution of the antigen or the target molecule, optimal detection reagents (antibodies, probes, primers) and their dilution, and optimal test conditions. It is also important to confirm the results of laboratory assays from time to time by biological assays, as this could reveal the emergence of pathogen variants that escape, for example, ELISA detection (Maury *et al.*, 1998).

7.2.2. Group testing of seeds

In large-scale testing of seeds, single-seed assays are not feasible. Instead, the working sample can be divided into groups of seeds (or seedlings) of a size in which one infected seed/seedling out of x healthy can be detected (see also assay sensitivity below), followed by a statistical analysis to determine with a certain probability the actual percentage of seed infection in the sample. When a representative sample (the working sample) of the seed lot is divided into N groups of n seeds, the most probable percentage of infection, P (in per cent) = $[1 - (Y/N)^{1/n}] \times 100$, can be estimated as a function of the number of assay-negative groups (Y) (Maury and Kheterpal, 1997). Example: if $N = 10$, $n = 50$ and $Y = 7$, we get $[1 - 0.7^{0.02}] \times 100 = 0.71\%$ as the most probable number (or the maximum likelihood estimate, MLE). A graphical method for estimating per cent seed infection is shown in Fig. 7.1. It shows the optimal range of group sizes for determining from 0.001% to 10% seed infection (see below). The figures in the left column correspond to $1 - Y$ in per cent.

The precision of the MLE method depends on the number of groups, which determine the confidence limits. Table 7.1 shows the confidence limits for an infection level of 0.11% estimated from different numbers of groups of 1000 seeds. The gain in precision must be balanced against the effort required to do the additional tests (Roberts *et al.*, 1993; Ridout and Roberts, 1997).

Figure 7.1 provides a tool based on the MLE formula for determining the per cent infection using different group sizes, but provides no confidence intervals. According to Roberts *et al.* (1993) and as informed by S.J. Roberts, Warwick,

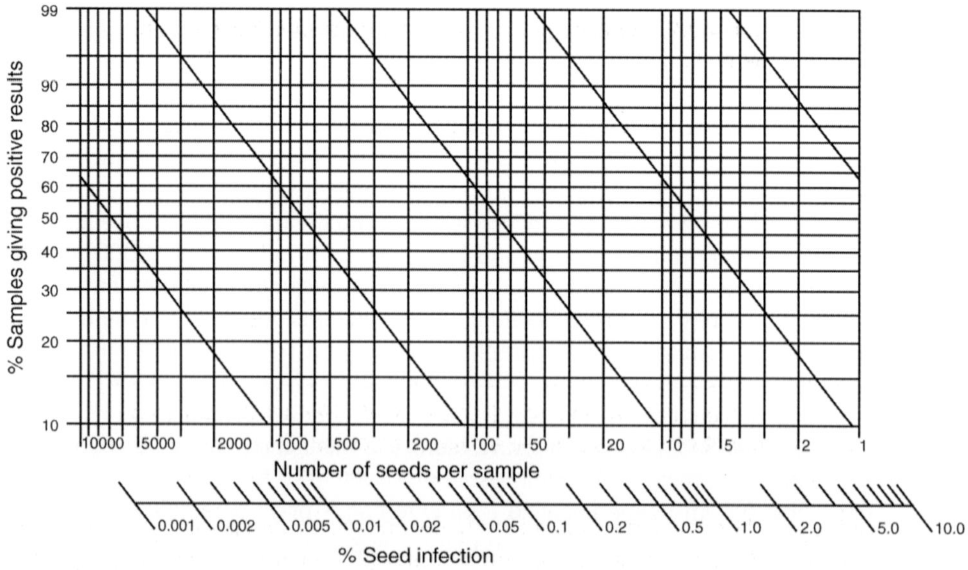

Fig. 7.1. Graphical method for estimating per cent seed infection. Reproduced from Taylor *et al.* (1993) with permission from authors and the copyright holder © ISTA.

Table 7.1. Effect of number of groups, *N*, on 95% confidence limits for a sample size, *n*, of 1000 seeds.

Number of groups, *N*	Number of positive	Estimated infection level (%)	95% confidence limits	
			Lower	Upper
3	2	0.11	0.02	0.38
6	4	0.11	0.03	0.27
12	8	0.11	0.05	0.21
24	16	0.11	0.06	0.18
48	32	0.11	0.07	0.16

Table reconstructed from Roberts *et al.* (1993) with permission from the authors and the copyright holder © ISTA.

UK (2004, personal communication), a computer program to estimate the proportion of infected seeds from the results of an assay, together with confidence intervals and a goodness-of-fit test, can be downloaded from www.planthealth.co.uk\downloads. On the website there is also an enquiry form, and additional information can be requested by email to enquiries@planthealth.co.uk.

Determination of the optimum group size is more difficult. If all groups in an assay are positive or all are negative, little information is obtained. An estimate

Table 7.2. Optimum group size, *n*, for determining infection level, θ.

θ (%)	Size, *n*
0.1	1593
0.5	318
1	159
5	31
10	15
25	6

Table reconstructed from Roberts *et al.* (1993) with permission from the authors and the copyright holder © ISTA.

of the infection level can only be made if some groups are positive and some are negative. Most information from such an assay is obtained when the proportion of negative samples is between 15 and 35%, which is when the expected mean number of infected seeds per group is between 1 and 2 (Roberts *et al.*, 1993). Examples for a range of different infection levels are given in Table 7.2 and these same values can be obtained approximately from Fig. 7.1 in the region where *Y* values are between 15 and 35%.

It is problematic that determination of the optimum group size is dependent on the proportion of infected seeds, which is usually unknown before testing. One way to overcome this is to define a 'range of interest' (e.g. 0.1 to 5.0%). The group size(s) can thereby be optimized for this range, where precise estimates of infection levels outside this range are of little interest. Another alternative is the use of a two- or three-stage (sequential) test design, where the results of each stage are used to establish group size for the next stage or, in the extreme case, whether to test at all. The choice will depend on the aims of the assay and especially the timescales necessary to obtain results; for more information see Roberts *et al.* (1993).

Group size and assay sensitivity

As can be seen from Table 7.2 and Fig. 7.1, the group size required for use in the MLE method grows with decreasing levels of infection to be detected. For example, to determine 5% seed infection, the optimal group size is 31, and for 0.5% it is 318.

Detection of one infected seed out of a group of *n* seeds if *n* is large, say 1000 or 2000, is usually not possible by ELISA, but may be possible in some cases by PCR. For such large groups, therefore, each group must be divided into subgroups of a size where one infected seed out of *x* seeds is detectable. The maximum size of such subgroups thus depends on the assay sensitivity for the virus–host concerned. The means of determining size of composite samples is described for ELISA in Sample preparation, Section 5.2.2 and in Section 5.2.8. A similar approach must be taken for NAH and PCR assays.

As mentioned previously, the higher the number of groups that are tested, the higher the level of confidence. For practical and economic reasons, a compromise must be made between the number of groups being tested and the level of confidence (Roberts *et al.*, 1993; Ridout and Roberts, 1997). More details on design and interpretation of seed-health assays can be found in, for example, Geng *et al.* (1983), Roberts *et al.* (1993) and Ridout and Roberts (1997). A statistical approach developed by Geng *et al.* (1983) involves a *non-tolerable* level (Int) and a lower *tolerable* level of infection (It), where the Int is determined by the pathogen's potential of causing economic losses in crop production. The method describes how to determine the number of tests required where the tolerance level, the sample size and the sensitivity of the assay are given (Hamilton, 1983). This approach apparently reduces the number of seed groups to be tested compared to those required when assessing the seed infection rate as described above (Roberts *et al.*, 1993; Masmoudi *et al.*, 1994; Ridout and Roberts, 1997). However, the statistical basis used for the approach by Roberts *et al.* (1993) and Ridout and Roberts (1997) is considered analytically more correct than that of Geng *et al.* (1983) (S.J. Roberts, Warwick, UK, 2004, personal communication).

7.3. Tolerance Levels of Infection, and Pathogens in Germplasm

Tolerance levels in quality control

Many seed-transmitted viruses, present even at low levels in seeds sown, can cause significant losses in the crop, depending on the presence of vectors, their vectoring efficiency and their number. Thus, tolerance levels of 0.1% seed infection, or lower in some localities, are required for controlling many seed-transmissible, aphid-borne viruses. The need for setting realistic tolerance levels per virus–host system concerned has already been discussed in Sections 1.4.1 and 3.2.2. As discussed above, a determination of low rates of seed infection makes it necessary for higher numbers of seeds to be tested.

Inoculum thresholds for viruses are discussed by Stace-Smith and Hamilton (1988), and relationships between seed-health assay results, tolerance standards and disease risks are presented by Roberts (1999).

Plant quarantine and gene banks

In contrast to that needed for quality control of seed, the tolerance of seed-transmitted diseases in quarantine must be zero. The general problems of excluding exotic seed-transmitted viral diseases by quarantine are treated in Sections 1.4.2 and 3.2.2, which include some that are related to exchanges of germplasm.

Of the germplasm handled by the 13 international agricultural research centres (IARCs), about 95% is conserved and exchanged in the form of true seeds. Between 1987 and 1991, approximately 750,000 samples of germplasm were distributed and, from one IARC (the International Crops Research Institute for the Semi-Arid Tropics (ICRISAT), India), almost one million seed samples were dispatched to 153 countries during 1974–1997 (Diekmann, 1997). One of the problems of the great diversity of germplasm is that different lines can be expected to have different levels of resistance to pests and pathogens. Another problem is that germplasm collections usually contain seeds from regions of the world that are not only the centres of origin of the crop, but may also be centres for genetic diversity of crop-specific pathogens. Seeds from these regions may therefore be contaminated by pathogenic strains or pathotypes exotic to the recipient location (Diekmann, 1997). The contamination of pea germplasm collections with PSbMV is one such example (Section 1.4.2).

Among the 1000-plus registered plant germplasm banks around the world, those maintained at the IARCs and those situated in developed countries have fewer problems with management of pathogen contamination than those located in developing countries. Curators responsible for the plant genetic resources (PGRs) in developing countries are often unaware of the potential phytosanitary problems that may be associated with the germplasm they receive from plant collectors and other sources (F.J. Morales, International Center for Tropical Agriculture (CIAT), Colombia, 2001, personal communication, and E. Frison, International Plant Genetic Resources Institute (IPGRI), Rome, 2004, personal communication).

According to these sources, the following factors result in international dissemination of pathogen-infected seed lots: (i) most germplasm bank curators lack formal training in the discipline of plant protection; (ii) seed-testing facilities are inadequate at many germplasm banks in developing countries; and (iii) poor funding at many germplasm banks precludes hiring plant- and seed-health specialists; all of which cause perpetuation of seed-borne pathogens during germplasm seed increases, loss of valuable PGRs and unrecognized international distribution of seed-borne pathogens.

The above-mentioned sources conclude that there is an urgent need for support from recipients in developed countries to help germplasm banks in the Third World to manage their PGRs, since the dissemination of seed-borne pathogens and the loss of PGRs are global concerns affecting both developing and industrialized countries.

7.4. Standardization and Cost of Tests

Seed-health tests differ from country to country and the results from one country may not be accepted by another. There is, therefore, a need for standard, or at least generally accepted, methods. Efforts to develop such methods have been made since 1958 by the Plant Disease Committee (now the Seed Health Committee, SHC) under ISTA. These efforts have resulted in a number of working sheets for seed-health testing for important fungi, bacteria and a single one for viruses, compiled in the ISTA *Handbook on Seed Health Testing* (Anon., 1981). Of these, 18 methods have been validated (by 2004) and moved from the ISTA *Handbook on Seed Health Testing* to the 'Annexe to Chapter 7, Seed Health Testing Methods' in ISTA's *International Rules for Seed Testing* (info: http://www.seedtest.org). So far, no up-to-date standard method for virus detection has been described or validated (only a single working sheet from 1981 for biological testing of lettuce for LMV appears in the above-mentioned ISTA *Handbook*; see Section 4.4.3).

Concurrently with the development and improvement of detection techniques, the ISTA SHC and its working groups are updating existing working sheets and preparing new ones. One of the tasks of the SHC is to organize comparative seed-health testing methods/results in a number of laboratories (see below), the results of which form an essential basis for preparing new working sheets. The organization of the SHC is described on http://www.seedtest.org/en/tcom-shc.html.

In 1993, new efforts to establish standard seed-health testing methods were initiated by an international consortium of seed industry and seed-health testing plant pathologists from The Netherlands, France, the USA, Japan and Israel by founding the International Seed Health Initiative (ISHI). ISHI works in collaboration with ISTA for the validation of seed-health testing methods, so as to ensure that methods are scientifically sound, reliable and robust. They also cooperate with national and international regulatory and accreditation authorities. ISHI has three divisions: vegetable crops (ISHI-Veg), herbage (ISHI-H) and field crops (ISHI-F). The seven International Technical Groups (ITG) under ISHI-Veg are working on important seed-borne diseases for each of their crops, and test methods for at least three pathogens updated/prepared by ISHI have been validated by ISTA (see ISHI, http://www.worldseed.org/ishis.htm). For detecting methods for tobamoviruses on tomato and pepper seed that have been comparatively tested by ISHI, see II. Tobamoviruses on tomato and pepper seed . . ., Section 4.4.3.

7.4.1. Comparative testing

Neergaard (1970) gives the following definitions of the requirements that will have to be fulfilled in order to adopt a procedure for routine seed-health testing:

1. The procedure must give results within a reasonable time limit and with a limited requirement of labour.
2. The results must enable a fair estimate of the amount of seed-borne inoculum, as a basis for evaluation of the planting value of the seed.
3. The results must be reproducible.

Neergaard (1970) further points out that, obviously, any agreement in methods can be obtained only by comparison of procedures and by checking mutually the interpretation of results. Both requirements are of equal importance. Accordingly, the line of action has been, *first*, to organize comparative tests for specific organisms, basing the tests on defined procedures thoroughly prescribed in all details, including the conditions of recording. *Secondly*, to check the results obtained independently by the participants in joint laboratory working parties where duplicate samples, set up according to the common methods that have been used, can be studied and the possible causes of discrepancies discussed.

Guidelines for comparative seed-health testing (organization, design, procedures and analysis of results) earlier issued by ISTA were replaced in 2003 with the *ISTA Handbook of Method Validation for Seed Testing* (Sheppard, 2003), published as a CD-ROM. The handbook presents guidelines for method validation through traditional multi-laboratory comparative or ring tests and performance validation of 'rapid test kits'.

Standard reagents

Test reagents for use in standardized methods, whether antisera, probes for NAH or primers for PCR, should be proved effective for detection of all strains or isolates of a particular virus. Polyclonal antisera are typically capable of detecting all strains of a given virus, but have some limitations (Section 5.1.2) compared to Mabs, which can be produced with selectable specificity, high sensitivity and in practically unlimited quantities, once hybridomas have been prepared. For testing purposes, the use of Mabs with selected specificity (broad-spectrum Mabs or pooled, strain-specific Mabs) would be an essential element for obtaining uniform results in different laboratories (Matthews, 1991; Van Regenmortel and Dubs, 1993; Barker and Torrance, 1997; Maury and Khetarpal, 1997; Hull, 2002).

Uniform probes and primers can also be produced in unlimited quantities. However, as emphasized by Candresse *et al.* (1998), only those primers thoroughly tested for sensitivity and detection of all pathogen variants are suited as standard reagents for broad-specificity detection (Section 6.2.7).

Standard reference material

In comparative tests, it is important to provide testers with adequate reference material for both positive and negative controls. Aspects of providing samples that are homogeneous and do not deviate over time are discussed for bacterial assays by van der Bulk and Taylor (1997) and van der Bulk *et al.* (1999). Such standardized reference material with guaranteed homogeneity and stability should ideally be available in the same way as standard reagents and reagent kits. For seed-transmitted viruses, dehydrated infected plant tissue would probably be the best form (see Section 4.5). However, because such samples are infective, they constitute a potential phytosanitary risk, though limited unless users propagate the material.

Standard procedures

In order to produce acceptable results in a seed-health testing laboratory, not only are standard reagents and standard reference material required, but several other factors must also be considered. Among these are: overall quality control measures; the organization of the laboratory; laboratory staff with necessary education and training; appropriate, detailed and approved methods and procedures for seed sampling and testing in order to obtain accurate and reproducible results; calibration of sampling, measuring and testing equipment; periodical monitoring of testing materials (water and chemicals, etc.); and maintenance of staff skills and of record-keeping systems for test results, test reports, audit reports, certificates, calibration and repair of equipment, etc. (Hutchins, 1997). Such a quality management system must be documented for ISTA accreditation of the laboratory.

Use of standardized reagents and protocols providing an internationally recognized standard of performance should lead to safe movement of germplasm and to greater confidence among plant breeders and plant health authorities (Barker and Torrance, 1997).

7.4.2. Cost of tests

The approximate price for ELISA tests ranges from US$0.60 to 1.80 per test sample, corresponding to duplicate wells of 100 μl each. About half of the price per sample includes IgG, conjugate and controls (reagent set) purchased from a phytodiagnostic test reagent supplier (Appendix 3), and the other half is for additional reagents and consumables required (price information: Adgen Phytodiagnostics, Neogen Europe Ltd, UK, 2004, personal communication). Lower costs per test sample can be obtained by using some of the EIA variants described in Chapter 5, and most probably also by purchasing crude antisera for in-laboratory preparation of IgG and conjugate.

The price for an NAB test per sample is higher, approximately three times the above-mentioned price for ELISA (O'Donnell, 1999), depending on the type of test and chemical costs. If commercial extraction kits are used, the price is yet higher (see also Chapter 6, pp. 159, 175 and 218). For NAB tests, the extra set-up costs and the increased level of skill of staff required should also be taken into account, compared to ELISA. On the other hand, the high sensitivity of PCR, which is just one of its advantages, will allow larger composite samples to be tested, resulting in fewer assays being needed (O'Donnell, 1999).

None of the above prices includes staff time for sample preparation or running the tests.

A list of commercial suppliers of phytodiagnostic test reagent *sets*, containing antibodies, conjugate and controls, or test *kits*, which contain all test ingredients (including solid support, buffers, substrate, etc.), can be found in Appendix 3, which also contains information on suppliers of laboratory equipment and materials.

References

AAB Descriptions of Plant Viruses (n.d.) Nos. 1–409. Available at: http://www.dpvweb.net

Anon. (1981) *Handbook on Seed Health Testing Working Sheets.* International Seed Testing Association, Zurich, Switzerland.

Anon. (2004a) *Crop Protection Compendium,* 2004 edn (CD-ROM). CAB International, Wallingford, UK.

Anon. (2004b) *International Rules for Seed Testing,* 2004 edn. International Seed Testing Association, Zurich, Switzerland.

Barker, H. and Torrance, L. (1997) Importance of biotechnology for germplasm health and quarantine. In: Callow, J.A., Ford-Lloyd, B.V. and Newbury, H.J. (eds) *Biotechnology and Plant Genetic Resources.* CAB International, Wallingford, UK, pp. 235–254.

Brunt, A.A., Crabtree, K., Dallwitz, M.J., Gibbs, A.J. and Watson, L. (eds) (1996) *Viruses of Plants.* CAB International, Wallingford, UK.

Candresse, T., Hammond, R.W. and Hadidi, A. (1998) Detection and identification of plant viruses and viroids using polymerase chain reaction (PCR). In: Hadidi, A., Khetarpal, R.K. and Koganezawa, H. (eds) *Plant Virus Disease Control.* APS Press, St Paul, Minnesota, pp. 399–416.

Diekmann, M. (1997) Activities at the international agricultural research centres to control pathogens in germplasm. In: McGee, D.C. (ed.) *Plant Pathogens and the Worldwide Movement of Seeds*. APS Press, St Paul, Minnesota.

Diener, T.O. and Raymer, W.B. (1971) Potato spindle tuber 'virus'. *AAB Descriptions of Plant Viruses* No. 66. Available at: http://www.dpvweb.net

Geng, S., Campbell, R.N., Carter, M. and Hills, F.J. (1983) Quality-control programs for seedborne pathogens. *Plant Disease* 67, 236–242.

Griffin, R.L. (1991) *Using the Transmission Electron Microscope in the Biological Sciences*. Ellis Horwood, London.

Hamilton, R.I. (1983) Certification schemes against seed-borne viruses in leguminous hosts, present status and further areas for research and development. *Seed Science and Technology* 11, 1051–1062.

Hull, R. (2002) *Matthews' Plant Virology*, 4th edn. Academic Press, London.

Hutchins, J.D. (1997) ISTA accreditation in seed health testing laboratories. In: Hutchins, J.D. and Reeves, J.C. (eds) *Seed Health Testing – Progress Towards the 21st Century.* CAB International, Wallingford, UK, pp. 219–226.

Kryczynski, S., Paduch-Cichal, E. and Skrzeczkowski, L.J. (1988) Transmission of three viroids through seed and pollen of tomato plants. *Journal of Phytopathology* 121, 51–57 (Abstr.).

McGee, D. (1997) Epidemiological approach to disease management through seed technology. *Annual Review of Phytopathology* 44, 445–466.

Masmoudi, K., Duby, C., Suhas, M., Guo, J.Q., Guyot, L., Olivier, V., Taylor, J. and Maury, Y. (1994) Quality control of pea seed for pea seed-borne mosaic virus. *Seed Science and Technology* 22, 407–414.

Mathur, S.B. and Kongsdal, O. (2003) *Common Laboratory Seed Health Testing Methods for Detecting Fungi*, 1st edn. International Seed Testing Association, Zurich, Switzerland.

Matthews, R.E.F. (1991) *Plant Virology*, 3rd edn. Academic Press, San Diego, California.

Maury, Y. and Khetarpal, R.K. (1997) Quality control of seed for viruses: present status and future prospects. In: Hutchins, J.D. and Reeves, J.C. (eds) *Seed Health Testing – Progress towards the 21st Century.* CAB International, Wallingford, UK, pp. 243–252.

Maury, Y., Duby, C. and Khetarpal, R.K. (1998) Seed certification for viruses. In: Hadidi, A., Khetarpal, R.K. and Koganezawa, H. (eds) *Plant Virus Disease Control.* APS Press, St Paul, Minnesota, pp. 237–248.

Mumford, R.A., Walsh, K. and Boonham, N. (2000) A comparison of molecular methods for the routine detection of viroids. *Bulletin OEPP/EPPO Bulletin* 30, 431–435.

Neergaard, P. (1970) Seed pathology, international co-operation and organization. *Proceedings of the International Seed Testing Association* 35, 19–42.

O'Donnell, K.J. (1999) Plant pathogen diagnostics: present status and future developments. *Potato Research* 42, 437–447.

Randles, J.W. and Rodriguez, M.J.B. (2003) Coconut cadang-cadang viroid. In: Hadidi, A., Flores, R., Randles, J.W. and Semancik, J.S. (eds) *Viroids.* CSIRO Publishing, Collingwood, Victoria, Australia, and Science Publishers, Enfield, New Hampshire, pp. 233–241.

Richardson, M.L. (1990) *An Annotated List of Seed-borne Diseases*, 4th edn. International Seed Testing Association, Zurich, Switzerland.

Ridout, M.S. and Roberts, S.J. (1997) Improving quality control procedures for seedborne pathogens by testing sub-samples of seeds. *Seed Science and Technology* 25, 195–202.

Roberts, S.J. (1999) Thresholds, standards, tests, transmission and risks. In: *Proceedings of 3rd ISTA Seed Health Symposium, Ames, Iowa, USA, 16–19 August 1999.* ISTA, Zurich, Switzerland, pp. 20–24.

Roberts, S.J., Phelps, K., Taylor, J.D. and Ridout, M.S. (1993) Design and interpretation of seed health assays. In: Sheppard, J.W. and Langerak, C.J. (eds) *Proceedings of the 1st ISTA PDC Symposium on Seed Health Testing*, Ottawa,

Canada. ISTA, Zurich, Switzerland, pp. 115–125.

Roenhorst, J.W., Butôt, R.P.T., van der Heijden, K.A., Hooftman, M. and Zaayen, A. (2000) Detection of *Chrysanthemum stunt viroid* and *Potato spindle tuber viroid* by return-polyacrylamide gel electrophoresis. *Bulletin OEPP/EPPO Bulletin* 30, 453–456.

Schumacher, J., Meyer, N., Riesner, D. and Weidemann, H.-L. (1986) Diagnostic procedure for detection of viroids and viruses with circular RNAs by 'return' gel electrophoresis. *Journal of Phytopathology* 115, 332–343.

Sheppard, J.W. (ed.) (2003) *ISTA Handbook of Method Validation for Seed Testing.* International Seed Testing Association, Zurich, Switzerland.

Singh, R.P. and Dhar, A.K. (1998) Detection and management of plant viroids. In: Hadidi, A., Khetarpal, R.K. and Koganezawa, H. (eds) *Plant Virus Disease Control.* APS Press, St Paul, Minnesota, pp. 428–447.

Singh, R.P. and Nie, X. (2002) Nucleic acid hybridization for plant virus and viroid detection. In: Khan, J.A. and Dijkstra, J. (eds) *Plant Viruses as Molecular Pathogens.* Food Products Press/Haworth Press, New York, pp. 443–469.

Singh, R.P., Boucher, A. and Seabrook, J.E.A. (1988) Detection of the mild strains of potato spindle tuber viroid from single true potato seed by return electrophoresis. *Phytopathology* 78, 663-667 (Abstr.).

Stace-Smith, R. and Hamilton, R.I. (1988) Inoculum thresholds of seedborne pathogens: viruses. *Phytopathology* 78, 875–880.

Taylor, J.D., Phelps, K. and Roberts, S.J. (1993) Most probable number (MPN) method: origin and application. In: Sheppard, J.W. and Langerak, C.J. (eds) *Proceedings of the 1st ISTA PDC Symposium on Seed Health Testing.* Ottawa, Canada. ISTA, Zurich, Switzerland, pp. 106–114.

Torok, V.A. and Randles, J.W. (2001) *Tobacco mosaic virus* RNA as an internal control for duplex RT-PCR assay of pea germplasm. *Australasian Plant Pathology* 30, 227–230.

van der Bulk, R.W. and Taylor, J.D. (1997) Microbiological reference materials for quality control in seed health testing. In: Hutchins, J.D. and Reeves, J.C. (eds) *Seed Health Testing – Progress towards the 21st Century.* CAB International, Wallingford, UK, pp. 233–241.

van der Bulk, R.W., Langerak, C.J., Roberts, S.J. and Lyons, N.F. (1999) Microbial reference materials in seed health test standardization. In: *Proceedings of 3rd ISTA Seed Health Symposium, Ames, Iowa, USA, 16–19 August 1999.* ISTA, Zurich, Switzerland, pp. 133–135.

Van Regenmortel, M.H.V. and Dubs, M.-C. (1993) Serological procedures. In: Matthews, R.E.F. (ed.) *Diagnosis of Plant Virus Diseases.* CRC Press, Boca Raton, Florida, pp. 159–214.

Wold, A. (1986) Foreword. In: *ISTA Handbook on Seed Sampling.* International Seed Testing Association, Zurich, Switzerland.

Appendix 1: List of Seed-transmitted Viruses and Viroids

List of Seed-transmitted Viruses and Viroids.[a]

Name	Acronym	Family	Genus
'Conventional' viruses			
Alfalfa mosaic virus	AMV	*Bromoviridae*	*Alfamovirus*
Apple stem grooving virus	ASGV	*Flexiviridae*	*Capillovirus*
Arabis mosaic virus	ArMV	*Comoviridae*	*Nepovirus*
Arracacha virus B	AVB	*Comoviridae*	*Nepovirus*
Artichoke Italian latent virus	AILV	*Comoviridae*	*Nepovirus*
Artichoke latent virus	ArLV	*Potyviridae*	*Potyvirus*
Artichoke yellow ringspot virus	AYRSV	*Comoviridae*	*Nepovirus*
Asparagus virus 2	AV-2	*Bromoviridae*	*Ilarvirus*
Barley stripe mosaic virus	BSMV	–	*Hordeivirus*
Bean common mosaic necrosis virus	BCMNV	*Potyviridae*	*Potyvirus*
Bean common mosaic virus	BCMV	*Potyviridae*	*Potyvirus*
Blackeye cowpea mosaic strain	BCMV-BlCM	*Potyviridae*	*Potyvirus*
Peanut stripe strain	BCMV-PSt	*Potyviridae*	*Potyvirus*
Bean pod mottle virus	BPMV	*Comoviridae*	*Comovirus*
Bean yellow mosaic virus	BYMV	*Potyviridae*	*Potyvirus*
Blackgram mottle virus	BMoV	*Tombusviridae*	*Carmovirus*
Blueberry leaf mottle virus	BLMoV	*Comoviridae*	*Nepovirus*
Blueberry shock virus	BlShV	*Bromoviridae*	*Ilarvirus*

Continued

Name	Acronym	Family	Genus
'Conventional' viruses (*Contd.*)			
Broad bean mottle virus	BBMV	*Bromoviridae*	*Bromovirus*
Broad bean stain virus	BBSV	*Comoviridae*	*Comovirus*
Broad bean true mosaic virus	BBTMV	*Comoviridae*	*Comovirus*
Broad bean wilt virus 2	BBWV-2	*Comoviridae*	*Fabavirus*
Cassava green mottle virus	CsGMV	*Comoviridae*	*Nepovirus*
Cassia yellow spot virus	CasYSV	*Potyviridae*	*Potyvirus*
Cherry leaf roll virus	CLRV	*Comoviridae*	*Nepovirus*
Cherry rasp leaf virus	CRLV	*Comoviridae*	*Nepovirus*
Chicory yellow mottle virus	ChYMV	*Comoviridae*	*Nepovirus*
Clover yellow mosaic virus	ClYMV	*Flexiviridae*	*Potexvirus*
Cocoa necrosis virus	CoNV	*Comoviridae*	*Nepovirus*
Coffee ringspot virus	CoRSV	*Rhabdoviridae*	*Nucleorhabdovirus*
Cowpea aphid-borne mosaic virus	CABMV	*Potyviridae*	*Potyvirus*
Cowpea green vein banding virus	CGVBV	*Potyviridae*	*Potyvirus*
Cowpea mild mottle virus	CPMMV	*Flexiviridae*	*Carlavirus*
Cowpea mosaic virus	CPMV	*Comoviridae*	*Comovirus*
Cowpea mottle virus	CPMoV	*Tombusviridae*	*Carmovirus*
Cowpea severe mosaic virus	CPSMV	*Comoviridae*	*Comovirus*
Crimson clover latent virus	CCLV	*Comoviridae*	*Nepovirus*
Cucumber green mottle mosaic virus	CGMMV	–	*Tobamovirus*
Cucumber mosaic virus	CMV	*Bromoviridae*	*Cucumovirus*
Cycas necrotic stunt virus	CNSV	*Comoviridae*	*Nepovirus*
Desmodium mosaic virus	DesMV	*Potyviridae*	*Potyvirus*
Dulcamara mottle virus	DuMV	–	*Tymovirus*
Eggplant mosaic virus	EMV	–	*Tymovirus*
Elm mottle virus	EMoV	*Bromoviridae*	*Ilarvirus*
Epirus cherry virus	EpCV	–	*Ourmiavirus*
Foxtail mosaic virus	FoMV	*Flexiviridae*	*Potexvirus*
Fragaria chiloensis latent virus	FClLV	*Bromoviridae*	*Ilarvirus*
Grapevine Bulgarian latent virus	GBLV	*Comoviridae*	*Nepovirus*
Grapevine fanleaf virus	GFLV	*Comoviridae*	*Nepovirus*
Guar symptomless virus	GSLV	*Potyviridae*	*Potyvirus*
Hibiscus latent ringspot virus	HLRSV	*Comoviridae*	*Nepovirus*
Hippeastrum mosaic virus	HiMV	*Potyviridae*	*Potyvirus*
Hop mosaic virus	HpMV	*Flexiviridae*	*Carlavirus*
Humulus japonicus latent virus	HJLV	*Bromoviridae*	*Ilarvirus*

Hydrangea mosaic virus	HdMV	*Bromoviridae*	*Ilarvirus*
Indian peanut clump virus	IPCV	–	*Pecluvirus*
Lettuce mosaic virus	LMV	*Potyviridae*	*Potyvirus*
Lucerne Australian latent virus	LALV	*Comoviridae*	*Nepovirus*
Lucerne transient streak virus	LTSV	–	*Sobemovirus*
Lychnis ringspot virus	LRSV	–	*Hordeivirus*
Maize dwarf mosaic virus	MDMV	*Potyviridae*	*Potyvirus*
Melon necrotic spot virus	MNSV	*Tombusviridae*	*Carmovirus*
Melon rugose mosaic virus	MRMV	–	*Tymovirus*
Mulberry ringspot virus	MRSV	*Comoviridae*	*Nepovirus*
Nicotiana velutina mosaic virus	NVMV	–	Unassigned
Olive latent virus 1	OLV-1	*Tombusviridae*	*Necrovirus*
Parsley latent virus	PaLV	–	Unassigned
Pea early-browning virus	PEBV	–	*Tobravirus*
*Pea enation mosaic virus 1**	PEMV-1	*Luteoviridae*	*Enamovirus*
*Pea enation mosaic virus 2**	PEMV-2	–	*Umbravirus*
Pea mild mosaic virus	PMiMV	*Comoviridae*	*Comovirus*
Pea seed-borne mosaic virus	PSbMV	*Potyviridae*	*Potyvirus*
Peach rosette mosaic virus	PRMV	*Comoviridae*	*Nepovirus*
Peanut clump virus	PCV	–	*Pecluvirus*
Peanut mottle virus	PeMoV	*Potyviridae*	*Potyvirus*
Peanut stunt virus	PSV	*Bromoviridae*	*Cucumovirus*
Pelargonium zonate spot virus	PZSV	–	Unassigned
Pepper mild mottle virus	PMMoV	–	*Tobamovirus*
Pepper ringspot virus	PepRSV	–	*Tobravirus*
Plum pox virus	PPV	*Potyviridae*	*Potyvirus*
Potato virus T	PVT	*Flexiviridae*	*Trichovirus*
Potato virus U	PVU	*Comoviridae*	*Nepovirus*
Prune dwarf virus	PDV	*Bromoviridae*	*Ilarvirus*
Prunus necrotic ringspot virus	PNRSV	*Bromoviridae*	*Ilarvirus*
Raspberry bushy dwarf virus	RBDV	–	*Idaeovirus*
Raspberry ringspot virus	RpRSV	*Comoviridae*	*Nepovirus*
Red clover vein mosaic virus	RCVMV	*Flexiviridae*	*Carlavirus*
Rubus Chinese seed-borne virus	RCSV	*Comoviridae*	*Nepovirus*
Satsuma dwarf virus	SDV	*Comoviridae*	*Nepovirus*
Soil-borne wheat mosaic virus	SBWMV	–	*Furovirus*
Southern bean mosaic virus	SBMV	–	*Sobemovirus*
Southern cowpea mosaic virus	SCPMV	–	*Sobemovirus*
Sowbane mosaic virus	SoMV	–	*Sobemovirus*

Continued

Name	Acronym	Family	Genus
'Conventional' viruses (*Contd.*)			
Soybean mosaic virus	SMV	*Potyviridae*	*Potyvirus*
Spinach latent virus	SpLV	*Bromoviridae*	*Ilarvirus*
Squash mosaic virus	SqMV	*Comoviridae*	*Comovirus*
Strawberry latent ringspot virus	SLRSV	*Comoviridae*	*Nepovirus*
Subterranean clover mottle virus	SCMoV	–	*Sobemovirus*
Sugarcane mosaic virus	SCMV	*Potyviridae*	*Potyvirus*
Sunn-hemp mosaic virus	SHMV	–	*Tobamovirus*
Telfairia mosaic virus	TeMV	*Potyviridae*	*Potyvirus*
Tobacco mosaic virus	TMV	–	*Tobamovirus*
Tobacco rattle virus	TRV	–	*Tobravirus*
Tobacco ringspot virus	TRSV	*Comoviridae*	*Nepovirus*
Tobacco streak virus	TSV	*Bromoviridae*	*Ilarvirus*
Tomato aspermy virus	TAV	*Bromoviridae*	*Cucumovirus*
Tomato black ring virus	TBRV	*Comoviridae*	*Nepovirus*
Tomato bushy stunt virus	TBSV	*Tombusviridae*	*Tombusvirus*
Tomato mosaic virus	ToMV	–	*Tobamovirus*
Tomato ringspot virus	ToRSV	*Comoviridae*	*Nepovirus*
Turnip yellow mosaic virus	TYMV	–	*Tymovirus*
Urdbean leaf crinkle virus	ULCV	–	Unassigned
Wheat streak mosaic virus	WSMV	*Potyviridae*	*Tritimovirus*
White clover mosaic virus	WClMV	*Flexiviridae*	*Potexvirus*
Zucchini yellow mosaic virus	ZYMV	*Potyviridae*	*Potyvirus*
Cryptic viruses			
Alfalfa cryptic virus 1	ACV-1	*Partitiviridae*	*Alphacryptovirus*
Alfalfa cryptic virus 2	ACV-2	*Partitiviridae*	*Betacryptovirus*
Beet cryptic virus 1	BCV-1	*Partitiviridae*	*Alphacryptovirus*
Beet cryptic virus 2	BCV-2	*Partitiviridae*	*Alphacryptovirus*
Beet cryptic virus 3	BCV-3	*Partitiviridae*	*Alphacryptovirus*
Carnation cryptic virus 1	CCV-1	*Partitiviridae*	*Alphacryptovirus*
Carnation cryptic virus 2	CCV-2	*Partitiviridae*	*Alphacryptovirus*
Carrot temperate virus 1	CteV-1	*Partitiviridae*	*Alphacryptovirus*
Carrot temperate virus 2	CteV-2	*Partitiviridae*	*Betacryptovirus*
Carrot temperate virus 3	CteV-3	*Partitiviridae*	*Alphacryptovirus*
Carrot temperate virus 4	CteV-4	*Partitiviridae*	*Alphacryptovirus*
Cucumber cryptic virus	CuCV	*Partitiviridae*	*Alphacryptovirus*
Fescue cryptic virus	FCV	*Partitiviridae*	*Alphacryptovirus*

Garland chrysanthemum temperate virus	GCTV	*Partitiviridae*	*Alphacryptovirus*
Hop trefoil cryptic virus 1	HTCV-1	*Partitiviridae*	*Alphacryptovirus*
Hop trefoil cryptic virus 2	HTCV-2	*Partitiviridae*	*Betacryptovirus*
Hop trefoil cryptic virus 3	HTCV-3	*Partitiviridae*	*Alphacryptovirus*
Mibuna temperate virus	MTV	*Partitiviridae*	*Alphacryptovirus*
Poinsettia cryptic virus	PnCV	*Partitiviridae*	*Alphacryptovirus*
Radish yellow edge virus	RYEV	*Partitiviridae*	*Alphacryptovirus*
Red clover cryptic virus 2	RCCV-2	*Partitiviridae*	*Betacryptovirus*
Red pepper cryptic virus 1	RPCV-1	*Partitiviridae*	*Alphacryptovirus*
Red pepper cryptic virus 2	RPCV-2	*Partitiviridae*	*Alphacryptovirus*
Rhubarb temperate virus	RTV	*Partitiviridae*	*Alphacryptovirus*
Ryegrass cryptic virus	RGCV	*Partitiviridae*	*Alphacryptovirus*
Santosai temperate virus	STV	*Partitiviridae*	*Alphacryptovirus*
Spinach temperate virus	SpTV	*Partitiviridae*	*Alphacryptovirus*
Vicia cryptic virus	VCV	*Partitiviridae*	*Alphacryptovirus*
White clover cryptic virus 1	WCCV-1	*Partitiviridae*	*Alphacryptovirus*
White clover cryptic virus 2	WCCV-2	*Partitiviridae*	*Betacryptovirus*
White clover cryptic virus 3	WCCV-3	*Partitiviridae*	*Alphacryptovirus*
Viroids			
Australian grapevine viroid	AGVd	*Pospiviroidae*	*Apscaviroid*
Avocado sunblotch viroid	ASBVd	*Avsunviroidae*	*Avsunviroid*
Chrysanthemum stunt viroid	CSVd	*Pospiviroidae*	*Pospiviroid*
Citrus exocortis viroid	CEVd	*Pospiviroidae*	*Pospiviroid*
Coconut cadang-cadang viroid	CCCVd	*Pospiviroidae*	*Cocadviroid*
Coleus blumei viroid 1	CbVd-1	*Pospiviroidae*	*Coleviroid*
Coleus blumei viroid 2	CbVd-2	*Pospiviroidae*	*Coleviroid*
Coleus blumei viroid 3	CbVd-3	*Pospiviroidae*	*Coleviroid*
Grapevine yellow speckle viroid 1	GYSVd-1	*Pospiviroidae*	*Apscaviroid*
Grapevine yellow speckle viroid 2	GYSVd-2	*Pospiviroidae*	*Apscaviroid*
Hop stunt viroid	HSVd	*Pospiviroidae*	*Hostuviroid*
Potato spindle tuber viroid	PSTVd	*Pospiviroidae*	*Pospiviroid*

[a]The list is updated but is probably not complete. Based on data from Mink (1993), Brunt *et al.* (1996), *AAB Plant Virus Descriptions* (n.d.), Hadidi *et al.* (2003), and different recent publications. Virus names in italics are those of species and those not in italics are of tentative species (Fauquet and Mayo, 1999).
*Two viruses in obligate symbiosis.
Names, acronyms and assignments to families and genera are those used in the compilation of plant virus names by Fauquet and Mayo (1999) and updating (Adams *et al.*, 2004). With permission from Fauquet and Mayo (1999) and the copyright holder © Springer-Verlag KG.

References

AAB Plant Virus Descriptions (n.d.) Nos. 1–409. Available at: http://www.dpvweb.net

Adams, M.J., Antoniw, J.F., Bar-Joseph, M., Brunt, A.A., Candresse, T., Foster, G.D., Martelli, G.P., Milne, R.G. and Fauquet, C.M. (2004) The new plant virus family *Flexiviridae* and assessment of molecular criteria for species demarcation. *Archives of Virology* 149, 1045–1060.

Brunt, A.A., Crabtree, K., Dallwitz, M.J., Gibbs, A.J. and Watson, L. (eds) (1996) *Viruses of Plants.* CAB International, Wallingford, UK.

Fauquet, M.C. and Mayo, M.A. (1999) Abbreviations for plant virus names – 1999. *Archives of Virology* 144, 1249–1273.

Hadidi, A., Flores, R., Randles, J.W. and Semancik, J.S. (eds) (2003) *Viroids.* CSIRO Publishing, Collingwood, Victoria, Australia, and Science Publishers, Enfield, New Hampshire.

Mink, G.J. (1993) Pollen and seed-transmitted viruses. *Annual Review of Phytopathology* 31, 375–402.

Appendix 2: Reagents, Solutions and Buffers

Note 1: It is assumed that the following procedures are carried out in a laboratory with the installations required for safe work, by persons familiar with the principles of good laboratory practice and safety precautions and that all waste materials are disposed of in an appropriate way and in accordance with local safety regulations.
Note 2: Chemicals and reagents of analytical grade should be used in all recipes.

Serological Testing Methods

Enzyme-linked immunosorbent assay

PBS buffer: 0.01 M phosphate buffered saline, pH 7.4, 1 litre

KH_2PO_4	0.2 g
$Na_2HPO_4 \cdot 12H_2O$	2.9 g
KCl	0.2 g
NaCl	8.5 g
Deionized H_2O up to	990 ml
NaN_3 2% stock solution*	10 ml

Adjust pH to 7.4 with diluted HCl or NaOH, if needed.**
A 10 × concentrated PBS – *without* NaN_3 – can be prepared as stock solution and kept at 5°C. As precipitate may be formed at 5°C, the 10 × buffer must be heated before dilution. (Note that the pH of the stock solution will increase approx. 0.3 pH units when diluted ten times.)
*NaN_3 stock solution, see below.
****Caution:** HCl is a strong acid; handle with care – see below. NaOH is a strong base; handle with care. NaN_3 is highly toxic; handle with care.

NaN_3 stock solution

Sodium azide (NaN_3) is a very effective preservative for buffers and solutions. Instead of adding the salt directly, prepare a 2% stock solution of NaN_3 (e.g. Merck 106688) in distilled water and add from this 1 ml per 100 ml of buffer etc. to give a final (standard) concentration of 0.02%.

Caution: NaN_3 is a highly toxic substance, which should be handled with great care. In addition, the white salt reacts heavily with acids and metals, developing toxic gases and explosive compounds, respectively. Handle in a fume cupboard and wear appropriate protective clothing, including dust mask.

Coating buffer: 0.05 M carbonate buffer, pH 9.6, 1 litre

Na_2CO_3	1.59 g
$NaHCO_3$	2.92 g
Deionized H_2O up to	990 ml
NaN_3 2% stock solution*	10 ml

If needed, the solution is adjusted to pH 9.6 with 0.1 M HCl or NaOH.** After storing, check pH before use.
*See above.
****Caution:** HCl is a strong acid; see below how to handle. NaOH is a strong base; handle with care. NaN_3 is highly toxic; handle with care.

Extraction buffer for DAS- and TAS-ELISA

Add per litre of PBS containing 0.02% NaN_3:

Tween-20	0.5 ml
Polyvinylpyrrolidone (MW 10,000 – 40,000)	20 g
Ovalbumin, BSA* or dry skimmed milk powder	2 g
Sodium sulphite (anhydrous)**	1.3 g

*Bovine serum albumin.
****Caution:** Sodium sulphite (Na_2SO_3) is harmful and should be handled with care.

Substrate buffer: diethanolamine, 10% (v/v), pH 9.8, 1 litre

Mix in the following sequence:

Deionized H_2O	200 ml
Diethanolamine ($C_4H_{11}NO_2$)*	100 ml
5 M HCl** *adjust with HCl to pH 9 (approx 2 ml)*	
Deionized H_2O up to	1000 ml

After storing, check pH and adjust, if needed, before use.
***Caution:** Diethanolamine is a harmful reagent. Protect skin, eyes and respiratory tract; work in fume cupboard. Store diethanolamine at 20°C to avoid solidification.
****Caution:** HCl is a strong acid; handle with care. *Always* add the acid to water, never the opposite. Wear safety glasses and proper lab clothing.

Dot immunobinding assay

Before starting the work, please see the **Notes** at the beginning of this appendix.

TBS: 0.02 M TRIS-HCl, pH 7.5, containing 0.5 M NaCl, 1 litre

TRIS base*	2.42 g
NaCl	29.2 g
Deionized H_2O	800 ml
*Dissolve and adjust the pH to 7.5 with diluted HCl** in fume cupboard*	
Deionized H_2O up to	1000 ml

*TRIS(hydroxymethyl)aminomethane
**__Caution:__ Strong acid, see above.

NBT stock solution

Nitro blue tetrazolium (Sigma™ N6876 or similar)	10 mg
Ethanolamine buffer (see below)	10 ml

Store the solution at 4°C.

BCIP stock solution

5-Bromo-4-chloro-3-indolylphosphate (e.g. Sigma™ B8503)	24 mg
Methanol*	4 ml
Acetone*	2 ml

Store this solution below 0°C.
***Caution:** Methanol and its vapour are extremely toxic with risk of permanent injuries; avoid contact to skin, eyes and respiratory tract. Both methanol and acetone are highly flammable.

$MgCl_2$, 2 M stock solution

$MgCl_2 \cdot 6H_2O$	4.06 g (1.9 g if water-free)
Deionized H_2O	10 ml

Substrate buffer: 0.1 M ethanolamine buffer, pH 9.6, 500 ml

Deionized H_2O	100 ml
Ethanolamine*	3.15 ml
*Mix and adjust the pH to 9.6 with 1 M HCl** (approx. 20 ml)*	
Deionized H_2O up to	500 ml

***Caution:** Harmful, irritant; protect eyes, skin and respiratory tract.
**__Caution:__ Strong acid, see ELISA above.

Nucleic Acid-based Testing Methods

Before starting the work, please see the **Notes** at the beginning of this appendix.

RNase-free tools and solutions

The robust and powerful enzymes RNases occur everywhere in the environment and may destroy RNA during its isolation unless tools and solutions are: (i) made RNase-free by treating with DEPC (diethylpyrocarbonate) and autoclaving prior to use; and (ii) kept RNase-free during their use. **Caution:** DEPC is a mutagen; use gloves, work in fume cupboard.

- *DEPC-treated H_2O:* Add DEPC to distilled (d H_2O) or double-distilled water (dd H_2O) at a concentration of 0.1% (v/v) and incubate at 37°C for 1 h, followed by autoclaving for 15 min at 15 p.s.i.
- *Glassware and plastic ware:* Fill with a solution of 0.1% DEPC in H_2O, let stand for 1 h at 37°C, followed by autoclaving for 15 min at 15 p.s.i.
  Use a special set of tools, including automatic pipetting devices, when handling RNA.
- *Solutions:* Use RNase-free glassware and chemicals, and DEPC-treated d H_2O. Wherever possible, treat solutions with 0.1% DEPC for at least 1 h at 37°C and then autoclave for 15 min at 15 p.s.i.

Note: Buffers containing TRIS (tris(hydroxymethyl)aminomethane) are decomposed by DEPC treatment – see procedures below for TRIS buffer preparation.

For details of the above procedures, see Sambrook and Russell (2001).

Non-isotopic nucleic acid spot hybridization

RNase precautions: *see above.*

GIT buffer

Guanidinium isothiocyanate*	4 M
TRIS-HCl, pH 7.5	100 mM
$MgCl_2$	25 mM
Na_2-EDTA, stock solution, see below	25 mM

Adjust to pH 7.5 with HCl or NaOH and autoclave.
Caution: HCl is a strong acid and NaOH a strong base; see p. 248, bottom.
This buffer cannot be DEPC-treated; prepare from RNase-free chemicals and use DEPC-treated, sterilized water and glassware.
***Caution:** Guanidinium isothiocyanate is toxic; handle in fume cupboard, and wear protective clothing.

EDTA stock solution, 0.5 M
Suspend Na_2-EDTA salt in dd H_2O using a magnetic stirrer. Dissolving of the salt requires adjustment of the pH to 8.0 by adding NaOH (pellets). After DEPC treatment and sterilization by autoclaving, store at room temperature.

SSC 20× (standard saline citrate buffer, 20 ×)

NaCl	3 M
Na citrate	0.30 M

Adjust to pH 7 with HCl or NaOH and autoclave.
Caution: HCl is a strong acid and NaOH a strong base; handle with care.

SSC 30× (standard saline citrate buffer, 30 ×)

NaCl	4.5 M
Na citrate	0.45 M

Adjust to pH 7 with HCl or NaOH and autoclave.
Caution: HCl is a strong acid and NaOH a strong base; handle with care.

Sodium phosphate buffer, 0.5 M, pH 7.2, 1 litre

Distilled H_2O	500 ml
$Na_2HPO_4 \cdot 7H_2O$	134 g
Dissolve the salt in water, and then add:	
H_3PO_4, 85% (concentrated phosphoric acid)*	4 ml
Adjust pH to 7.2 with diluted NaOH or diluted phosphoric acid	
Distilled H_2O up to	1000 ml

Sterilize by autoclaving.
***Caution:** Handle the concentrated phosphoric acid with care. *Always* add the acid to water, never the opposite. Wear safety glasses and proper lab clothing.

'Hyb solution' for prehybridization and hybridization, 500 ml

Use RNase-free components and stock solutions. Combine the following stock solutions in the order shown:

	Amount	*Final conc.*
Formamide, deionized, 100% **Caution:** harmful	250 ml	50% v/v
SSC, 30 ×	83 ml	5 ×
Sodium phosphate buffer, 0.5 M, pH 7.2	50 ml	50 mM
Blocking solution, 10% (see below)	100 ml	2% v/v
Yeast tRNA (10 mg/ml)	2.5 ml	50 µg/ml
N-lauroylsarcosine, 10%	5 ml	0.1% v/v
Sodium dodecyl sulphate (SDS)*, *adding, see below*	35 g	7% w/v
Sterile dd H_2O^* up to	500 ml	

*Pour the solution, containing the ingredients, *except* SDS and H_2O, into an Erlenmeyer flask that contains 35 g SDS (**Caution:** SDS is harmful, do not inhale, wear face guard). Heat and stir until the SDS dissolves. Then adjust to 500 ml with sterile dd H_2O.
Store at –20°C, but heat to 65°C before use.

Maleic acid buffer

Maleic acid	100 mM
NaCl	150 mM

Adjust to pH 7.5 with NaOH (**Caution:** strong base), autoclave and store at room temperature.

Blocking stock solution, 10% w/v

Blocking reagent, powder (Roche Diagnostics)	10 g
Maleic acid buffer	100 ml

Dissolve by shaking and heating. Autoclave and store in aliquots at –20°C.

Detection buffer

TRIS-HCl	100 mM, pH 9.5
NaCl	100 mM
$MgCl_2$	50 mM

Note, this buffer cannot be DEPC-treated; prepare from RNase-free chemicals and use DEPC-treated sterilized water and glassware. Sterilize by autoclaving and store at room temperature.
Discard the solution if a precipitate appears after a long storage period.

Polymerase chain reaction

See RNase-free tools and solutions above.

Nucleic acid extraction, protocol I

Extraction buffer

TRIS-HCl	50 mM, pH 8.5
Na_2-EDTA, 0.5 M stock solution*	10 mM
NaCl	200 mM

Note, this buffer cannot be DEPC-treated; prepare from RNase-free chemicals and use DEPC-treated sterilized water and glassware. Sterilize by autoclaving and store at room temperature.
Discard the solution if a precipitate appears after a long storage period.
*See p. 250.

PHENOL, ACID (FOR RNA VIRUSES): saturated with 0.1 M citrate buffer, ~pH 4.5 (e.g. Sigma P4682). Store in a dark bottle at 4°C (Seal and Coates, 1998).

or

PHENOL, NEUTRAL (FOR DNA VIRUSES): saturated with TE (e.g. Sigma P4557). Store in a dark bottle at 4°C (Seal and Coates, 1998).

Caution: Phenol is highly toxic; handle in fume cupboard, wearing protective clothing, gloves and face shield.

Do not use phenol solutions that have oxidized and have changed from colourless to light pink. Dispose of waste according to laboratory regulations.

CHLOROFORM. **Caution:** Highly toxic, avoid contact and inhalation. Wear protective clothing and work in a fume cupboard. Store in the dark at room temperature.

TE buffer, pH 7.4

TRIS-HCl	10 mM, pH 7.4
Na_2-EDTA, 0.5 M stock solution*	1 mM

Adjust to pH 7.4 with diluted HCl**.
Note, this buffer cannot be DEPC-treated; prepare from RNase-free chemicals and use DEPC-treated sterilized water and glassware. Sterilize by autoclaving and store at room temperature.
*See p. 250.
****Caution:** strong acid, handle with care.

ETHANOL, 70%. Dilute 95% ethanol with sterile H_2O; store at –20°C.

Nucleic acid extraction, protocol II A

Extraction buffer

TRIS-HCl	200 mM, pH 8.5
Lithium dodecyl sulphate*	15 g/l
LiCl*	375 mM
Sodium deoxycholate*	10 g/l
Igepal Ca-630 (Sigma I8896)	1% v/v (formerly Nonidet P-40)
Na_2-EDTA, 0.5 M stock solution**	10 mM

Note, this buffer cannot be DEPC-treated; prepare from RNase-free chemicals and use DEPC-treated sterilized water and glassware. Sterilize by autoclaving and store at room temperature.
***Caution:** Lithium salts are harmful to eyes and skin, especially lithium chloride. Sodium deoxycholate is also harmful, and highly flammable. Wear safety glasses, face guards and protective lab clothing when handling these compounds.
**See p. 250.

Nucleic acid extraction, protocol II B

Grinding buffer

Guanidine hydrochloride*	6 M
Sodium acetate	0.2 M, pH 5.2

Na_2-EDTA, 0.5 M stock solution**	25 mM
Potassium acetate	1 M
Polyvinylpyrrolidone (PVP) 40	2.5% w/v

Sterilize by autoclaving.
***Caution:** Guanidine hydrochloride is toxic; handle in fume cupboard, and wear protective clothing.
**See p. 250.

SE buffer, pH 7.4

NaCl	140 mM
KCl	2 mM
KH_2PO_4	2 mM
$Na_2HPO_4{\cdot}2H_2O$	8 mM
Adjust the pH to 7.4 with NaOH or HCl if needed and add further:*	
Tween 20	0.05% v/v
PVP 40	2% w/v
Ovalbumin	0.2% w/v
Bovine serum albumin (BSA)	0.5% w/v
NaN_3 2% stock solution**	0.02% w/v **Caution:** highly toxic.

***Caution:** NaOH is a strong base and HCl a strong acid; handle with care.
**NaN_3 2% stock solution, see p. 248; add 1 ml per 100 ml buffer.

Sodium iodide–sodium sulphite solution

Na_2SO_3*	0.75 g (0.15 M)
Sodium iodide	36 g (6 M)
Sterile H_2O	40 ml

Dissolve the Na_2SO_3 in the 40 ml water, then add and dissolve the 36 g NaI. Store the solution in the dark at 4°C. Sterilize by filtration.
***Caution:** Na_2SO_3 is harmful, handle with care.

Silica slurry, pH 2

Silica particles (Sigma S5631 or similar)	60 g
Distilled H_2O	2 × 500 ml
HCl, diluted*	

1. Suspend the silica particles in 500 ml distilled water, mix well and allow to settle for 24 h.
2. Discard the upper 470 ml of the supernatant, add another 500 ml distilled water, mix well and let the suspension settle for 5 h.
3. Discard the upper 440 ml of the suspension, and adjust the remaining slurry to pH 2.0 with HCl.

4. Autoclave and store in a dark bottle at room temperature, or aliquot into 1.5 ml microcentrifuge tubes, which can be stored at 4°C for several months.
***Caution:** HCl is a strong acid, handle with care.

Wash buffer

TRIS-HCl	10 mM, pH 7.5
Na_2-EDTA, 0.5 M stock solution*	0.5 mM
NaCl	50 mM
Ethanol	50% v/v

Note, this buffer cannot be DEPC-treated; prepare from RNase-free chemicals and use DEPC-treated sterilized water and glassware. Sterilize by autoclaving.
*See p. 250.

Nucleic acid extraction, protocol III

Extraction solution: TRIS-HCl-KCl-EDTA, pH 8.4

TRIS-HCl	100 mM, pH 8.4
KCl	1 M
Na_2-EDTA, 0.5 M stock solution*	10 mM

Adjust to pH 8.4 with HCl or NaOH (**Caution:** strong acid and strong base, respectively; handle with care).
Note, this buffer cannot be DEPC-treated; prepare from RNase-free chemicals and use DEPC-treated sterilized water and glassware. Sterilize by autoclaving.
Discard the solution if a precipitate appears after a long storage period.
*See p. 250.

Agarose gel electrophoresis

DEPC treatment of buffers, chemicals and equipment for gel electrophoresis of DNA is not needed.

TAE buffer, 10× stock solution: 0.40 M TRIS-acetate, 10 mM EDTA, pH 8.0

Sterile H_2O	500 ml
TRIS base	48.4 g
Glacial acetic acid	11.4 ml
Na_2-EDTA, 0.5 M stock solution*	20 ml
Sterile H_2O up to	1000 ml

Sterilize by autoclaving and store at room temperature.
*See p. 250.

TBE buffer, 5 × stock solution: 0.45 M TRIS-borate, 10 mM EDTA, pH 8.0

Sterile H_2O	500 ml
TRIS base	54 g
Boric acid	27.5 g
Na_2-EDTA, 0.5 M stock solution*	20 ml
Sterile H_2O up to	1000 ml

Sterilize by autoclaving and store at room temperature.
*See p. 250.

Ethidium bromide, stock solution: 10 mg/ml in TE buffer, pH 7.4

Ethidium bromide*	100 mg (one tablet, if Sigma E2515)
TE buffer, pH 7.4**	10 ml

***Caution:** Though moderately toxic, ethidium bromide is a powerful mutagen; wear double layers of gloves and protective clothing when handling.
Store in the dark at room temperature.
**See p. 253.

Gel-loading buffer, 5 ml
Make two solutions:

Ficoll 400 (Sigma F2637)	750 mg
Sterile H_2O	2.5 ml

and

Bromophenol blue	12.5 mg
Xylene cyanol	12.5 mg
Sterile H_2O	2.5 ml

Mix the two solutions. Filter and store at 4°C.

References

Sambrook, J. and Russell, D. (2001) *Molecular Cloning: A Laboratory Manual*, 3rd edn. Cold Spring Harbor Laboratory Press, New York.

Seal, S. and Coates, D. (1998) Detection and quantification of plant viruses by PCR. In: Foster, G.D. and Taylor, S.C. (eds) *Plant Virology Protocols: From Virus Isolation to Transgenic Resistance*. Humana Press, Totowa, New Jersey, pp. 469–485.

Appendix 3: Suppliers of Laboratory Equipment and Materials

Equipment

An overview, description, comparison and suppliers of a variety of laboratory equipment can be found on the web from, for example: http://www.biocompare.com. Suppliers of plant growth equipment (screenhouses): e.g. Clovis Lande Associates Ltd, UK (http://www.clovis.co.uk/virology).

Chemicals and Reagents

Suppliers of laboratory chemicals and reagents mentioned in the text can be accessed on the web via Google or other search engines. Additional suppliers can be found via http://www.biocompare.com or the gateway http://www.bioinformatics.vg/index.shtml with many links.

Phytodiagnostic Test Reagents (Examples)

Some of the following suppliers also deal in various equipment and materials for laboratory assays for viruses, as well as indicator plant seed.

ADGEN DIAGNOSTICS
Neogen Europe Ltd, Cunningham Building, Auchincruive, Ayr KA6 5HW, UK.
http://www.neogeneurope.com

AGDIA
30380 County Road 6, Elkhart, Indiana 46514, USA.
http://www.agdia.com

AMERICAN TYPE CULTURE COLLECTION (ATCC)*
PO Box 1549 Manassas, Virginia 20108, USA.
http://www.atcc.org

BIOREBA AG
Christoph-Merian-Ring 7, CH-4153 Reinach BL 1, Switzerland.
http://www.bioreba.ch

DSMZ–German Collection of Microorganisms and Cell Cultures* **
Mascheroder Weg 1 b, 38124 Braunschweig, Germany.
http://www.dsmz.de

LOEWE Biochemica GmbH
Mühlweg 2a, D-82054 Sauerlach, Germany.
http://www.loewe-info.com/index

*Also supplier of virus and viroid isolates.
**Also supplier of indicator plant seed.

Index